SEDIMENTOLOGY
AND STRATIGRAPHY

Sedimentology and Stratigraphy

Gary Nichols

Department of Geology, Royal Holloway, University of London

Blackwell
Publishing

BLACKWELL PUBLISHING
350 Main Street, Malden, MA 02148-5020, USA
9600 Garsington Road, Oxford OX4 2DQ, UK
550 Swanston Street, Carlton, Victoria 3053, Australia

First published 1999 by Blackwell Science Ltd

9   2005

Library of Congress Cataloging-in-Publication Data

Nichols, Gary.
   Sedimentology and stratigraphy/ Gary Nichols.
      p. cm.
   Includes bibliographical references and index.
   ISBN 0-632-03578-1
   1. Sedimentation and deposition.   2. Geology, Stratigraphic.
   I. Title.
   QE571.N53   1998
   551.3'03—dc21   98–7562
                                          CIP

ISBN-13: 978-0-632-03578-6

A catalogue record for this title is available from the British Library.

Set by Setrite Typesetters, Hong Kong
Printed and bound in the United Kingdom
by TJ International, Padstow, Cornwall

The publisher's policy is to use permanent paper from mills that operate a sustainable forestry policy, and
which has been manufactured from pulp processed using acid-free and elementary chlorine-free practices.
Furthermore, the publisher ensures that the text paper and cover board used have met acceptable
environmental accreditation standards.

For further information on
Blackwell Publishing, visit our website:
www.blackwellpublishing.com

Contents

v

Preface

All course tutors have their own ideas about the content and organization of the courses they have to give to students, and the idea of writing a textbook for a course in sedimentology and stratigraphy at Royal Holloway, University of London, had some appeal. It was going to be a project for the future, but Simon Rallison of Blackwell Science provided the impetus and encouragement to get me started on the writing and keep me going (even though I was very slow at times).

The text for this book was written over a number of years, mostly in a very erratic manner of short bursts of activity separated by long periods of time when I was otherwise occupied with teaching, research and administration at Royal Holloway. The two most productive periods were times when I was in a more isolated environment and could give the book more of my attention. The first of these periods was somewhat extreme isolation in a tent 75° south in the Antarctic on days when the weather precluded fieldwork for the British Antarctic Survey. My commiserations to Karl Farkas and Dave Cantrill who, at different times, had to put up with me sitting next to them tapping away on my solar-powered Macintosh Powerbook for days on end. The second was in the more civilized environment of the Faculty of Science in Charles University, Prague, where I spent three months as a visiting professor. My thanks to David Uličný for arranging my post in Prague; David and other friends in the Czech Republic are also to be thanked for distracting me from writing and ensuring that I enjoyed myself whilst I was there.

The text

The objective of this book has been to provide students who are starting to study geology at university level with an introduction to sedimentology and stratigraphy. It is hoped that the text is accessible to those completely new to the subject but at the same time covers the technical jargon and terminology used in more advanced work. A balance also has to be struck between the need to treat the subject as a natural science and the recognition that not all students of the Earth sciences have strong backgrounds in physics and chemistry.

Arrangement of the text

This book does not contain a glossary as such, but where a technical term is first used it appears in *italics* and is accompanied by some form of definition. Cross-referencing within the text is made by placing the section or subsection number in parentheses: for example *(3.4.2)* indicates Chapter 3, Section 3.4.2.

Relationship to other texts

This text is intended to complement books in sedimentary petrology by concentrating on the environments of deposition instead of the composition of the sediments and sedimentary rocks: Maurice Tucker's book *Sedimentary Petrology* (1991) is recommended for readers who wish to take a more petrographic approach to the subject. For a more comprehensive treatment of sedimentary processes the reader is referred to the texts by Mike Leeder, *Sedimentology: Process and Product* (1982), and Philip Allen, *Earth Surface Processes* (1997). In providing an introduction to the principles of stratigraphy in the later chapters of the book it is hoped that the reader may use this as a forerunner to considering the stratigraphy of a particular geographic area: an example would be *A Dynamic Stratigraphy of the British Isles* by Roger Anderton *et al.* (1982). The book is also intended to serve as an introduction to further study of sedimentary environments based on texts such as *Sedimentary Environments: Processes, Facies and Stratigraphy*, edited by

Harold Reading (1996) or *Facies Models: Response to Sea Level Change*, edited by Roger Walker and Noel James (1992).

Illustrations

All line drawings in the text have been prepared for this publication. In cases where the diagram is a direct copy of one published elsewhere this has been acknowledged. However, there are many which are original but have undoubtedly been influenced by illustrations in other publications. All photographs, with the exception of Figures 8.2, 8.3 and 8.5 (courtesy of Neil Harbury), were taken by the author, the opportunities to capture the images having been provided by many research grants and opportunities for geological tourism around the world.

References

The references provided in the text only scrape the surface of the voluminous literature on the topics covered in this book. The intention has been to cite the key references which are important to consult for some of the original work on a subject and indicate where further information can be found. Many of the main texts listed are books, compilation works and review papers which provide the most important routes to finding out more on these subjects. In selecting references for further reading an emphasis has been placed on books and periodicals which should be relatively easy to find in libraries.

Acknowledgements

Many colleagues in the Geology Department at Royal Holloway, University of London and elsewhere provided me with ideas and inspiration which have directly and indirectly helped me with this book: my thanks to them all. Thanks also to Jan Alexander and Gary Hampson for the care and considerable effort they both put into reviewing a first manuscript of the text: any remaining errors or inadequacies are entirely my responsibility.

Gary Nichols

1 Introduction: sedimentology and stratigraphy

Sedimentology is the study of the processes of formation, transport and deposition of material which accumulates as sediment in continental and marine environments and eventually forms sedimentary rocks. Stratigraphy is the study of rocks to determine the order and timing of events in Earth history. The two subjects can be considered to form a continuum of scales of observation and interpretation. The study of sedimentary processes and products allows us to interpret the dynamics of depositional environments. The record of these processes in sedimentary rocks allows us to interpret the rocks in terms of environments. To establish lateral and temporal changes in these past environments a chronological framework is required. This time framework is provided by different aspects of stratigraphy and allows us to interpret sedimentary rocks in terms of dynamic evolving environments. The record of tectonic and climatic processes through geological time is held in these rocks along with evidence for the evolution of life on Earth. This chapter introduces the general themes of this book.

1.1 Sedimentology and stratigraphy in Earth sciences

The Earth sciences have traditionally been divided into sub-disciplines which focus on aspects of geology such as palaeontology, geophysics, mineralogy, petrology, geochemistry and so on. Within each of these specialized areas the science has evolved as new analytical techniques have been applied and innovative theories developed. At the same time as advances have been made in separate fields of the subject the importance of an integrated approach combining ideas and expertise from different disciplines has been recognized. Geology is a multidisciplinary science which can best be understood if different aspects are seen in relation to each other.

Sedimentology and stratigraphy are two of the main sub-disciplines of geology, often considered separately in the past but now increasingly combined in teaching, academic research and economic application. They can be considered together as a continuum of processes and products, in both space and time. Sedimentology may be concerned primarily with the formation of sedimentary rocks but as soon as these beds of rock are looked at in terms of their temporal and spatial relationships the study has become stratigraphic (Fig. 1.1). Similarly, if the stratigrapher wishes to interpret layers of rock in terms of environments of the past the research is sedimentological. It is therefore logical to consider sedimentology and stratigraphy together. In fact it could easily be argued that mineralogy should also be incorporated to cover the components of the rocks and that evolutionary palaeontology provides much of the time framework and is therefore also inextricably linked to stratigraphy. However, in order to divide up the Earth sciences into manageable sections a series of arbitrary lines have to be drawn and a text has to be limited to certain topics.

The first part of this book covers aspects of the processes of sedimentation and their products in different depositional environments. Sedimentary rocks are then considered in terms of their spatial and temporal relationships in stratigraphic successions in sedimentary basins. Plate tectonics, petrology and palaeontology are complementary topics in geology which may be considered in parallel to sedimentology and stratigraphy, although only a basic knowledge of these topics is assumed in this text.

Fig. 1.1 Conglomerate and sandstone beds (centre, left) exposed in northern Spain, interpreted as the deposits of an alluvial fan *(8.4)*: these are stratigraphically younger than the limestone beds in the background.

1.2 Stratigraphy and sedimentology

Use of the term 'stratigraphy' dates back to d'Orbigny in 1852, but the concept of layers of rocks, or strata, representing a sequence of events in the past is much older. In 1667 Steno developed the principle of superposition: 'in a sequence of layered rocks, any layer is older than the layer next above it'. Stratigraphy can be considered as the relationship between rocks and time, and the history of the Earth is recorded in layers of rock, albeit in a very incomplete way. The stratigrapher is concerned with the observation, description and interpretation of direct and tangible evidence in rocks to determine temporal and spatial relationships during Earth history.

Stratigraphy is enjoying a renaissance of attention in Earth sciences as a result of new ideas which have developed in recent years, particularly the concept of 'sequence stratigraphy'. Whilst both the nomenclature of stratigraphic units in different areas and the biostrati-graphic basis for their definition are still important, stratigraphy is now as often considered in terms of changes in environment during the development of sedimentary basins. It is also recognized as the key to understanding almost all Earth processes as stratigraphic analysis provides information about events through geological history. Geophysics may provide the physical basis for the behaviour of the lithosphere but the stratigraphic record provides evidence of the way the lithosphere has behaved through time.

'Sedimentology' has only existed as a distinct branch of the geological sciences for a few decades. It developed as the observational elements of physical stratigraphy became more quantitative and the layers of strata were considered in terms of the physical, chemical and biological processes which formed them. There were no great breakthroughs of the type which occurred when Earth structures began to be looked at in a different light as the plate tectonics theory developed. The concept of interpreting rocks in terms of modern processes which underpins modern sedimentology dates back to the 18th and 19th centuries ('the present is the key to the past'). Sedimentology developed as researchers paid more attention to the interpretation of sedimentary rocks and began to embrace sedimentary petrology, which they had previously considered more or less separately from stratigraphy. The subject now covers everything from the sub-microscopic analysis of grains to the palaeogeographic evolution of whole sedimentary basins.

1.3 See the world in just one grain of sand

The measurement of both space and time in sedimentology and stratigraphy involves 17 orders of magnitude (Fig. 1.2). At one extreme, the behaviour of the Earth in its orbit around the Sun controls the world-wide climate which influences sedimentary processes. At the other extreme, the properties of clay particles a micrometre across also determine the character of

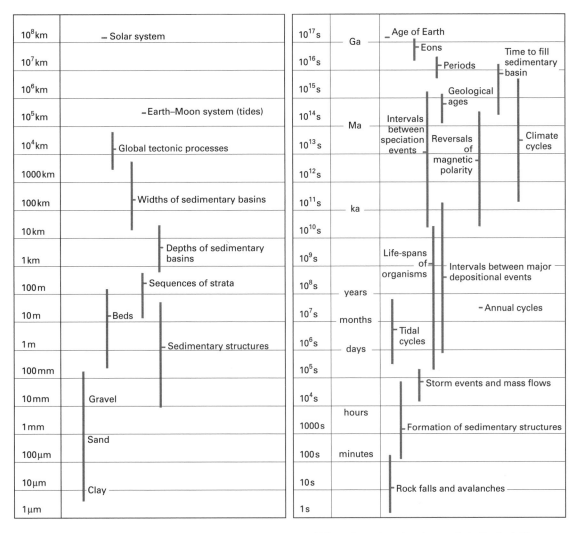

Fig. 1.2 Magnitudes of geological processes in space and time.

sedimentary rocks. The time-scale of stratigraphy is the entire history of the Earth, a period of 4½ billion years, yet a single sedimentation event can occur in seconds. To consider all this in a logical order the large-scale time and space issues could be considered first as the overall controlling factors, or one could start from the smallest elements and shortest-period events. It is largely a matter of personal preference and each approach has its advantages and disadvantages. Different scales interact and it is not possible to follow issues through in a strict order working in either direction.

The starting point taken in this book is the 'grain of sand'. The smallest elements — the particles of sand, pebbles, clay minerals, pieces of shell, algal filaments, chemical precipitates and other constituents which make up sediments — are considered first, together with the processes which move and deposit them. The materials are then considered in terms of depositional environments, the places where sediments accumulate into sedimentary rocks and become layers of stratigraphy through time. Climatic and tectonic processes control the large-scale patterns of stratigraphy as the rocks form the fill of sedimentary basins which are seen in the present day and in rocks throughout the world.

1.4 Processes and products

The nature of sedimentary material is very varied in

origin, size, shape and composition. Particles such as grains and pebbles may be derived from the erosion of older rocks or directly ejected from volcanoes. Organisms form a very important source of material, ranging from microbial filaments encrusted with calcium carbonate to whole or broken shells, coral reefs, bones and plant debris. Direct precipitation of minerals from solution in water also contributes to sediments in some situations.

Formation of a body of sediment involves either the transport of particles to the site of deposition or the chemical or biological growth of the material in place. Accumulation of sediments in place is largely influenced by the chemistry, temperature and biological character of the setting. The processes of transport which bring in material include movement by water, air, ice or mass flows. The type and velocity of the transporting medium, and the amount and size of the material carried, will determine the nature of the sediment which may then accumulate.

The processes of transport and deposition can be determined by looking at individual layers of sediment. The size, shape and distribution of particles all provide clues to the way in which the material was carried and deposited. The processes also involve the formation of structures in the sediment which are preserved in the rocks. Primary sedimentary structures such as ripples in sands can be seen forming today, either in natural environments or in a laboratory tank, and the conditions of flow velocity and water depth measured (Fig. 1.3). By recognizing the same size and shape of ripples in

sedimentary rocks it may be assumed that they formed under the same velocity and flow depth *(4.2)*.

By making observations of sedimentary rocks it is possible to make estimates (to varying degrees of accuracy) of the physical, chemical and biological conditions which existed at the time of sedimentation. These conditions may include the salinity, depth and flow velocity in lake or sea water, the strength and direction of the wind and the tidal range in a shallow marine setting. A fundamental assumption made in interpreting the processes of sedimentation from the character of sedimentary rocks is that the laws which govern physical and chemical processes have not changed through time.

By comparison of present-day processes and their products with the characteristics of sedimentary rocks, the physical, chemical and biological conditions under which the sediment formed can be determined.

1.5 Sedimentary environments and facies

The environment at any point on the land or under the sea is governed by the prevailing physical and chemical processes and the organisms which live under those conditions at that time. A depositional environment can therefore be characterized in terms of these processes. As an example, a fluvial (river) environment includes a channel confining the flow of fresh water which carries and deposits gravelly or sandy material on bars in the channel (Fig. 1.4). When the river floods, water spreads relatively fine sediment over the floodplain where it is

Fig. 1.3 A laboratory flume tank can be used to study flow over a bed of sand under controlled experimental conditions. Experiments have made it possible to establish quantitative relationships between flow conditions and features seen on the sand which may be preserved in the stratigraphic record.

Fig. 1.4 A modern sedimentary environment: a sandy river channel and vegetated floodplain (near Morondava, western Madagascar).

deposited in thin layers. Soils form and vegetation grows on the floodplain area. In a succession of sedimentary rocks (Fig. 1.5) the channel may be represented by a lens of sandstone or conglomerate which shows internal structures formed by deposition on the channel bars. The floodplain setting will be represented by thinly bedded mudrock and sandstone with roots and other evidence of soil formation.

In the description of sedimentary rocks in terms of depositional environments, the term *facies* is often used. A rock facies is a body of rock with specified characteristics which reflect the conditions under which it was formed (Reading & Levell 1996). Describing the facies of a sediment involves documenting all the characteristics of its lithology, texture, sedimentary structures and fossil content which can aid in determining the processes of formation. If sufficient facies information is available, an interpretation of the depositional environment can be made. The lens of sandstone may be shown to be a river channel if the floodplain deposits are found associated with it. However, channels filled with sand exist in other settings, including deltas, tidal environments and the deep sea floor. The recognition of a channel form on its own is hence not a sufficient basis to determine the depositional environment.

The depositional facies of sedimentary rocks can be used to determine the environmental conditions under which the sediments accumulated.

Fig. 1.5 Sedimentary rocks interpreted as the deposits of a river channel (the lens of sandstones beneath the figure's feet) scoured into mudstone deposited on a floodplain (the darker, thinly bedded strata below and to the side of the sandstone lens). Eocene rocks near Roda de Isabena in northern Spain.

1.6 Modern and ancient sedimentary environments

The combination of physical, chemical and biological processes acting in any place at any time is unique, and hence the products of those processes are infinitely variable. From a strictly objective scientific viewpoint the processes which determined the form of a sedimentary rock should be elaborated from first principles in order to establish the physical processes occurring in the environment, the chemistry of the water, and so on. For practical purposes we can consider a number of principal environments which have recognizable characteristics. These categories of environment consist of end members and points along a spectrum of depositional settings. The possibilities of variation from the 'typical' character of a particular environment are endless and situations intermediate between two settings are possible. For example, at what point is a pond in a floodplain environment considered to be a lake? The dangers of pigeon-holing must always be kept in mind: a succession of thin sandstone and mudstone beds might have the general character of deposition in a deep marine setting but the presence of desiccation cracks in the mudstone would be clear evidence of subaerial exposure, inconsistent with a formation in deep water.

The most straightforward way of considering the range of depositional environments is to start with mountain areas where weathering and erosion generate clastic detritus, and work down to the bottoms of the deep oceans. The character of the continental, coastal and shallow marine environments in between is influenced by the supply of clastic detritus, rainfall, temperature, biogenic productivity, topography on land and bathymetry in the sea. Some processes may be common to many different environments: settling from suspension of fine-grained material to form a layer of mud may occur on floodplains, in lakes, lagoons, sheltered bays, outer shelf settings and the deepest oceans. Other processes are unique to certain settings: the regular reversal of flow due to tidal action is unique to shallow marine and coastal environments. In general, combinations of processes can be considered characteristic of each depositional setting.

Associations of depositional processes can be considered characteristic of different depositional environments and allow a number of main categories of environment to be recognized.

1.7 Geographical distribution of environments and facies

Depositional environments are clearly of limited lateral extent. A river may pass into a delta with shallow marine conditions and deeper water further offshore. Along the shoreline from the delta there may be a beach and perhaps a lagoon behind it. At any point in time all of these may be places where sediment is accumulating, with many more similar or different depositional environments on land or in the sea in other parts of the world. The boundaries between depositional environments may be sharp, such as the edge of some lakes, or gradational, where the conditions progressively change with depth offshore in shallow marine settings. There is considerable variation in the dimensions and extent of these environments. A beach may be only a few tens of metres across but stretch tens of kilometres along the coastline. A region of aeolian sand dunes in a desert may cover tens of thousands of square kilometres.

The sediments deposited will show a lateral variation which reflects the changes in environment. For example, a moraine at the snout of a glacier will consist of a poorly sorted mixture of mud, sand and gravel, but the glacial outwash rivers flowing away from the glacier will be depositing better-sorted sands and gravels. A glacial lake nearby may be the location of deposition of mud and silt. Distinctly different sediments therefore will form in these sub-environments of the glacial setting at the same time and only a few metres to kilometres apart. In the stratigraphic record these contrasting sediments will occur side by side: a muddy, sandy conglomerate formed by the moraine, lenses and sheets of sandstone and conglomerate deposited by the outwash rivers, and laminated mudstone and siltstone which accumulated in the lake. A reconstruction of the palaeoenvironment can therefore include the geographical distribution of different depositional settings.

By looking at the lateral distribution of sedimentary facies in rocks of the same age we can reconstruct palaeoenvironments and an overall palaeogeography.

1.8 Changing environments and facies through time

The surface of the Earth is dynamic at all scales of space and time. The landscape is continuously modified by rock being eroded from one place and moved to another by gravity, water, wind and ice. Through time mountains are worn down and seas fill with sediment. New mountains are created by the movement of tectonic plates around the surface of the planet, and these same plate motions generate new areas for sediment to accumulate. These processes have been going on for thousands of millions of years. Pieces of crust move around the surface of the globe taking depositional environments with them, modifying them and sometimes changing them to areas of uplift and erosion. The plates pass through different climate belts as they move, and the world's climate changes over short and long time periods.

Depositional environments are therefore in a state of flux, although the rate of change with time may be slow enough to allow conditions to remain constant for millions of years. An area of continental sedimentation in river channels, overbank areas and lakes may become flooded by the sea and become a region of shallow marine sedimentation. A warm tropical shallow sea area with coral reefs may be uplifted, eroded and blanketed by desert sands. An advancing ice sheet forming during a period of climate cooling may transform a vegetated coastal swamp into an area of glacial moraines. The changes in depositional environment caused by tectonic and climatic processes are recorded in the sedimentary facies of rocks. Sediments accumulate and rocks formed in different environments pile up on top of each other to provide the stratigraphic record of these changes in environment (Fig. 1.6).

The stratigraphic record provided by sedimentary rocks can be interpreted in terms of changing depositional environments through geological time and these changes can ultimately be related to tectonic and climatic processes.

1.9 The stratigraphic record and geological time

In order to achieve the objective of interpreting sedimentary rocks in terms of a dynamic Earth a time framework is required. We need to know what was happening at certain points in time in order to reconstruct palaeoenvironments and palaeogeography. Some means of correlating rocks is therefore required which tells us which ones were formed at the same time. The order in which events occurred tells us how conditions in any area changed so we need to be able to determine the relative age of different rock units, which ones are older and which are younger. In order to understand the rates

at which geological processes have operated in the past some form of dating is required which will give our time framework a scale in years.

The relative age of rocks can be determined by simple stratigraphic relations in the first instance. In a simple, undeformed succession of beds, the upper beds are younger than the lower beds. Within these beds changes in fossil content can sometimes be observed. The form of organisms has changed through time by the process of evolution, and consequently particular types of fossil are characteristic of particular periods of time in Earth history. We can use the presence or absence of fossils to put the rocks in which the fossils occur in stratigraphic order. Rocks which contain the same characteristic fossils can be considered to be of approximately the same age. In some circumstances the rate of radioactive decay of isotopes of elements in rocks can be used to calculate an isotopic age in years for the rock. A combination of different stratigraphic techniques has been used to construct a stratigraphic column to which all rocks and events can be related and a geological time-scale which can provide absolute ages for those events.

The geological time-scale constructed from stratigraphic information present in rocks provides us with the temporal framework for events in Earth history.

1.10 Earth history, global tectonics, climate and evolution

Stratigraphy provides the record of Earth history and with it much of the evidence for the way the planet works as a physical, chemical and biological unit. Successions of sedimentary rocks indicate how regions of accumulation (sedimentary basins) have formed and filled. These records can be interpreted in terms of the behaviour of the lithosphere when subjected to the extensional and compressional forces of plate tectonics. Both the magnitudes and the rates of tectonic processes can be determined from the stratigraphic record. Geophysical observations and interpretations of the structure of the lithosphere, and of volcanic and seismic activity at plate boundaries, have been instrumental in understanding the dynamics of plate tectonics, but the stratigraphic record has provided the temporal framework for our understanding of the way the Earth works.

In addition to a record of the tectonic history, sedimentary rocks contain information about changes in local and global climate through geological time. Some

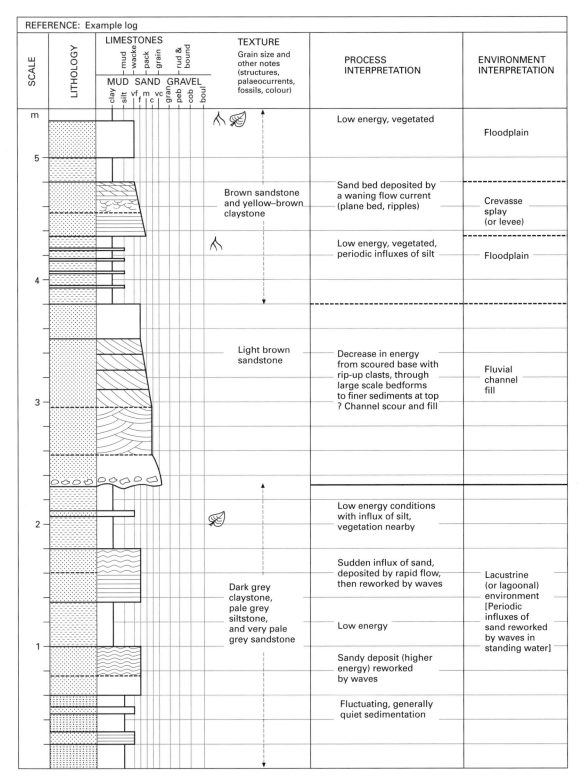

Fig. 1.6 A succession of sedimentary rocks interpreted in terms of depositional processes and environments. This representation of stratigraphy — a graphic sedimentary log — is discussed in Section 5.6.

sedimentary environments are very sensitive to climate, particularly temperature and precipitation. For example, coral reefs are believed to flourish only in relatively warm shallow seas and accumulations of evaporite minerals only form in places where the rate of evaporation exceeds the rate of precipitation. A more indirect indicator of climatic fluctuations comes from the record of evidence for sea level changes present in sedimentary rocks. One of the causes of sea level rise is the melting of polar continental ice caps which occurs when the global temperature rises. In this case the sedimentary record not only tells us about events of the past but also provides clues about the future when the global climate changes.

Fossils in sedimentary rocks are the record of past life on Earth. They provide the evidence for changes in life forms through time and hence much of the information on which theories of evolutionary processes are based. The record is very incomplete, but general trends are clear and events of speciation and extinction are documented by the appearance and disappearance of the fossils of particular species, genera and families. The debates about the causes of the extinction of major groups such as the dinosaurs are all based on interpretation of physical, chemical and biological evidence found in the stratigraphic record.

The stratigraphic record contained within rocks can be used to determine how plate tectonic processes operate over long time periods and how the Earth's climate has changed in the past, as well as providing clues to the processes of evolution of life.

2 Terrigenous clastic sediments: gravel, sand and mud

Four main groups of sedimentary rocks can be recognized on the basis of their composition: terrigenous clastic sediments, carbonate sediments, evaporites and volcaniclastic deposits. The general and specific composition determines the properties of the sediment and the character of the sedimentary rock which will be formed by lithification of the material. Classification schemes and a descriptive nomenclature have been developed to allow sediments and sedimentary rocks to be categorized and given names which clearly indicate the nature of the material. Description of sediment and sedimentary rocks includes an assessment of the mineralogy and/or biogenic origin of the constituents as well as a quantitative analysis of the size, shape and distribution of particles present. These lithological descriptions can be made from loose material or hand specimens and complemented by petrographic analysis using a microscope. Some information about the processes and conditions of deposition can be obtained from examination of sediments and rocks at this scale, and this will complement observations made from field exposures or subsurface data. This chapter mainly deals with terrigenous clastic sediments (muds, sands and gravels) and their lithified equivalents: these are composed of minerals and rock fragments largely derived from the weathering and erosion of older rocks.

2.1 The components of sediments and sedimentary rocks

Sediments and sedimentary rocks can be classified by either their constituents or their mode of origin, or a combination of the two. A convenient division of all sedimentary rocks is as follows (Fig. 2.1).

TERRIGENOUS CLASTIC MATERIAL

This is material which is made up of particles or *clasts* derived from pre-existing rocks. The clasts are principally detritus eroded from bedrock and are commonly made up largely of silicate minerals: the terms *detrital sediments* and *siliciclastic sediments* are also used for this material. Clasts range in size from clay particles measured in micrometres (μm), to boulders metres across. Sandstones and conglomerates make up 20–25% of the sedimentary rocks in the stratigraphic record and mudrocks are 60% of the total.

CARBONATES

By definition, a limestone is any sedimentary rock containing over 50% calcium carbonate ($CaCO_3$). In the natural environment the hard parts of organisms, mainly invertebrates such as molluscs, constitute a principal source of calcium carbonate. Limestones constitute 10–15% of the sedimentary rocks in the stratigraphic record.

EVAPORITES

These are deposits formed by the precipitation of salts out of water due to evaporation.

VOLCANICLASTIC SEDIMENTS

These are the products of volcanic eruptions or the result of the breakdown of volcanic rocks.

OTHERS

Other sediments and sedimentary rocks are sedimentary

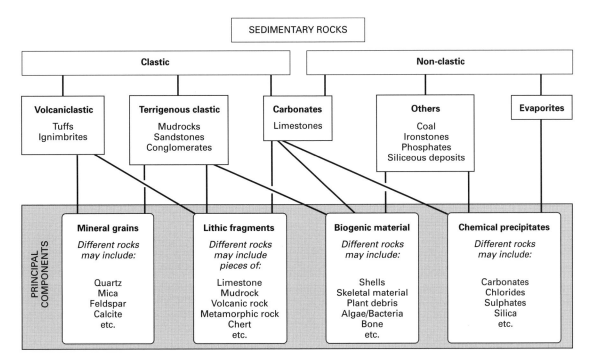

Fig. 2.1 Table of the principal constituents of sedimentary rocks.

ironstone, phosphate sediments, organic deposits (coals and oil shales) and cherts (siliceous sedimentary rocks). These are volumetrically less common than the others, making up about 5% of the stratigraphic record, but some are of considerable economic importance.

As with most classification systems, there are overlaps and grey areas in this scheme. Some limestone beds can form by the chemical precipitation of calcium carbonate out of water during evaporation, and could therefore be considered to be an evaporite deposit. In other cases the nomenclature may not seem to be entirely logical: a rock containing 51% quartz sand grains and 49% calcareous fragments would be termed a calcareous sandstone; however, with the proportions reversed (49% quartz and 51% calcareous fragments), the rock would be called a sandy limestone.

2.2 Classification and nomenclature of terrigenous clastic sediments and sedimentary rocks

A distinction can be drawn between sediments and sedi-mentary rocks. *Terrigenous clastic sediments* are loose aggregates of clastic material which becomes a *terri-genous clastic sedimentary rock* when the material is lithified (*lithification* is the process of 'turning into rock': *17.2.1*). Mud, silt and sand are all loose *aggregates*; the addition of the suffix '-stone' (mudstone, siltstone, sandstone) indicates that the material has been lithified and is now a solid rock. Loose gravel material is named according to its size as granule, pebble, cobble and boulder aggregates which may become lithified into conglomerate (sometimes with the size range added as a prefix, e.g. 'pebble conglomerate').

2.2.1 Terrigenous clastic sediments and sedimentary rocks

A threefold division on the basis of grain size is used as the starting point to classify and name terrigenous clastic sediments and sedimentary rocks: gravel and conglom-erate consist of clasts greater than 2 mm in diameter; sand-sized grains are between 2 mm and $\frac{1}{16}$ mm (63 μm) across; mud (including clay and silt) is made up of particles less than 63 μm in diameter. There are variants on this scheme and there are a number of ways of providing subdivisions within these categories, but sedimentologists tend to use the *Wentworth scale*

(Fig. 2.2) to define and name terrigenous clastic deposits.

2.2.2 The Udden–Wentworth grain size scale

Known generally as the 'Wentworth scale', this is the scheme in most widespread use for the classification of aggregate particulate matter (Udden 1914; Wentworth 1922). The divisions on the scale are made on the basis of factors of 2: for example, medium sand grains are 0.25–0.5 mm in diameter, coarse sand grains are 0.5–1.0 mm, very coarse sand 1.0–2.0 mm, and so on. It is therefore a logarithmic progression, but a logarithm to the 'base 2', as opposed to the 'base 10' of the more common log scales. This scale has been chosen because these divisions appear to reflect the natural distribution of sedimentary particles: in a simple way, it can be related to starting with a large block and repeatedly breaking it into two pieces.

Four basic divisions are recognized:
1 clay ($<4\,\mu m$);
2 silt ($4\,\mu m$ to $63\,\mu m$);
3 sand ($63\,\mu m$ to $2.0\,mm$);
4 gravel/aggregates ($>2.0\,mm$).

The *phi scale* is a numerical representation of the Wentworth scale. The Greek letter 'φ' (phi) is often used as the unit for this scale. Using the logarithm to base 2, the grain size can be denoted on the phi scale as:

$$\phi = -\log_2(\text{grain diameter in mm})$$

The negative is used because it is conventional to represent grain sizes on a graph as decreasing from right to left. Using this formula, a grain diameter of 1 mm is 0φ; increasing the grain size, 2 mm is –1φ, 4 mm is –2φ, and so on; decreasing the grain size, 0.5 mm is +1φ, 0.25 mm is 2φ, and so on.

2.3 Gravel and conglomerate

Clasts over 2 mm in diameter are divided into granules, pebbles, cobbles and boulders (Fig. 2.2). The name given to a consolidated gravel will depend on the dominant clast size: for example, if most of the clasts are between 64 mm and 256 mm in diameter the rock is called a cobble conglomerate. The term *breccia* is commonly used for conglomerate made up of clasts which are angular in shape *(2.6)*. In some circumstances it is prudent to specify that a deposit is a 'sedimentary breccia' as 'tectonic breccias' may form by the fragmentation of rock in fault zones by friction between bodies of rock moving past each other. Mixtures of rounded and angular clasts are sometimes termed *breccio-conglomerate*. Occasionally the noun *rudite* and the adjective *rudaceous* are used: these terms are synonymous with conglomerate and conglomeratic.

2.3.1 Composition of gravel and conglomerate

Further description of the nature of a gravel or conglomerate can be provided by considering the types of clast present. If all the clasts are of the same material (all of granite, for example), the conglomerate is considered to be *monomict*. A *polymict* conglomerate is one which contains clasts of many different lithologies, and sometimes the term *oligomict* is used where there are just two or three clast types present.

mm	phi	Name		
256	–8	Boulders		Gravel / Conglomerate
128	–7			
64	–6	Cobbles		
32	–5			
16	–4			
8	–3	Pebbles		
4	–2			
2	–1	Granules		
1	0	Very coarse sand		Sand / Sandstone
0.5	1	Coarse sand		
0.25	2	Medium sand		
0.125	3	Fine sand		
0.063	4	Very fine sand		
0.031	5	Coarse silt		Mud / Mudrock
0.0156	6	Medium silt		
0.0078	7	Fine silt		
0.0039	8	Very fine silt		
		Clay		

Fig. 2.2 The Udden–Wentworth scale of grain size classification.

Almost any lithology may be found as a clast in gravel and conglomerate. *Resistant lithologies*, those which are less susceptible to physical and chemical breakdown, have a higher chance of being preserved as a clast in a conglomerate. Factors controlling the resistance of a rock type include the minerals present and the ease with which they are chemically or physically broken down in the environment. Some sandstones break up into sand-sized fragments when eroded because the grains are weakly held together. The most important factor controlling the varieties of clast found is the bedrock being eroded in the source area. A gravel will be composed entirely of limestone clasts if the source area is made up only of limestone bedrock. Recognition of the variety of clasts can therefore be a means of determining the source (or provenance: *5.5*) of a conglomeratic sedimentary rock.

2.3.2 Texture of conglomerate

Conglomerate beds are rarely composed entirely of gravel-sized material. Between the granules, pebbles, cobbles and boulders, finer sand and/or mud will often be present: this finer material between the large clasts is referred to as the *matrix* of the deposit. If there is a high proportion (over 20%) of matrix, the rock may be referred to as a *sandy conglomerate* or *muddy conglomerate*, depending on the grain size of the matrix present (Fig. 2.3). An *intraformational conglomerate* is composed of clasts of the same material as the matrix and is formed as a result of reworking of lithified sediment soon after deposition.

The proportion of matrix present is an important factor in the *texture* of conglomeratic sedimentary rock—that is, the arrangement of different grain sizes within it *(2.6)*. A distinction is commonly made between conglomerates which are *clast-supported* (that is, with clasts touching each other throughout the rock) and those which are *matrix-supported* (in which most of the clasts are completely surrounded by matrix). The term *orthoconglomerate* is sometimes used to indicate that the rock is clast-supported, and *paraconglomerate* for a matrix-supported texture. These textures are significant when determining the mode of transport and deposition of a conglomerate (e.g. on alluvial fans: *8.4*).

The arrangement of the sizes of clasts in a conglomerate can also be important in interpretation of depositional processes. In a flow of water pebbles are moved more easily than cobbles, which in turn require less energy to move them than boulders. A deposit which is

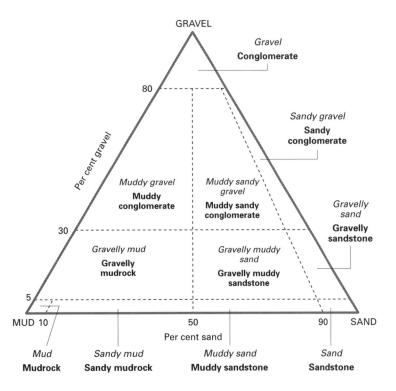

Fig. 2.3 Nomenclature used for mixtures of terrigenous clastic sediments and sedimentary rocks.

made up of boulders overlain by cobbles and then pebbles may be interpreted in some cases as having been formed from a flow which was decreasing in velocity. This sort of interpretation is one of the techniques used in determining the processes of transport and deposition of sedimentary rocks *(4.2)*.

2.3.3 Shapes of clasts

The shapes of clasts in gravel and conglomerate are determined by the fracture properties of the bedrock they are derived from and the history of transport (see clast sphericity and clast roundness: *2.6*). Rocks with equally spaced fracture planes in all directions form cubic or *equant* blocks which form spherical clasts when the edges are rounded off (Fig. 2.4). Bedrock which breaks up into slabs, such as a well-bedded limestone or sandstone, forms clasts with one axis shorter than the other two (Krumbein & Sloss 1951). This is termed an *oblate* or *discoid* form. Rod-shaped or *prolate* clasts are less common, forming mainly from metamorphic rocks with a strong linear fabric.

When discoid clasts are moved in a flow of water they are preferentially orientated and may stack up in a form known as *imbrication* (Fig. 2.5). These stacks are arranged in the pattern which is most stable in the flow,

Fig. 2.5 Imbrication produced by reorientation of pebbles in a flow (direction of flow from left to right).

which is with the discoid clasts dipping upstream. In this orientation, the water can flow most easily up the upstream side of the clast. When orientated dipping downstream, flow at the edge of the clast causes it to be reorientated. The direction of imbrication of discoid pebbles in a conglomerate can be used to indicate the direction of the flow which deposited the gravel.

2.4 Sand and sandstone

A *sand* may be defined as a sediment consisting primarily of grains in the size range 63 μm to 2 mm, and a *sandstone* defined as a sedimentary rock with grains of these sizes. This size range is divided into five intervals: very fine, fine, medium, coarse and very coarse (Fig. 2.2). It should be noted that this nomenclature refers only to the size of the particles. Although many sandstones contain mainly quartz grains, the term 'sandstone' carries no implication about the amount of quartz present in the rock and some sandstones contain no quartz at all. Similarly, the term *arenite*, which is a sandstone with less than 15% matrix, does not imply any particular clast composition. Along with the adjective *arenaceous* used to describe a rock as sandy, arenite has its etymological roots in the Latin word for sand, *arena*, also used to describe a stadium with a sandy floor.

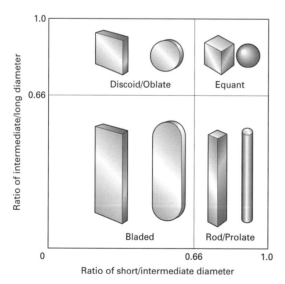

Fig. 2.4 The shape of clasts can be considered in terms of four end members: equant, rod, disc and blade. Equant and disc-shaped clasts are most common. (After Tucker 1991.)

2.4.1 Composition of sandstones

Sand grains are formed by the breakdown of pre-existing rocks by weathering and erosion *(6.5, 6.6)*, and from material which forms within the environments of transport and deposition. The breakdown products fall into two categories: *detrital mineral grains*, eroded from pre-existing rocks, and sand-sized pieces of rock, or *lithic fragments*. Grains which form within the depositional environment are principally biogenic in origin — that is, they are pieces of plant or animal — but there are some which are formed by chemical reactions.

2.4.2 Detrital mineral grains

A very large number of different minerals may occur in sands and in sandstones, and only the most common are described here.

QUARTZ

Quartz is the commonest mineral species found as grains in sandstone and siltstone. As a primary mineral it is the major constituent of granitic rocks, occurs in some igneous rocks of intermediate composition and is absent from basic igneous rock types. Metamorphic rocks such as gneisses formed from granitic material and many coarse-grained metasedimentary rocks contain a high proportion of quartz. Quartz is a very stable mineral which is resistant to chemical breakdown at the Earth's surface. Grains of quartz may be broken or abraded during transport, but with a hardness of 7 on Mohs' scale quartz grains remain intact over long distances and long periods of transport. In hand specimen quartz grains show little variation: coloured varieties such as smoky or milky quartz and amethyst occur but mostly quartz is seen as clear grains.

FELDSPAR

Most igneous rocks contain feldspar as a major component. Feldspar is hence very common and is released in large quantities when granites, andesites and gabbros, as well as some schists and gneisses, break down. However, feldspar is susceptible to chemical alteration during weathering and, being softer than quartz, tends to be abraded and broken up during transport. Feldspars are only commonly found in circumstances where the chemical weathering of the bedrock has not been too intense and the transport pathway to the site of deposition is relatively short. Potassium feldspars are more common as detrital grains than sodium- and calcium-rich varieties as they are chemically more stable when subjected to weathering *(6.7.4)*.

MICA

The two commonest mica minerals, biotite and muscovite, are relatively abundant as detrital grains in sandstone, although muscovite is more resistant to weathering. They are derived from granitic to intermediate composition igneous rocks and from schists and gneisses where they have formed as metamorphic minerals. The platy shape of mica grains makes them distinctive in hand specimen and under the microscope. Micas tend to be concentrated in bands on bedding planes and often have a larger surface area than the other detrital grains in the sediment. This is because a platy grain has a lower settling velocity than an equant mineral grain of the same mass and volume *(4.2.5)* so micas stay in temporary suspension longer than quartz or feldspar grains of the same mass.

HEAVY MINERALS

The common minerals found in sands have densities of around 2.6 or $2.7 \, \text{g cm}^{-3}$: quartz has a density of $2.65 \, \text{g cm}^{-3}$, for example. Most sandstones contain a small proportion, commonly less than 1%, of minerals which have a greater density. These *heavy minerals* have densities greater than $2.85 \, \text{g cm}^{-3}$ and are traditionally separated from the bulk of the lighter minerals by using a liquid of that density in which the common minerals will float but the small proportion of dense minerals will sink. These minerals are rarely seen in hand specimen and are occasionally seen in thin sections of sandstone. Usually it is possible to study them only after concentrating them by dense liquid separation. The reason for studying them is that they can be characteristic of a particular source area and are therefore valuable for studies of the sources of detritus *(5.5)*. Common heavy minerals include zircon, tourmaline, rutile, apatite, garnet and a range of other metamorphic and igneous accessory minerals.

MISCELLANEOUS MINERALS

Other minerals rarely occur in large quantities in sandstone. Most of the common minerals in igneous silicate rocks (e.g. olivine, pyroxenes and amphiboles) are all too readily broken down by chemical weathering. Oxides of iron are relatively abundant. Local concentrations of a particular mineral may occur when there is a nearby source.

2.4.3 Lithic fragments

The breakdown of pre-existing, fine- to medium-grained igneous, metamorphic and sedimentary rocks results in sand-sized fragments. Sand-sized lithic fragments are only found of fine- to medium-grained rocks because by definition the mineral crystals and grains of a coarser-grained rock type are the size of sand grains or larger. Determination of the lithology of these fragments of rock usually requires thin section examination to identify the mineralogy and fabric *(3.9)*.

Grains of igneous rocks such as basalt and rhyolite are susceptible to chemical alteration at the Earth's surface and are only commonly found in sands formed close to the source of the volcanic material. Beaches around volcanic islands such as Hawaii are black in places, being made up almost entirely of lithic grains of basalt. Sandstone of this sort of composition is rare in the stratigraphic record, but grains of volcanic rock types may be common in sediments deposited in basins related to volcanic arcs or rift volcanism (Chapter 23).

Fragments of schists and pelitic (fine-grained) metamorphic rocks can be recognized under the microscope by the strong aligned fabric which these lithologies possess: pressure during metamorphism results in mineral grains becoming reoriented or growing into an alignment perpendicular to the stress field. Micas most clearly show this fabric, but quartz crystals in a metamorphic rock may also display a strong alignment. Rocks formed by the metamorphism of quartz-rich lithologies break down into relatively resistant grains which can be incorporated into a sandstone.

Lithic fragments of sedimentary rocks are generated when pre-existing strata are uplifted, weathered and eroded. Sand grains can be reworked by this process and individual grains may go through a number of cycles of erosion and redeposition *(6.6)*. Finer-grained mudrock lithologies may break up to form sand-sized grains, although their resistance to further breakdown during transport is largely dependent on the degree of induration of the mudrock *(17.2)*. Pieces of limestone are commonly found as lithic fragments in sandstone although a rock made up largely of calcareous grains would be classified as a limestone *(3.1)*. One of the most common lithologies seen as sand grains is chert *(3.4)* which, being silica, is a resistant material.

2.4.4 Biogenic particles

Small pieces of calcium carbonate found in sandstone are commonly broken shells of molluscs and other organisms which have calcareous hard parts. Such fragments are common in sandstone deposited in shallow marine environments where these organisms are most abundant. If these calcareous fragments make up over 50% of the bulk of the rock it would be considered to be a limestone (the nature and occurrence of calcareous biogenic fragments is described in the next chapter: *3.1.2*). Fragments of bone and teeth may be found in sandstones from a wide variety of environments but are rarely common. Wood, seeds and other parts of land plants may be preserved in sandstone deposited in continental and marine environments.

2.4.5 Authigenic minerals

Minerals which grow crystals in a depositional environment are called *authigenic* minerals. They are distinct from all the detrital minerals which formed by igneous or metamorphic processes and were subsequently reworked into the sedimentary realm. Many carbonate minerals form authigenically, and another important mineral formed in this way is *glauconite*, a green iron silicate which forms in shallow marine environments. Glauconite is a useful indicator of the environment of deposition *(11.6.1)*. It forms preferentially when rates of sedimentation are slow, and hence is useful in stratigraphic analysis *(21.2.4)*, and because it forms within the depositional environment, a radiometric date from glauconite crystals can be used to determine the age of the deposit *(20.1)*.

2.4.6 Mineral and clast durability

The durability of a grain is a measure of its tendency to remain unaltered during erosion, transport and deposition. Minerals such as quartz and lithic fragments of chert are resistant because they are least affected by the physical and chemical processes at the surface of the Earth. Feldspars, micas, other rock-forming silicate minerals and lithic fragments tend to break down and are therefore less resistant.

2.4.7 Sandstone nomenclature and classification

Full description of a sandstone usually includes some

information concerning the types of grain present. Informal names such as 'micaceous sandstone' are used when the rock clearly contains a significant amount of a distinctive mineral, in this case mica. Terms such as 'calcareous sandstone' and 'ferruginous sandstone' may also be used to indicate a particular chemical composition, in these cases a noticeable proportion of calcium carbonate and iron, respectively. These names for a sandstone are useful and appropriate for field and hand specimen descriptions, but when a full petrographic analysis is possible with a thin section of the rock under a microscope, a more formal nomenclature is used. This is usually the Pettijohn (1975) classification scheme (Fig. 2.6).

The Pettijohn sandstone classification combines textural criteria (the proportion of muddy matrix) with compositional criteria (the percentages of the three commonest components of sandstone: quartz, feldspar and lithic fragments). The triangular plot has these three components as the end members to form a 'QFL' triangle, which is commonly used in clastic sedimentology. To use this scheme for sandstone classification, the relative proportions of quartz, feldspar and lithic fragments must first be determined by visual estimation or by counting grains under a microscope: other components, such as mica or biogenic fragments, are disregarded. The third dimension of the classification diagram is used to display the texture of the rock, the relative proportions of clasts and matrix. In a sandstone

the matrix is the silt and clay material which was deposited with the sand grains. The second stage is therefore to measure or estimate the amount of muddy matrix: if the amount of matrix present is less than 15% the rock is called an *arenite*; between 15% and 75% it is a *wacke*; and if most of the volume of the rock is fine-grained matrix it is classified as a mudstone *(2.5)*.

Quartz is the most common grain type present in most sandstones, so this classification emphasizes the presence of other grains. Only 25% feldspar need be present for the rock to be called a *feldspathic arenite*, *arkosic arenite* or *arkose* (these three terms are interchangeable when referring to sandstone rich in feldspar grains). By the same token, 25% of lithic fragments in a sandstone make it a *lithic arenite* by this scheme. Over 95% of quartz must be present for a rock to be classified as a *quartz arenite*; sandstone with intermediate percentages of feldspar or lithic grains is called subarkosic arenite and sublithic arenite. Wackes are similarly divided into quartz wacke, feldspathic (arkosic) wacke and lithic wacke, but without the subdivisions. If a grain type other than the three main components is present in significant quantities (at least 5% or 10%), a prefix may be used such as 'micaceous quartz arenite': note that such a rock would not necessarily contain 95% quartz as a proportion of *all* the grains present, just 95% of the quartz, feldspar and lithic fragments when they are added together.

The term *greywacke* is sometimes used for a

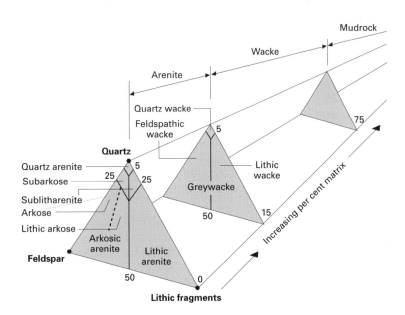

Fig. 2.6 The Pettijohn classification of sandstones, often referred to as a 'Toblerone plot'. (After Pettijohn 1975.)

sandstone which might also be called a feldspathic or lithic wacke. Greywackes are typically mixtures of rock fragments, quartz and feldspar grains with a matrix of clay- and silt-sized particles.

2.5 Clay, silt and mudrock

Fine-grained terrigenous clastic sedimentary rocks tend to receive less attention than any other group of deposits despite the fact that they are volumetrically the most common *(2.1)* of all sedimentary rock types. The grain size is generally too small for optical techniques of mineral determination, and until scanning electron microscopes (SEMs) and X-ray diffraction analysis techniques were developed *(2.5.4)* little was known about the constituents of these sediments. In the field mudrocks do not often show the clear sedimentary and biogenic structures seen in coarser clastic rocks and limestone. Exposure is commonly poor because they do not generally form steep cliffs, and soils readily form supporting vegetation which covers the outcrop. This group of sediments therefore tends to be overlooked but, as will be seen in later sections concerning depositional environments and stratigraphy, they can provide as much information as any other sedimentary rock type.

2.5.1 Definitions of terms in mudrocks

Clay is a textural term to define the finest grade of clastic sedimentary particles, those less than 4 µm in diameter. Individual particles are not discernible to the naked eye and can only just be resolved with a high-power optical microscope. *Clay minerals* are a group of phyllosilicate minerals which are the main constituents of clay-sized particles. *Silt* is the name given to material consisting of particles between 4 and 62 µm in diameter (Fig. 2.2). This size range is subdivided into coarse, medium, fine and very fine. The coarser grains of silt are just visible to the naked eye or with a hand lens. Finer silt is most readily distinguished from clay by touch, as it will feel 'gritty' if a small amount is ground between teeth, whereas clay feels smooth.

When clay- and silt-sized particles are mixed in unknown proportions as the main constituents in unconsolidated sediment we call this material *mud*. The general term *mudrock* can be applied to any indurated sediment made up of silt and/or clay. If it can be determined that most of the particles (over two-thirds) are clay-sized the rock may then be called a *claystone*,

and if silt is the dominant size a *siltstone*; mixtures of more than one-third of each component are referred to as *mudstone* (Folk 1974; Blatt *et al.* 1980). The term *shale* is sometimes applied to any mudrock (e.g. by drilling engineers) but it is best to use this term only for mudrocks which show a *fissility*, that is a strong tendency to break in one direction, parallel to the bedding. (Note the distinction between shale and slate: the latter is a term used for fine-grained metamorphic rocks which break along one or more cleavage planes.)

2.5.2 Silt and siltstone

The mineralogy and textural parameters of silt are more difficult to determine than for sandstone because of the small particle size. Only coarser silt grains can be easily analysed using optical microscope techniques. Resistant minerals are most common at this size because other minerals will often have been broken down chemically before they are physically broken down to this size. Quartz is the most common mineral seen in silt deposits. Other minerals occurring in this grade of sediment include feldspars, muscovite, calcite and iron oxides, amongst many other minor components. Silt-sized lithic fragments are only abundant in the 'rock flour' formed by glacial erosion *(7.2.1)*.

In aqueous currents silt remains in suspension until the flow slows down to be sluggish or almost stationary. Silt deposition is therefore characteristic of low-velocity flows or standing water with little wave action *(4.2.4)*. Silt-sized particles can remain in suspension in air as dust for long periods and may be carried high into the atmosphere. Strong persistent winds can carry silt-sized dust thousands of kilometres and deposit it as laterally extensive sheets (Pye 1987). Wind-blown silt forming *loess* deposits appears to have been important during glacial periods *(7.3.4, 24.7.4)*.

2.5.3 Clay minerals

Clay minerals commonly form as breakdown products of feldspars and other silicate minerals. They are phyllosilicates with a layered crystal structure similar to that of micas, and compositionally they are aluminosilicates. The layers are made up of silica with aluminium and magnesium ions, with oxygen atoms linking the sheets (Fig. 2.7). Two patterns of layering occur, one with two layers (the *kandite group*) and the other with three layers (the *smectite group*). Of the many different clay minerals

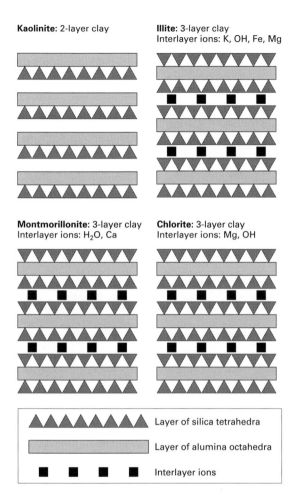

Kaolinite: 2-layer clay

Illite: 3-layer clay
Interlayer ions: K, OH, Fe, Mg

Montmorillonite: 3-layer clay
Interlayer ions: H₂O, Ca

Chlorite: 3-layer clay
Interlayer ions: Mg, OH

Layer of silica tetrahedra

Layer of alumina octahedra

Interlayer ions

Fig. 2.7 The structure of clay minerals. (After Tucker 1991.)

which occur in sedimentary rocks the four most common (Tucker 1991) are considered here (Fig. 2.7).

Kaolinite is the commonest member of the kandite group and is generally formed in soil profiles in warm, humid environments where acid waters intensely leach bedrock lithologies such as granite. Clay minerals of the smectite group include the expandable or *swelling clays* such as *montmorillonite* which can absorb water within their structure. Montmorillonite is a product of more moderate temperature conditions in soils with neutral to alkaline pH. It also forms under alkaline conditions in arid climates. Another three-layer clay mineral is *illite* which is related to the white mica muscovite. It is the most common clay mineral in sediments forming in soils in temperate areas where leaching is limited. *Chlorite* is a three-layer clay mineral which forms most commonly

in soils with moderate leaching under fairly acidic groundwater conditions and in soils in arid climates. Montmorillonite, illite and chlorite all form as a weathering product of volcanic rocks, particularly volcanic glass.

2.5.4 Clay mineral petrography

Identification and interpretation of clay minerals requires a higher-technology approach than is needed for coarser sediment. There are two principal techniques: scanning electron microscopy and X-ray diffraction pattern analysis (Tucker 1988). An image from a sample under an SEM is generated from secondary electrons produced by a fine electron beam which scans the surface of the sample. Features only micrometres across can be imaged by this technique, providing much higher resolution than is possible under an optical microscope. It is therefore used for investigating the form of clay minerals and their relationship to other grains in a rock. The distinction between clay minerals deposited as detrital grains and those formed diagenetically *(17.4.2)* within the sediment can be most readily made using an SEM.

An X-ray diffractometer operates by firing a beam of X-rays at a powder of a mineral or disaggregated clay and determining the angles at which the radiation is diffracted by the crystal lattice. The pattern of intensity of diffracted X-rays at different angles is characteristic of particular minerals and can be used to identify the mineral(s) present. Analysis by X-ray diffractometer is a relatively quick and easy method of semi-quantitatively determining the mineral composition of a fine-grained sediment. It is also used to distinguish certain carbonate minerals *(3.1.1)* which have very similar optical properties.

2.5.5 Clay particle properties

Because of their small size and platy shape, clay minerals remain in suspension in quite weak fluid flows and only settle out when the flow is very sluggish or the fluid still. Clay particles are therefore present as suspended load in most currents of water and air, and are only deposited when the flow ceases.

Once they come into contact clay particles tend to stick together—they are *cohesive*. This cohesion can be considered to be partly due to a thin film of water between two small platy particles having a strong surface tension effect (in much the same way as two plates of glass can be held together by a thin film of water between them) but is also a consequence of an electrostatic effect between clay minerals charged due to incomplete bonds in the mineral structure. As a result of these cohesive properties, clay minerals in suspension tend to *flocculate* and form small aggregates of individual particles. These flocculated groups have a greater settling velocity than individual clay particles and will be deposited out of suspension more rapidly. Flocculation is enhanced by saline water conditions and a change from fresh to saline water deposition (e.g. at the mouth of a delta or in an estuary: *12.1, 12.7*) results in the flocculation of clay particles. Once clay particles are deposited the cohesion makes them resistant to remobilization in a flow (Fig. 4.6). This allows deposition and preservation of fine sediment in areas which experience intermittent flows.

2.6 Description of the textures of terrigenous clastic sedimentary rocks

The shapes of clasts, their degree of sorting and the proportions of clasts and matrix are all aspects of the *texture* of the material. A number of terms are used in the petrographic description of the texture of terrigenous clastic sediments and sedimentary rocks.

CLASTS AND MATRIX

The fragments which make up a sedimentary rock are called 'clasts'. They may range in size from silt through sand to gravel (granules, pebbles, cobbles and boulders). A distinction is usually made between the clasts and the matrix, the latter being finer-grained material which lies between the clasts. There is no absolute size range for the matrix: the matrix to a sandstone may be silt- and clay-sized material, whereas the matrix of a conglomerate may be sand, silt or clay.

SORTING

This is a description of the distribution of clast sizes present: a well-sorted sediment is composed of clasts which fall for the most part in one class on the Wentworth scale (e.g. medium sand); a poorly sorted deposit contains a wide range of clast sizes. Sorting is a function of the origin and transport history of the detritus. With increased transport distance or repeated agitation of a sediment, the different sizes tend to become separated. A visual estimate of the sorting may be made

Very well sorted	'Standard deviation' < 0.35
Well sorted	= 0.35–0.5
Moderately well sorted	= 0.5–0.71
Moderately sorted	= 0.71–1.0
Poorly sorted	= 1.0–2.0
Very poorly sorted	> 2.0

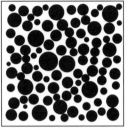

'Standard deviation' = 0.35

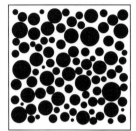

'Standard deviation' = 0.5

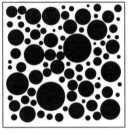

'Standard deviation' = 1.0

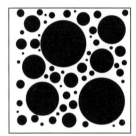

'Standard deviation' = 2.0

Fig. 2.8 Sorting estimate comparison chart. (After Harrell 1984; courtesy of the Society for Sedimentary Geology.)

by comparison with a chart (Fig. 2.8) or calculated from grain size distribution data *(2.7)*.

CLAST ROUNDNESS

During transport of sediment the individual clasts will repeatedly come into contact with each other and stationary objects, resulting in abrasion. Sharp edges tend to be chipped off first, smoothing the surface of the clast. A progressive rounding of the edges occurs with prolonged agitation of the sediment, and hence the roundness is a function of the transport history of the material. Roundness is normally estimated visually (Fig. 2.9), but may also be calculated from the cross-sectional shape of a clast.

CLAST SPHERICITY

In describing individual clasts, the dimensions can be considered in terms of closeness to a sphere (Fig. 2.9).

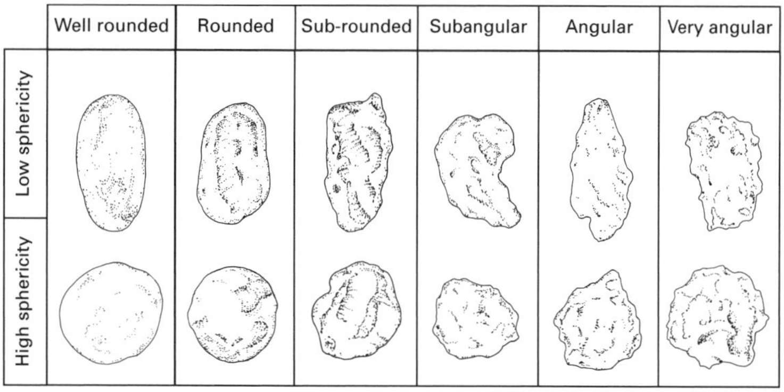

	Well rounded	Rounded	Sub-rounded	Subangular	Angular	Very angular
Low sphericity						
High sphericity						

Fig. 2.9 Roundness and sphericity estimate comparison chart. (After Pettijohn *et al.* 1987.)

Discoid or needle-like clasts have a low sphericity. Sphericity is an inherited feature — that is, it depends on the shapes of the fragments which formed during weathering. A slab-shaped clast will become more rounded during transport and become disc-shaped, but will generally retain its form, with one axis much shorter than the other two.

FABRIC

If a rock has a tendency to break in a certain direction, or shows a strong alignment of elongate clasts, this is described as the *fabric* of the rock. Mudrock which breaks in a platy fashion is considered to have a shaly fabric (and may be called a shale), and sandstone which similarly breaks into thin slabs is sometimes referred to as being 'flaggy'. Fabrics of this type are due to an anisotropy in the arrangement of particles: a rock with an isotropic fabric would not show any preferred direction of fracture because it consists of evenly and randomly orientated particles.

2.7 Granulometric and clast shape analysis

Quantitative assessment of the percentages of different grain sizes in clastic sediments and sedimentary rocks is called *granulometric analysis*. These data and measurements of the shape of clasts can be used in the description and interpretation of clastic sedimentary material (see Lewis & McConchie 1994).

2.7.1 Techniques in granulometric analysis

The techniques used will depend on the grain size of the material examined. Gravels are normally assessed by direct measurement in the field. A quadrant is laid over the loose material or on a surface of the conglomerate, and each clast measured within the area of the quadrant. The size of quadrant required will depend on the approximate size of the clasts: a metre square is appropriate for pebble- and cobble-sized material.

A sample of unconsolidated sand is collected or a piece of sandstone disaggregated by mechanical or chemical breakdown of the cement. The sand is then passed through a stack of sieves which have meshes at intervals of half or one unit on the ϕ scale *(2.2.2)*. All the sand which passes through the $500 \mu m$ ($\phi = 1$) mesh sieve but is retained by the $250 \mu m$ ($\phi = 2$) mesh sieve will have the size range of medium sand. By weighing the contents of each sieve the distribution by weight of different size fractions can be determined.

It is not practicable to sieve material finer than coarse silt, so the proportions of clay- and silt-sized material are determined by other means. Most laboratory techniques employed in the granulometric analysis of silt- and clay-size particles are based on settling velocity relationships predicted by Stokes' law *(4.2.5)*. A variety of methods using settling tubes and pipettes have been devised (Krumbein & Pettijohn 1938; Lewis & McConchie 1994), all based on the principle that particles of a given grain size will take a predictable period of time to settle a certain distance in a water-filled tube. Samples are siphoned off at time intervals, dried and weighed to determine the proportions of different clay and silt size ranges. These settling techniques do not fully take into account the effects of grain shape or density on settling velocity and care must be taken in comparing the results of these analyses with grain size distribution data obtained from more sophisticated techniques such as the Coulter counter, which determines grain size on the basis of the electrical properties of grains suspended in a fluid.

The results from the analyses are plotted in one of three forms: either a histogram of the weight percentages of each of the size fractions, a frequency curve or a cumulative frequency curve (Fig. 2.10). Note in each case that the coarse sizes plot on the left and the finer material on the right of the graph. Each provides a graphic representation of the grain size distribution, enabling a value for the mean grain size and *sorting* (standard deviation from a normal distribution) to be calculated. Other values which can be calculated are the *skewness* of the distribution, an indicator of whether the grain size histogram is symmetrical or is skewed to a higher percentage of coarser or finer material; and the *kurtosis*, a value which indicates whether the histogram has a sharp peak or a flat top (Pettijohn 1975; Lewis & McConchie 1994).

2.7.2 Use of granulometric analysis results

The grain size distribution is determined to some extent by the processes of transport and distribution. Glacial sediments are normally very poorly sorted, river sediments moderately sorted and both beach and aeolian deposits are often well sorted. The reasons for these differences are discussed in later chapters. In most circumstances the general sorting characteristics can be assessed in a qualitative way, and there are many other features such as sedimentary structures which would allow the deposits of different environments to be distinguished. A quantitative granulometric analysis is

therefore often unnecessary and may not provide much more information than is evident from other, quicker observations.

Moreover, determination of environment of deposition from granulometric data can be misleading under circumstances where material has been reworked from older sediment. A river transporting material eroded from an outcrop of older sandstone formed in an aeolian environment will deposit very well-sorted material. The grain size distribution characteristics would indicate deposition by aeolian processes, but the more reliable field evidence would better reflect the true environment of deposition from sedimentary structures and facies associations *(5.2)*.

Granulometric analysis provides quantitative information when a comparison of the character is required from sediments deposited within a known environment, such as a beach or along a river. It is therefore most commonly used in the analysis and quantification of present-day processes of transport and deposition.

2.7.3 Clast shape analysis

Attempts have been made to relate the shape of pebbles to the processes of transport and deposition. Analysis is carried out by measuring the longest, shortest and intermediate axes of a clast and calculating an index for its shape (approaching a sphere, a disc or a rod: Fig. 2.4). Whilst there may be some circumstances where clasts are sorted according to their shape, the main control on the shape of a pebble is the shape of the material eroded from the bedrock in the source area. If a rock breaks up into cubes after transport the rounded clasts will be spherical, and if the bedrock is thinly bedded and breaks

Fig. 2.10 Histogram, frequency distribution and cumulative frequency curves of grain size distribution data.

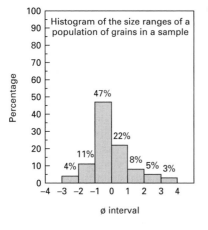

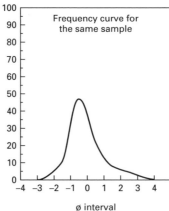

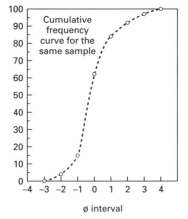

up into slabs the resulting clasts will be discoid. No amount of rounding of the edge of a clast will change its fundamental dimensions. Clast shape analysis is therefore most informative about the character of the rocks in the source area and provides little information about the depositional environment.

2.8 Maturity of terrigenous clastic material

A terrigenous clastic sediment or sedimentary rock can be described as having a certain degree of *maturity*. This refers to the extent to which the material has changed when compared to the starting material of the bedrock from which it was derived. Maturity can be measured in terms of texture and composition. Normally a compositionally mature sediment is also texturally mature but there are exceptions—for example, on a beach around a volcanic island where only mineralogically unstable components (basaltic rock and minerals) are available but the texture reflects an environment where there has been prolonged movement and grain abrasion by the action of waves and currents.

2.8.1 Textural maturity

The texture of sediment or sedimentary rock can be used to indicate something about the erosion, transport and depositional history. The determination of the *textural maturity* of a sediment or sedimentary rock can best be represented by a flow diagram (Fig. 2.11). Using this particular scheme for assessing maturity, any sandstone

Fig. 2.11 Flow diagram of the determination of the textural maturity of a terrigenous clastic sediment or sedimentary rock.

that is classified as a wacke would be considered texturally immature. Arenites can be subdivided on the basis of the sorting and shape of the grains. If sorting is moderate to poor the sediment is considered to be sub-mature, whilst well-sorted or very well-sorted sands are considered mature if the individual grains are angular to sub-rounded, and supermature if rounded to well rounded. The textural classification of the maturity is independent of composition of the sands. An assessment of the textural maturity of a sediment is most useful when comparing material derived from the same source as it may be expected that the maturity will increase as the amount of energy that has been input increases. For example, maturity often increases downstream in a river, and once the same sediment reaches a beach the high wave energy will further increase the maturity. Care must be taken when comparing sediments from different sources as they are likely to have started with different grain size and shape distributions and are therefore not directly comparable.

2.8.2 Mineralogical maturity

A distinction may be drawn between mineralogical maturity, which is strongly influenced by the composition of the source rock area, and textural maturity, which is more related to the history of transport and deposition. *Mineralogical* or *compositional maturity* is a measure of the proportion of resistant or stable minerals present in the sediment. The proportion of highly resistant clasts (such as quartz and siliceous lithic fragments) in a sandstone is compared to the amount of less resistant clast types present (such as feldspars, most other mineral types and lithic clasts) when assessing compositional maturity. A sandstone is compositionally mature if the

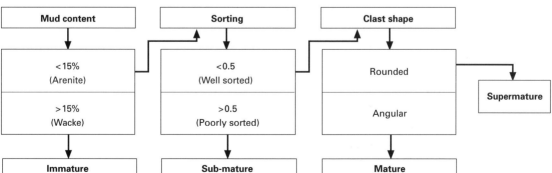

proportion of quartz grains is very high and it is a quartz arenite according to the Pettijohn classification scheme (Fig. 2.6); if the ratio of quartz, feldspar and lithic fragments means that the composition falls in the lower part of the triangle it is a mineralogically immature sediment.

2.8.3 Cycles of sedimentation

Mineral grains and lithic clasts eroded from an igneous rock, such as granite, are transported by a variety of processes (Chapter 4) to a point where they are deposited to form an accumulation of clastic sediment. Material formed in this way is referred to as a *first cycle* deposit because there has been one cycle of erosion, transport and deposition. Once this sediment has been lithified into sedimentary rock, it may subsequently be uplifted by tectonic processes and be subject to erosion, transport and redeposition. The redeposited material is considered to be a *second cycle* deposit as the individual grains have gone through two *cycles of sedimentation*. Clastic sediment may go through many cycles of sedimentation, and each time the mineralogical and textural maturity of the clastic detritus increases. The only clast types which survive repeated weathering, erosion, transport and redeposition are resistant minerals such as quartz and lithic fragments of chert. Certain heavy minerals such as zircon *(2.4.2)* are also extremely resistant and the degree to which zircon grains are rounded may be used as an index of the number of cycles of sedimentation to which material has been subjected.

2.9 Terrigenous clastic sediments: summary

Terrigenous clastic gravels, sands and muds are widespread modern sediments and are found abundantly as conglomerate, sandstone and mudrock in successions of sedimentary rocks. They are composed chiefly of the products of the breakdown of bedrock *(6.5)* and may be transported by a variety of processes *(4.1)* to depositional environments *(5.1)*. The main textural and compositional features of sand and gravel can be readily determined in the field and in hand specimen. This makes it possible to determine much about the origin and history of the material without requiring sophisticated laboratory techniques. Investigation of mudrocks depends on sub-microscopic and chemical analysis of the material. Sedimentary structures formed in clastic sediments *(4.3)* provide further information about the conditions under which the material was deposited; this information is the key to the palaeoenvironmental analysis discussed in later chapters of this book.

Further reading

Blatt, H. (1982) *Sedimentary Petrology*, 564 pp. W.H. Freeman, New York.
Blatt, H., Middleton, G.V. & Murray, R.C. (1980) *Origin of Sedimentary Rocks*, 2nd edn, 782 pp. Prentice Hall, Englewood Cliffs, NJ.
Chamley, H. (1989) *Clay Sedimentology*, 623 pp. Springer-Verlag, Berlin.
Leeder, M.R. (1982) *Sedimentology, Process and Product*, 344 pp. Unwin Hyman, London.
Lewis, D.G. & McConchie, D. (1994) *Analytical Sedimentology*, 197 pp. Chapman & Hall, London.
Pettijohn, F.J. (1975) *Sedimentary Rocks*, 3rd edn, 718 pp. Harper and Row, New York.
Pettijohn, F.J., Potter, P.E. & Siever, R. (1987) *Sand and Sandstone*, 553 pp. Springer-Verlag, New York.
Tucker, M.E. (1991) *Sedimentary Petrology*, 2nd edn, 260 pp. Blackwell Scientific Publications, Oxford.

3 Biogenic, chemical and volcanogenic sediments

In areas where there is not a large supply of clastic detritus other processes are important in the accumulation of sediments. The hard parts of plants and animals, ranging from microscopic algal to vertebrate bones, make up deposits in many different environments. Of greatest significance are the many organisms which build shells and structures of calcium carbonate in life and leave behind these hard parts when they die as calcareous sediments which form limestone. Chemical processes also play a part in the formation of limestone but are most important in the generation of evaporites which are precipitated out of waters concentrated in salts. Volcaniclastic sediments are largely the products of primary volcanic processes of generation of ashes and deposition of them subaerially or under water. In areas of active volcanism these deposits can swamp out all other sediment types. Of the miscellaneous deposits which do not fit into the four main categories, most are primarily of biogenic origin (siliceous sediments, phosphates and carbonaceous deposits) whilst ironstones are chemical deposits. The final section of this chapter provides some guidance notes on the description of sedimentary rocks in hand specimen and under the microscope.

3.1 Limestone

Limestones are the second most abundant sedimentary rocks after terrigenous clastic rocks. They form from material made up of calcium carbonate deposited in a range of environments (Tucker & Wright 1990). Many limestones are composed of *biomineralized* calcium carbonate, formed as part of a living organism. In addition to this *biogenic* material there are also purely chemical precipitates and some deposits formed from a combination of biological and chemical processes (Fig. 2.1, Table 3.1).

3.1.1 Mineralogy

Mineralogically, calcium carbonate is either *calcite* (the trigonal crystal form) or *aragonite* (the orthorhombic crystal form). Aragonite is not stable at Earth surface temperatures and pressures and with time it recrystallizes to calcite. Other ions, principally magnesium, may substitute for calcium in the lattice of calcite crystals, and two forms of calcite are recognized — *low magnesium calcite* (with less than 4% magnesium) and *high magnesium calcite* (which may have 11–19% magnesium present). Of these two forms the low magnesium calcite is more stable, and primary high magnesium calcite eventually recrystallizes. Strontium may substitute for calcium in both calcite and aragonite, although in small quantities (less than 1%); it is important because of the use of strontium isotopes in dating rocks *(20.1.2)*. *Dolomite* is a different mineral, a calcium magnesium carbonate which forms almost exclusively as a replacement of calcite and aragonite *(17.5.2)*.

3.1.2 Biomineralized constituents of limestone

The constituents of calcium carbonate deposits range in size from mud particles a few micrometres across to large structures built by organisms such as coral colonies within a reef.

Skeletal fragments in carbonate sediments are whole or broken pieces of the hard body parts of organisms which have calcium carbonate minerals as part of their structure. Many of these are familiar organisms such as

Table 3.1 The principal components of carbonate rocks.

Skeletal fragments
Molluscs (cephalopods, bivalves, gastropods, etc.)
Brachiopods
Echinoids
Crinoids
Corals
Foraminifera

Algae
Red algae (encrusting)
Green algae (planktonic)
Yellow-green algae (sea grasses and lake algae)

Bioherms and biostromes
Corals
Bryozoa
Cyanobacteria (stromatolites)

Non-skeletal carbonate
Ooids
Pisoids
Peloids (faecal pellets)
Intraclasts
Lime mud

bivalves and gastropods which have hard shells which may accumulate as whole entities or be broken up into fragments which can nevertheless still be recognized as part of a particular animal.

The shells of *molluscs* (which include bivalves, gastropods and cephalopods) have a characteristic layered structure of fine crystals. The mineral is most commonly aragonite, and because of recrystallization the structure may not be seen in skeletal fragments in a sedimentary rock. Only certain molluscs — notably oysters, scallops and the guards of belemnites — have original calcite skeletons which are preserved in their original form. *Brachiopods* are also shelly organisms with an overall body morphology similar to bivalves. They are not common today but were very abundant in the Palaeozoic and Mesozoic. Their shells are made up of low magnesium calcite and in them a two-layer structure of fibrous crystals may be completely preserved.

Another group of shelly organisms, the *echinoids* (sea urchins), can be easily recognized because they construct their hard body parts out of whole low magnesium calcite crystals. Individual plates of echinoids are preserved in carbonate sediments. *Crinoids* (sea lilies) belong to the same phylum as echinoids (the Echino-

dermata) and are similar in the sense that they too construct their body parts out of whole calcite crystals, with the discs which make up the stem of a crinoid forming sizeable accumulations in Carboniferous sediments.

Foraminifera are small, single-celled marine animals which range from a few tens of micrometres to tens of millimetres in diameter. In life they are either floating (*planktonic*) or live on the sea floor (*benthic*), and most modern and ancient forms have hard outer parts (tests) made up of high or low magnesium calcite. Both modern sediments and ancient limestone beds have been found with huge concentrations of Foraminifera such that they may form the bulk of the sediment.

Some of the largest calcium carbonate biogenic structures are built by *corals* which may be in the form of colonies many metres across; other corals are solitary. Calcite seems to have been the main crystal form in Palaeozoic corals, with aragonite crystals making the skeleton in younger corals. Hermatypic corals have a symbiotic relationship with algae which require clear, warm, shallow marine waters. These corals form more significant build-ups than the less common, ahermatypic corals which do not have algae and can exist in colder, deeper water. Another group of colonial organisms which may contribute to carbonate deposits are the *Bryozoa*. These single-celled protozoans are seen mainly as encrusting organisms today but in the past they formed large colonies. Their structure is made up of aragonite, high magnesium calcite or a mixture of the two. Structures built by colonial organisms may be *bioherms* if they form mounds or *biostromes* if they form sheet-like bodies.

Algae and microbial organisms are an important source of biogenic carbonate and are the most important contributors of fine-grained sediment in many carbonate environments. Three types of alga are carbonate-producers. *Red algae* (Rhodophyta) are otherwise known as the coralline algae. Some forms are found encrusting surfaces such as shell fragments and pebbles. They have a layered structure and are effective at binding soft substrate. The *green algae* (Chlorophyta) have calcified stems and branches, often segmented, which contribute fine rods and grains of calcium carbonate to the sediment when the organisms die. *Nannoplankton*, planktonic algae which belong to the *yellow-green algae*, are extremely important contributors to marine sediments in parts of the stratigraphic record. This group, the Chrysophyta, includes *coccoliths* which are spherical bodies a few tens of micrometres across made up of plates. Coccoliths are an important constituent of pelagic limestone, including chalk *(15.5.1)*.

Cyanobacteria are now classified separately from algae. The algal mats formed by these organisms are more correctly called bacterial or microbial mats. In addition to sheet-like mats, columnar and domal forms are also known. The filaments and sticky surfaces of cyanobacteria act as traps for fine-grained carbonate, and as the structure grows it forms biostromes or bioherms called *stromatolites (13.4.3)*. *Oncoids* are irregular concentric structures, millimetres to centimetres across, formed of layers bound by cyanobacteria found as clasts within carbonate sediments. Other cyanobacteria bore into the surface of skeletal debris and alter the original structure of a shell into a fine-grained micrite (*micritization*).

3.1.3 Other constituents of limestone

A variety of other types of grain also occur commonly in carbonate sediments and sedimentary rocks (Fig. 3.1). *Ooids* are spherical bodies of calcium carbonate less than 2 mm in diameter. They have an internal structure of concentric layers which suggests that they form by the precipitation of calcium carbonate around the surface of the sphere. At the centre of an ooid lies a nucleus which may be a fragment of other carbonate material or a clastic sand grain. Accumulations of ooids form shoals in shallow marine environments today *(14.5)* and are components of limestone throughout the Phanerozoic. A

rock made up of carbonate ooids is an *oolitic limestone*. The origin of ooids has been the subject of much debate, and the present consensus is that they form by chemical precipitation out of agitated water saturated in calcium carbonate in warm waters (Tucker & Wright 1990). It is likely that bacteria also play a role in the process, especially in less agitated environments. Concentrically layered carbonate particles over 2 mm across are called *pisoids*. Pisoids are often more irregular in shape but are otherwise similar in form and origin to ooids. *Oncoids* are similar to pisoids and ooids but have an internal structure consisting of irregular, overlapping laminae of micrite.

Some round particles made up of fine-grained calcium carbonate found in sediments do not show any concentric structure and have apparently not grown in water in the same way as ooids or pisoids. These *peloids* are commonly the *faecal pellets* of marine organisms such as gastropods and may be very abundant in some carbonate deposits, mostly as particles less than a millimetre across.

Intraclasts are fragments of calcium carbonate material which has been partly lithified and then broken up and reworked to form a clast which is incorporated into the sediment. This commonly occurs where lime mud (see below) dries out by subaerial exposure in a mudflat and is then reworked by a current. A conglomerate of flakes of carbonate mud can be formed in this way. Other settings where clasts of lithified calcium carbonate occur are associated with reefs where the framework of the reef is broken up by wave or storm action *(14.7.2)* and redeposited.

Fine-grained calcium carbonate particles less than 4 μm across (cf. clay: *2.5*) are referred to as *lime mud*, *carbonate mud* or *micrite*. This fine material may be the result of purely chemical precipitation from water saturated in calcium carbonate, or of the breakdown of skeletal fragments, or have an algal or bacterial origin. The small size of the particles usually makes it impossible to determine the source. Lime mud is found in many carbonate-forming environments and can be the main constituent of limestone.

3.1.4 Classification of limestones

It is possible to classify most limestones in the same manner as terrigenous clastic rocks by using the sizes of the particles present as the principal criterion. The terms *calcilutite*, *calcarenite* and *calcirudite* are used in some

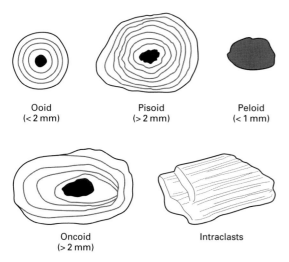

Ooid
(< 2 mm)

Pisoid
(> 2 mm)

Peloid
(< 1 mm)

Oncoid
(> 2 mm)

Intraclasts

Fig. 3.1 Non-skeletal components of carbonate sediments.

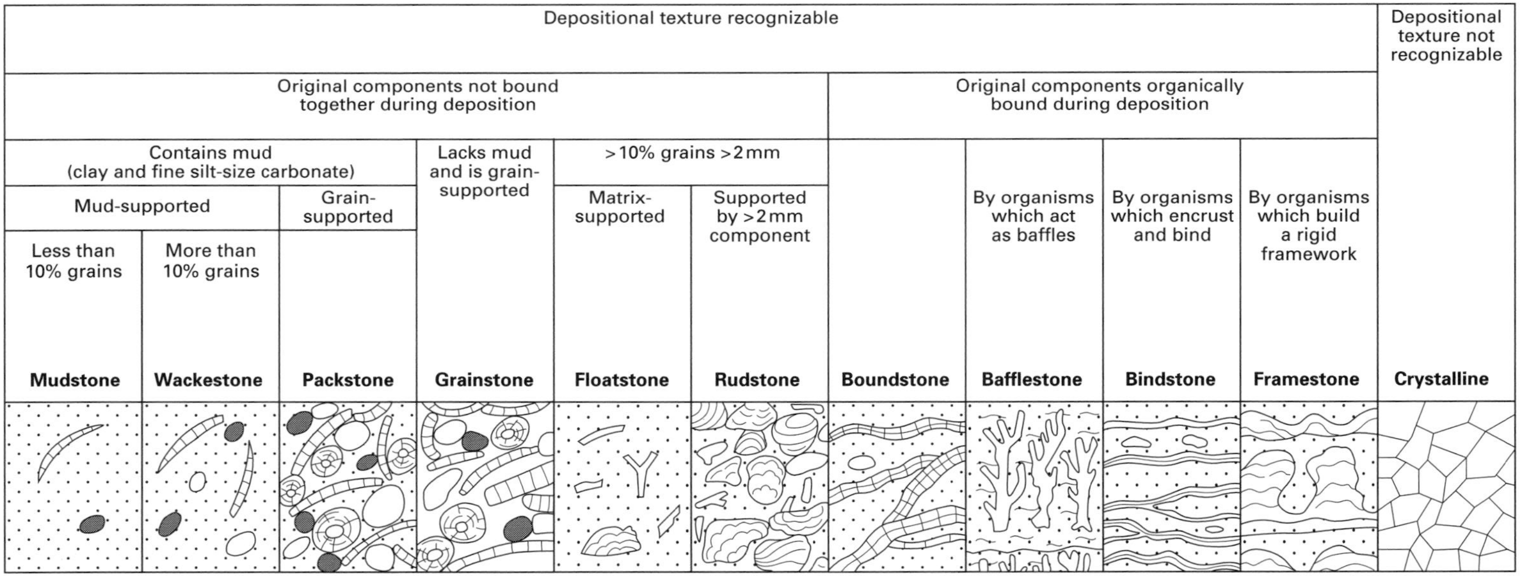

Fig. 3.2 The Dunham classification of carbonate sedimentary rocks (Dunham 1962; AAPG ©1962, reprinted by permission of the American Association of Petroleum Geologists) with modifications by Embry and Klovan (1971). This scheme is the most commonly used for description of limestones in the field and in hand specimen.

Table 3.2 Classification of carbonate rocks according to grain size.

Grain size	Name for carbonate rock
>2 mm	Calcirudite
63 μm–2 mm	Calcarenite
<63 μm	Calcilutite

circumstances to describe limestone made up principally of mud-sized material, sandy detritus and gravelly material, respectively (Table 3.2).

Other classification schemes for carbonates have developed which more usefully reflect the different ways in which a limestone may form. The most widely used scheme for the description of limestone in the field, in hand specimen and in thin section is the *Dunham classification* (Fig. 3.2). The criteria used in this classification scheme are principally textural—that is, the proportion of carbonate mud present and the framework of the rock—but the nature of the grains or framework material also forms part of the classification. The first stage in using the Dunham classification is to determine whether the fabric is matrix- or clast-supported. Matrix-supported limestone is divided into *carbonate mudstone* (less than 10% clasts) and *wackestone* (more than 10% clasts). If the limestone is clast-supported it is termed a *packstone* if there is mud present or a *grainstone* if there is no matrix. A *boundstone* has an organic framework such as a coral colony. The original scheme (Dunham 1962) did not include the subdivision of boundstone into *bafflestone, bindstone* and *framestone* which describes the type of organisms which build up the framework. These categories, along with the addition of *rudstone* (clast-supported limestone conglomerate) and *floatstone* (matrix-supported limestone conglomerate), were added by Embry and Klovan (1971)—see also James and Bourque (1992). Note that the terms 'rudstone' and 'floatstone' are used for carbonate *intraformational conglomerate*—that is, they are made up of material deposited in an adjacent part of the same environment and then redeposited (e.g. the breaking up of the front of a reef: *14.7.2*). These should be distinguished from conglomerate made up of clasts of limestone eroded from an older bedrock and deposited in a quite different setting, for example in a river or on an alluvial fan (*8.4*).

Using this combination of textural and compositional criteria, the name given to a limestone in the Dunham scheme provides information about the likely conditions under which the sediment formed: a coral boundstone forms under quite different conditions than a foraminiferal wackestone (*14.6, 14.7*). The *Folk classification* (Fig. 3.3) is an alternative scheme for description in thin section (Folk 1959). The sediment is described in terms of the nature of the main framework grains (ooids, bioclasts, intraclasts, etc.) and the material between the grains, which may be micrite or sparry cement. A name under this scheme provides more information about the diagenetic history of the rock (*17.5*) but less about the processes of deposition.

3.1.5 Environments of deposition of carbonate sediment

Carbonate sediments are largely the product of biogenic and biochemical processes. The hard parts of large organisms and carbonate precipitation associated with algae and bacteria provide large amounts of calcareous sediment, especially in shallow warm seas. They may build up in any location where there is a supply of biogenic carbonate and a restriction on the amount of clastic detritus. Most limestone beds formed as deposits in coastal and shallow marine environments (*13.4, 14.5*), although carbonate deposition also occurs in caves, springs, soils (*9.7*), lakes (*10.3.4*) and deep marine settings (*15.5.1*). Many of the organisms which make up carbonate rocks occur in specific environments (e.g. corals, benthic organisms, different types of algae) making it possible to determine quite precisely the environment of deposition of a limestone on the basis of its biogenic constituents. Even greater detail is available from microscopic examination of limestone.

3.2 Volcanic and volcaniclastic rocks

Volcanic eruptions are the most obvious and spectacular examples of the formation of both igneous and sedimentary rocks on the Earth's surface. During eruption volcanoes produce a range of materials, from molten rock, which forms a lava flowing from fissures in the volcano, to fine particulate material, which is ejected from the vent as volcaniclastic ash which falls as a sediment far away from the site of eruption (Cas & Wright 1987). Lavas and ash may occur in any depositional environment near a volcano and hence volcanic and volcaniclastic units may be found in association with a wide variety of sedimentary rocks. The location of volcanoes can be related to the plate tectonic setting, mainly in the vicinity of plate margins and other areas

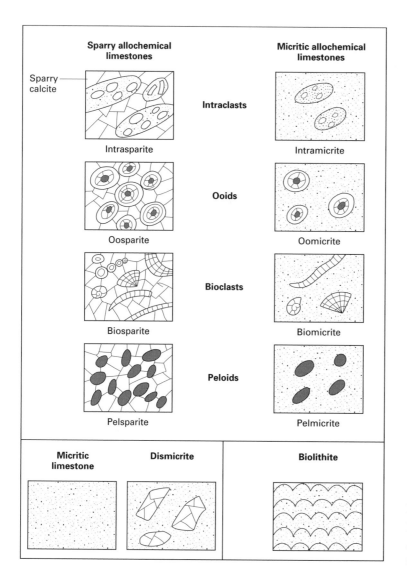

Fig. 3.3 The Folk classification scheme for limestones (Folk 1959, 1962; AAPG ©1959, reprinted by permission of the American Association of Petroleum Geologists) which is sometimes used in the description of limestone rocks in thin section.

of high heat flow in the crust. The presence of beds formed by volcanic processes can be an important indicator of the tectonic setting in which the sedimentary succession formed. Volcanic rocks are also of considerable value in stratigraphy as they may often be dated radiometrically, providing an absolute time constraint on the sedimentary succession *(20.1).*

3.2.1 Types of volcanic rock

Lava flows from craters or fissures result in sheets of volcanic rock when the molten magma cools and soli-

difies. These sheets may be tens of centimetres to tens of metres thick and cover areas kilometres to hundreds of kilometres across. The rock is crystalline, made up of interlocking crystals of minerals formed from the silicate melt, the molten rock in the magma chamber of the volcano. Lava cools relatively quickly, giving individual crystals little time to grow. Volcanic rocks are therefore characterized by small crystals, often too small to see with the naked eye. However, larger crystals may be present, formed by slower crystallization in the magma chamber and then carried out with the rest of the magma. The minerals present depend on the chemistry of the magma. A relatively low proportion of SiO_2 results in the crystallization of minerals such as olivine, pyroxene and plagioclase feldspar and the rock would be petrographically basaltic. With a higher percentage of SiO_2 the rock is compositionally rhyolitic (the fine-grained equivalent of a granite) containing quartz, micas and potassium feldspars.

The composition of the magma affects the style of eruption. Basaltic magmas tend to form volcanoes which produce large volumes of lava, but small amounts of volcanic ash. Mauna Loa in Hawaii is an example of a lava-dominated, basaltic volcano. In contrast, the Mount St Helens eruption in the USA involved more silicic magma and was much more explosive, with large amounts of the molten rock being ejected from the volcano as particulate matter. The particles ejected are known as *pyroclastic* material, also collectively referred to as *tephra*. Note that the term 'pyroclastic' is used for material ejected from the volcano as particles and 'volcaniclastic' refers to any deposit which is mainly composed of volcanic detritus (see also 'epiclastic': *6.5.4*). Pyroclastic material may be individual crystals, pieces of volcanic rock (lithic fragments), or *pumice*, the highly vesicular, chilled 'froth' of the molten rock. The size of this pyroclastic debris ranges from fine dust a few micrometres across to pieces which may be several metres across.

3.2.2 Nomenclature of volcaniclastic rocks

The textural classification of volcaniclastic deposits is a modification of the Wentworth scheme (Table 3.3). Coarse material (over 64 mm) is divided into *volcanic blocks* which were solid when erupted and *volcanic bombs* which were partly molten and have cooled in the air; consolidated into a rock, these are referred

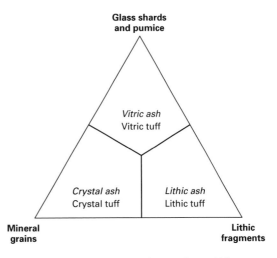

Fig. 3.4 Volcanic components and nomenclature. (After Pettijohn *et al.* 1987.)

to as *volcanic breccia* and *agglomerate*, respectively. Granule- to pebble-sized particles (2–64 mm) are called *lapilli* and form a *lapillistone*. Sand-, silt- and clay-grade tephra is *ash* when unconsolidated and *tuff* upon lithification. Coarse ash/tuff is sand-sized and fine ash/tuff silt- and clay-grade material. Compositional descriptions hinge on the relative proportions of crystals, lithic fragments and 'vitric' material, which is fragments of volcanic glass formed when the molten rock cools very rapidly, sometimes forming pumice (Fig. 3.4). The processes of transport and the environments of deposition of volcaniclastic sediments are considered further in Chapter 16.

3.3 Evaporite minerals

These are minerals formed by precipitation out of solution as ions become more concentrated when water evaporates. On average sea water contains $35\,g\,l^{-1}$ (grams per litre) of dissolved ions (Table 3.4). The chemistry of lake waters is variable, often with the same principal ions as sea water but in different proportions. The combination of anions and cations into minerals occurs as they become concentrated and the water saturated with respect to a particular compound. The least soluble compounds are precipitated first. Calcium carbonate is first precipitated out of sea water, followed by calcium sulphate and sodium chloride as the waters become more concentrated; potassium and magnesium

Table 3.3 The classification of volcanic sedimentary rocks.

Unconsolidated		Consolidated
Bombs Blocks	>64 mm	Agglomerate Volcanic breccia
Lapilli	2–64 mm	Lapillistone
Coarse ash	0.06–2 mm	Coarse tuff (Volcanic sandstone)
Fine ash	<0.06 mm	Fine tuff (Volcanic mudstone)

Table 3.4 The proportions of the principal ions in sea water of normal salinity and 'average' river water. (From Krauskopf 1979.)

Dissolved species	Sea water		River water (average)
	ppm	% of total	
Cl^-	18 000	55.05	7.8
Na^+	10 770	30.61	6.3
SO_4^{2-}	2715	7.68	11.2
Mg^{2+}	1290	3.69	4.1
Ca^{2+}	412	1.16	15.0
K^+	380	1.10	2.3
HCO_3^-	140	0.41	58.4
Br^-	67	0.19	0.02
H_3BO_3	26	0.07	0.1
Sr^{2+}	8	0.03	0.09
F^-	1.3	0.005	0.09
H_4SiO_4	1	0.004	13.1

chlorides precipitate out of very concentrated sea water. The order of precipitation of evaporite minerals from sea water and the relative amount are shown in Fig. 3.5.

The most commonly encountered evaporite minerals in sedimentary rocks are forms of calcium sulphate, as either *gypsum* or *anhydrite*. Calcium sulphate is precipitated from sea water once evaporation has concentrated the water to 19% of its original volume. Gypsum is the hydrous form of the mineral. It precipitates at the surface under all but the most arid conditions but may become dehydrated to anhydrite on burial (*17.6*). Anhydrite has no water in the crystal structure and forms

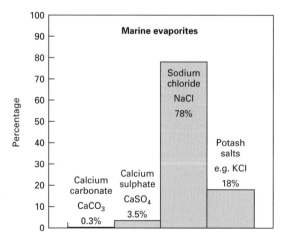

Fig. 3.5 The proportions of minerals precipitated by the evaporation of sea water of average composition. (Data from Krauskopf 1979.)

either by direct precipitation in arid shorelines (*13.5*) or as a result of alteration of gypsum. It may become hydrated to gypsum if water is introduced. Primary gypsum occurs as elongate crystals of *selenite* when it forms from precipitation out of water. If it forms as a result of the rehydration of anhydrite it has a fine crystalline form in nodules of *alabaster*. Gypsum also occurs as a fibrous form in secondary veins.

Halite precipitates out of sea water once it has been concentrated to less than 10% of its original volume. It may occur as thick crystalline beds or as individual crystals which have a distinctive cubic symmetry, sometimes with a stepped crystal face (a *hopper crystal*). The high solubility of sodium chloride means that it is only preserved in rocks in the absence of fresh groundwater which would dissolve it. Surface exposures of halite can be found in some arid regions where it is not removed by rainwater.

Other evaporite minerals are less common. The magnesium and potassium chlorides which form in the final stages of evaporation of sea water are so soluble that they are rarely preserved. Different evaporite minerals may occur in saline lakes (*10.4*), depending on the chemistry of the lake waters. They are mainly sulphates and carbonates of sodium and magnesium such as trona ($Na_2CO_3.NaHCO_3.2H_2O$), mirabilite ($Na_2SO_4.10H_2O$) and epsomite ($MgSO_4.7H_2O$).

3.4 Cherts

Cherts are fine-grained siliceous sedimentary rocks. They are hard, compact rocks which are made up of silt-sized quartz crystals (*microquartz*) and *chalcedony*, a form of silica which is made up of radiating fibres a few tens to hundreds of micrometres long. Beds of chert form either as primary sediments or by diagenetic processes.

On sea and lake floors the siliceous skeletons of microscopic organisms may accumulate to form a *siliceous ooze*. These organisms are diatoms in lakes and these may also accumulate in marine conditions, although radiolaria are more commonly the main components of marine siliceous oozes. Radiolarians are *zooplankton* (microscopic animals with a planktonic lifestyle) and diatoms are *phytoplankton* (free-floating plants and algae). Upon consolidation these oozes form beds of chert. The opaline silica of the diatoms and radiolaria is metastable and recrystallizes to chalcedonic silica or microquartz. Cherts formed from oozes are often thin-bedded with a layering caused by variations in the proportions of clay-sized material present. They are most common in deep ocean environments *(15.5.2)*.

Some cherts are diagenetic *(17.3.1)*, formed by the replacement of another mineral by waters rich in silica flowing through the rock. They commonly replace limestone (e.g. as *flint* in chalk) and sometimes occur in mudstone. They are in the form of nodules or irregular layers and can hence easily be distinguished from primary cherts. *Jasper* is chert with a strong red coloration due to the presence of haematite.

3.5 Phosphates

Sedimentary phosphate deposits are referred to as *phosphorites*. Phosphorus is a common element which is essential to life forms and present in all living matter. Mineralogically phosphorites are composed of a calcium phosphate, carbonate hydroxyl fluorapatite. High concentrations of sedimentary phosphorites are not common and they are found most frequently associated with shallow marine continental shelf deposits *(11.6.2)*. Phosphatic material in the form of bone, teeth and fish scales also occurs dispersed in many clastic and biogenic sedimentary rocks.

3.6 Sedimentary ironstone

The metal iron is a common element in sediments, occurring in traces in almost all deposits. Sedimentary rocks which contain at least 15% of the metal are referred to as *ironstone* and they attract attention because of their economic significance. The iron may be in the form of oxides, hydroxides, carbonate, sulphides or silicates (Berner 1971) (Table 3.5).

Iron is transported as hydroxides in colloidal suspension or it is attached to clay minerals and organic particles. Deposition occurs when the chemistry of the

Table 3.5 Common sedimentary ironstone minerals.

Oxides	Haematite	Fe_2O_3
	Magnetite	Fe_3O_4
Hydroxides	Goethite	$FeO.OH$
	Limonite	$FeO.OH.H_2O$
Carbonate	Siderite	$FeCO_3$
Sulphide	Pyrite	FeS_2
Silicates	Glauconite	$KMg(FeAl)(SiO_3)_6.3H_2O$
	Chamosite	$(Fe_5Al)(Si_3Al)O_{10}(OH)_8$

environment favours the precipitation of iron minerals. If there is a well-oxygenated environment *haematite*, an iron oxide, is the most common mineral formed, although under less oxidizing conditions the iron hydroxide *goethite* forms. Haematite is red to black in colour whilst the hydroxide is yellow to pale brown. It is likely that goethite forms as a precursor to haematite in desert environments, giving desert sands their yellowish colour. Further oxidation to haematite to give these sands the red colour seen in some ancient desert deposits may well be a post-depositional process.

Under reducing conditions the type of iron mineral formed depends on the availability of sulphide or sulphate ions. In sulphur-rich settings iron sulphide (*pyrite*) is common, occurring either as golden crystals or more commonly as finely disseminated particles which give the sediment a black colour. Fine-grained pyrite is found in reducing, organic-rich environments such as tidal mudflats and fetid lakes. In the absence of sulphides or sulphates the iron carbonate *siderite* may be precipitated: appropriate conditions for the formation of siderite occur mainly in non-marine muddy environments such as lakes and marshes. The authigenic mineral glauconite *(2.4.5)* is an iron silicate, as is *chamosite* which is a mineral found in some ironstone beds as ooids, although the mineral may be diagenetic in origin, occurring as a replacement of calcium carbonate.

3.7 Carbonaceous (organic) deposits

Sedimentary materials with a distinct proportion of organic matter are termed *carbonaceous* as they are rich in carbon (not to be confused with sediments rich in *carbonate* which, if calcium carbonate is the main constituent, are *calcareous*). Organic matter normally decomposes on the death of the organism and is only preserved if it partially breaks down to stable compounds. This only occurs under conditions of limited oxygen availability, referred to as *anaerobic*. Environments where this may happen are waterlogged mires, swamps and bogs, stratified lakes *(10.2.2)* and marine waters with restricted circulation.

3.7.1 Modern organic-rich deposits

Accumulation of organic material most obviously occurs in soils as *humus*, but because soils are commonly well oxygenated due to the burrowing activities of organisms there is little long-term preservation of this material in a soil profile. The wet conditions of mires, bogs and swamps are more favourable for the preservation of organic matter as they are more anaerobic settings and thick accumulations of *peat* may form. The composition of peat depends on the plant ecosystem, which may range from mosses in cool upland areas to trees in lowland fens and swamps. Peats are forming at present in a wide range of climatic zones, from sub-polar boggy regions in Siberia and Canada to mangrove swamps in the tropics (McCabe 1984; Hazeldine 1989). Thick peat deposits are most commonly associated with river floodplains *(9.3)* but also occur in the upper parts of deltas *(12.1)* and associated with coastal plains *(13.2.4)*. An accumulation made up of organic material with only small amounts of clastic detritus can only occur in places where there is little or no clastic input. Thick layers of pure peat will not form in environments which are regularly flooded by fresh or sea water carrying suspended sediment.

Not all accumulations of organic matter are made up of partly decomposed large plants. The remains of planktonic algae living in lakes and seas concentrate at the bottom of the water under anaerobic conditions. This aquatic organic material is called *sapropel* and may include spores and the finely comminuted detritus of larger plants.

3.7.2 Ancient organic-rich deposits

A deposit is considered to be organic-rich (carbonaceous) if it contains a proportion of organic material which is significantly higher than the averages of 2% for mudrock, 0.2% for limestone and 0.05% for sandstone.

The organic material may be present either because it was deposited with the sediment (as in the case of coal and oil shale) or because a fluid hydrocarbon has migrated from elsewhere to be concentrated in a porous sediment or rock. The latter are *hydrocarbon reservoirs* containing oil or natural gas which may be exploited if present in economic quantities *(17.8.2)*.

If over two-thirds of the material is solid organic matter it may be called a *coal*. Most economic coals have less than 10% non-organic, non-combustible material, often referred to as 'ash'. The carbonaceous material in a coal is not homogeneous and different types of organic matter can be recognized, known as *coal macerals* (Table 3.6) (McCabe 1984). The proportions of these macerals can be used to define a series of coal lithotypes (Table 3.6). *Humic coals* formed from the series of processes which convert peat into lignite and then coal *(17.8.2)* are composed mainly of *vitrain*, which is a shiny black form of coal. *Durain* has a dull lustre, and coals composed of alternating shiny and dull layers are called *clarain*. Soft, friable coaly material formed of inertinite macerals are called *fusain*: this occurs in a variety of clastic sedimentary rocks as well as in pure coals and in many cases can be clearly recognized as being fossil charcoal. *Sapropelic coals* are accumulations of algae, spores and fine plant material formed under water and buried. The coalification of carbonaceous matter into macerals and coal lithotypes takes place as a series of post-depositional bacteriological, chemical and physical processes *(17.8.2)*.

Mudrocks which contain a high proportion of organic material which can be driven off as a liquid or gas by

Table 3.6 Coal macerals and coal lithotypes.

Coal maceral groups

Vitrinite	Woody material (trunks, branches, leaves, roots)
Exinite	Spores, cuticles, resin, algae
Inertinite	Oxidized plant material including fusinite (charcoal)

Coal lithotypes

Humic coals: formed from macroscopic plant parts, organic matter accumulated in a soil profile as peat

Vitrain	Mainly vitrinite	Black, bright lustre
Durain	Exinite/inertinite	Grey/black, dull
Clarain	Finely layered vitrain and durain	
Fusain	Mainly fusinite	Black, friable and soft

Sapropelic coals: formed mainly from microscopic plant parts, mainly algal matter in water

Cannel coal	Fine particles	Black, dull, 'greasy'
Boghead coal	Alginite (exinite)	Black/brown, dull, 'greasy'

heating are called *oil shales*. The organic material is usually the remains of algae which have broken down during diagenesis *(17.8.2)* to form *kerogen*, long-chain hydrocarbons which form *petroleum* (natural oil and gas) when they are heated. Oil shales are therefore important *source rocks* of the hydrocarbons which ultimately form concentrations of oil and gas, although not all source rocks have a sufficiently high carbonaceous content to be called oil shales. The environments in which oil shales are formed must be anaerobic to prevent oxidization of the organic material; suitable conditions are found in lakes and certain restricted shallow marine environments (Eugster 1985).

3.8 The description of sedimentary rocks in hand specimen

The following points should be considered when a sedimentary rock is described in the field or in hand specimen. Additional, more detailed information can be obtained from thin sections of the rock (see below).

GENERAL PROPERTIES OF THE ROCK

This should include a description of the colour, the degree of consolidation or how well cemented the rock is and, if it is well lithified, how easily it breaks along parallel fractures (its *fissility*) and its fracture characteristics (e.g. the conchoidal fracture of a chert).

CONSTITUENTS AND TEXTURE

A more complete examination of the constituents can be made with a thin section of the rock, but a number of observations can be made from a hand specimen. If the clasts are large enough the mineral grains and lithic fragments present should be described and their proportions estimated. In sandstone and conglomerate, textural characteristics such as the grain size and sorting, roundness and sphericity of clasts can be determined. In limestone the nature of the clasts (ooids, bioclasts, intraclasts, etc.) and the presence of any framework-forming organisms are important. If a distinct matrix can be recognized in any sedimentary rock, its type and proportion should be noted.

SEDIMENTARY STRUCTURES AND FABRIC

Notes and sketches on this should include the type and spacing of any lamination, the scale and shape of cross lamination, cross bedding, sole structures, lineations, bioturbation, and so on (see the following chapters for descriptions of these features). Features such as grading, preferred orientation of grains and the relationship between the grains and the matrix should also be considered.

This information can be used to give the rock a name and to make some interpretations about the origin and depositional environment of the rock using information from Chapters 6 to 16. Note that interpretations should be made with care because the environment of deposition cannot always be determined from hand specimen and the depositional context, determined from field relationships, is almost always needed.

3.9 Examination of sedimentary rocks under the microscope

By examining sedimentary rocks under a petrographic microscope it is possible to determine many more details of the composition and texture than is possible by looking at a hand specimen (Cox *et al.* 1974; Adams *et al.* 1984). A thin slice of the rock (normally 30 μm thick) is mounted on to a microscope slide to make a *thin section*. At this thickness most silicate and carbonate minerals are transparent whilst many metal oxides and sulphides are opaque. The petrographic microscope can be used to determine a number of properties of the mineral grains which allow them to be identified as particular mineral types. For transparent minerals a source of light passed through a polarizing filter is directed through the thin section from below. A second polarizing filter orientated perpendicularly to the first can be inserted between the thin section and the eyepiece.

The main mineral properties considered are:
1 the shape of the mineral, although this may have been modified by erosion during transport;
2 the number of cleavage planes present, if any, and angles between them;
3 the refractive index of transparent minerals, which is qualitatively measured by determining the relief between the mineral and the cement used to stick the rock slice to the slide: minerals with a high refractive index have sharp edges, a high relief;
4 the colour in thin section and changes in that colour when the grains are rotated in the polarized light (this is called *pleochroism*);
5 the position, relative to the mineral outline, at which

the mineral is dark when both polarizers are inserted: the angle between this position and a particular crystal face, usually the longest edge, is the *extinction angle*;

6 the colours due to the distortion of polarized light as it passes through the mineral which is seen when both polarizers are used: the hue and intensity of these *birefringence* colours are useful characteristics in mineral identification.

In addition to the mineral identification, thin section examination of sedimentary rocks allows a number of textural properties of the rock to be determined.

1 The sorting, roundness and sphericity of sand grains *(2.6)* can be determined more readily in thin section than in hand specimen.

2 The fabric of the rock can be seen more clearly—that is, any tendency for grains to be aligned in a particular direction, layering on a fine scale, and so on.

3 Any post-depositional, diagenetic features *(17.2)* can be recognized more easily in thin section.

Further reading

Kendall, A.C. (1992) Evaporites. In: *Facies Models: Response to Sea Level Change* (eds R.G. Walker & N.P. James), pp. 375–409. Geological Association of Canada, St Johns, Newfoundland.

Leeder, M.R. (1982) *Sedimentology, Process and Product*, 344 pp. Unwin Hyman, London.

Lewis, D.G. & McConchie, D. (1994*) Analytical Sedimentology*, 197 pp. Chapman & Hall, London.

Scoffin, T.P. (1987) *Carbonate Sediments and Rocks*, 274 pp. Blackie, Glasgow.

Tucker, M.E. (1991) *Sedimentary Petrology*, 2nd edn, 260 pp. Blackwell Scientific Publications, Oxford.

Tucker, M.E. & Wright, V.P. (1990) *Carbonate Sedimentology*, 482 pp. Blackwell Scientific Publications, Oxford.

4   Processes of transport and sedimentary structures

Biological build-ups such as reefs, shell banks and microbial mats are created in place with no transport of material. Similarly, precipitation of evaporite minerals in lakes, lagoons and along coastlines does not involve any movement of particulate matter. However, almost all other sedimentary deposits are created by the transport of material. Movement of detritus may be due purely to gravity, but more commonly it is the result of flow in water, air, ice or dense mixtures of sediment and water. The interaction of the sedimentary material with the transporting medium results in the development of sedimentary structures, some of which are due to bedforms built up in the flow whilst others are erosional. These sedimentary structures are preserved in the rocks and provide a record of the processes occurring at the time of deposition. If the physical processes occurring in different modern environments are known and if the sedimentary rocks are interpreted in terms of those same processes it is possible to infer the probable environment of deposition. A knowledge of these processes and their products is therefore fundamental to sedimentology. In this chapter the main physical processes occurring in depositional environments are discussed. The nature of the deposits resulting from these processes and the main sedimentary structures formed by the interaction of the flow medium and the detritus are introduced. Many of these features occur in a number of different sedimentary environments and should be considered in the context of the environments in which they occur.

4.1 Transport media

GRAVITY

The simplest case of sediment transport does not significantly involve a surrounding medium as it is that of particles falling by gravity off a cliff or down a slope. *Rock falls* generate piles of sediment at the base of slopes, typically consisting mainly of coarse debris which is not subsequently reworked by other processes. These accumulations are seen as *screes* along the sides of valleys in mountainous areas. They build up as *talus cones* with a surface at the *angle of rest* of the gravel, the maximum angle at which the material is stable without clasts falling further downslope. This angle varies with the shape of the clasts and distribution of clast sizes, but is typically between 30 and 35 degrees from the horizontal. Scree deposits are localized in mountainous areas *(6.6.1)* and occasionally along coasts: they are rarely preserved in the stratigraphic record.

WATER

Transport of particles in water is by far the most significant of all transport mechanisms. Water flows on the land surface in channels and as overland flow. Currents in seas are driven by wind, tides and oceanic circulation. These flows may be strong enough to carry coarse material along their base and finer material in suspension. Material may be carried in water hundreds or thousands of kilometres before being deposited as sediment. The mechanisms by which water moves this material are considered below.

AIR

After water, air is the next most important transporting medium. Wind blowing over the land can pick up dust and sand and carry it large distances. The capacity of the wind to transport material is limited by the low density of air. As will be seen in Section 4.2.6, the density contrast between the fluid medium and the clasts is critical to the effectiveness of the medium in moving sediment.

ICE

Water and air are clearly fluid media, but we can also consider ice as a fluid because over long time periods it moves across the land surface, albeit very slowly. Ice is therefore a rather high-viscosity fluid which is capable of transporting large amounts of clastic debris. Movement of detritus by ice is significant in and around polar ice caps and in mountainous areas with permanent or semi-permanent glaciers *(7.2, 7.3)*. The volume of material moved by ice has been very great at times of extensive glaciation.

DENSE SEDIMENT AND WATER MIXTURES

When there is a very high concentration of sediment in water the mixture forms a *debris flow (4.6.1)*, which can be thought of as a slurry with a consistency similar to that of wet concrete. These dense mixtures move by gravity over land or under water, behaving in a different way than sediment dispersed in a water body. More dilute mixtures may also move under gravity in water as *turbidity currents (4.6.2)*. These gravity-driven flow mechanisms are important as means of transporting coarse material into the deep oceans.

4.2 The behaviour of fluids and particles in fluids

A brief introduction to *fluid dynamics*, the behaviour of moving fluids, is provided in this section to give some physical basis to the discussion of sediment transport and the formation of sedimentary structures in later sections. More comprehensive treatments of sedimentary fluid dynamics are provided in Leeder (1982), J.R.L. Allen (1985, 1994) and P.A. Allen (1997).

4.2.1 Laminar and turbulent flow

A fluid in motion can move in two distinct ways. In *laminar flows*, all molecules within the fluid move parallel to each other in the direction of transport. In a heterogeneous fluid almost no mixing occurs during laminar flow. In *turbulent flows*, molecules in the fluid move in all directions but with a net movement in the

transport direction. Heterogeneous fluids are thoroughly mixed in turbulent flows.

The distinction between laminar and turbulent motion was first documented by O. Reynolds in the late 19th century. He performed experiments on flow through a tube, and noticed that plotting the flow rate against the pressure drop between the inlet and outlet did not produce a straight-line graph. The greater loss of pressure at higher flow rates can be attributed to increased friction between particles in turbulent flow. Experiments using threads of dye in tubes show that the lines of flow are parallel at low flow rates, but at higher flow velocities the dye thread breaks up as the fluid is mixed up by the turbulent motion (Fig. 4.1).

It is possible to assign a parameter called the *Reynolds number (Re)* to flows. This is a dimensionless quantity which indicates the extent to which a flow is laminar or turbulent. The Reynolds number is obtained by relating the following factors: the velocity of flow (u), the ratio between the density of the fluid and the viscosity of the fluid (v, the fluid kinematic viscosity) and a 'characteristic length' (l, the diameter of a pipe or depth of flow in an open channel). The equation defining the Reynolds number is:

$$Re = ul / v$$

Fluid flow in pipes and channels is found to be laminar when the Reynolds number is low (less than 500) and turbulent at higher values (greater than 2000). With increased velocity the flow is more likely to be

Laminar flow

At all points in flow all molecules are moving downstream

Turbulent flow

At any point in the flow a molecule may be moving in any direction, but the net flow is downstream

Fig. 4.1 Turbulent and laminar fluid flows.

turbulent and a transition from laminar to turbulent flow in the fluid occurs. Fluids with low kinematic viscosity, such as air, are turbulent at low velocities so all natural air flows which can carry particles in suspension are turbulent. Water flows are only laminar at very low velocities or very shallow water depths, so turbulent flows are much more common in aqueous sediment transport and deposition processes. Laminar flow occurs in some debris flows, in moving ice and in lava flows, all of which have kinematic viscosities orders of magnitude greater than that of water.

Most flows in water and air which are likely to carry significant volumes of sediment are turbulent. The behaviour of particles in these flows will now be considered.

4.2.2 Transport of particles in a fluid

Particles of any size are moved in a fluid by one of three mechanisms (Fig. 4.2). First, they can move by *rolling* along at the bottom of the air or water flow without losing contact with the bed surface. Second, they can move in a series of jumps, periodically leaving the bed surface and being carried short distances within the body of the fluid before returning to the bed again; this is known as *saltation*. Finally, turbulence within the flow can produce sufficient upward motion to keep particles in the moving fluid more or less continually; this is known as *suspension*.

A number of factors control the motion of particles in a turbulent fluid. First, as flow velocity is increased the higher kinetic energy within the fluid results in particles leaving the bed surface and moving by saltation. Second, the increase in turbulence also provides sufficient upward force to keep particles in suspension. Third, particles with greater mass require more energy to lift them in saltation and to keep them in suspension. Finally, particles with large surface areas relative to their mass (for example, platy minerals such as mica) have lower settling velocities (they take longer to sink) and hence can be kept in (permanent or temporary) suspension more easily.

At low current velocities only fine particles (clays) and low-density particles are kept in suspension, with sand-size particles moving by rolling and some saltation. At higher flow rates all silt and some sand may be kept in suspension, with granules and fine pebbles saltating and coarser material rolling.

These processes are essentially the same in air and water, but in air higher velocities are required to move particles of a given size because of the lower density and viscosity of air compared to water (Table 4.1). A further consequence of the lower viscosity of air is that as saltating grains land the cushioning effect of the fluid medium is relatively slight, and the grains have sufficient momentum to knock impacted grains up into the free stream of flow. This effect is less pronounced in water because the friction between the moving grain and the fluid depletes the energy before landing. Particulate

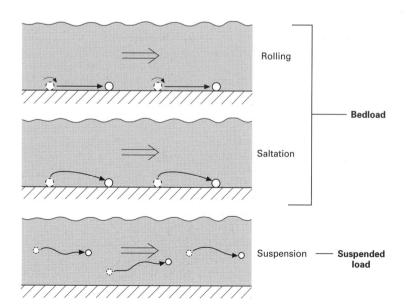

Fig. 4.2 Mechanisms of transport of particles in a flow: rolling and saltation (bedload); and suspension (suspended load).

Table 4.1 Density and viscosity of fluid transport media.

Substance	Density (kg m^{-3})	Viscosity (N s m^{-2})
Air	1.3	1.78×10^{-5}
Water	1000	1×10^{-3}
Debris flows	1500 to 2600	1×10^2 to 1×10^3

matter carried by a flow is normally considered in terms of *bedload* (rolling and saltating particles) and *suspended load* (material in suspension), also sometimes referred to as *washload* (Fig. 4.2).

4.2.3 Entraining particles in a flow

It is not immediately obvious why a particle which is sitting at the base of a flow (e.g. in the bed of a river) should do anything other than be moved along by frictional drag. Indeed, frictional drag between flowing water and objects in the flow is the main mechanism by which coarse material is transported as the rolling component of bedload. However, some particles move up from the base of the flow and are temporarily entrained within it before being deposited again further downflow. These are the saltating particles. The flow is unable to maintain these grains in suspension because they fall back down again, so what makes them move upwards in the first place? The answer lies in the *Bernoulli effect*, the phenomenon which allows birds and aircraft to fly and yachts to sail 'close to the wind'.

The Bernoulli effect can best be explained by considering the flow of a fluid (air, water or any fluid medium) in a tube which is narrower at one end than at the other (Fig. 4.3). The cross-sectional area of the tube is less at one end than at the other, but in order to maintain a constant transport of the fluid along the tube the same amount must go in one end and out of the other in a given time period. In order to get the same amount of fluid through a smaller gap it must move at a greater velocity through the narrow end. This effect is familiar to anyone who has squeezed and constricted the end of a garden hose: the water comes out as a faster jet when the end of the hose is partly closed off.

The next thing to consider is the conservation of mass and energy along the length of the tube. The variables involved can be presented in the form of the *Bernoulli equation*:

$$\text{Total energy} = \rho g h + \rho u^2/2 + P$$

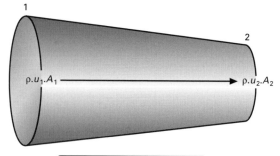

Mass of fluid at '1' = mass at '2'
$\rho . u_1 . A_1 = \rho . u_2 . A_2$
$u_1 . A_1 = u_2 . A_2$
Area A_1 has decreased to A_2
Velocity u_1 must increase to u_2

Bernoulli's equation
Total energy = $0.5 \rho u^2 + \rho g h + P$
If u increases P must decrease
= Pressure drop

Fig. 4.3 The Bernoulli effect illustrated by a fluid passing through a tapered cylinder.

where ρ is the density of the fluid, u the velocity, g the acceleration due to gravity, h the height difference and P the pressure. The three terms in this equation are potential energy ($\rho g h$), kinetic energy ($\rho u^2/2$) and pressure energy (P). This equation assumes no loss of energy due to frictional effects, so in reality the relationship is:

$$\rho g h + \rho u^2/2 + P + E_{\text{loss}} = \text{constant}$$

The potential energy is constant because there is no difference in level between where the fluid is starting from and where it is ending up. Kinetic energy is changed as the velocity of the flow is increased or decreased. If the total energy in the system is to be conserved, there must be some change in the final term, the pressure energy. Pressure energy can be thought of as the energy which is stored when a fluid is compressed: a compressed fluid (as in a canister of a compressed gas) has a higher energy than an uncompressed one.

Returning to the flow in the tapered tube, in order to balance the Bernoulli equation, the pressure energy must be reduced to compensate for an increase in kinetic energy caused by the constriction of the flow at the end of the tube. This means that there is a reduction in pressure at the narrower end of the tube.

Transferring these ideas to a flow along a channel, a

clast in the bottom of the channel will reduce the cross-section of the flow over it. The velocity over the clast will be greater than upstream and downstream of it and in order to balance the Bernoulli equation there must be a reduction in pressure over the clast. This reduction in pressure provides a temporary *lift force* which moves the clast off the bottom of the flow (Middleton & Southard 1978). The clast is then temporarily entrained in the moving fluid before falling under gravity back down on to the channel base in a single saltation event (Fig. 4.4).

4.2.4 Grain size and flow velocity

The fluid velocity at which a particle becomes entrained in the flow can be referred to as the *critical velocity*. If the forces acting on a particle in a flow are considered then a simple relationship between the critical velocity and mass of the particle would be expected. The drag force required to move a particle along in a flow will increase with mass, as will the lift force required to bring it up into the flow. At moderate flow velocities sand grains may saltate, granules roll and pebbles remain unmoved, but as the flow velocity increases the forces acting on these particles increase and finer sand may be in suspension, granules saltating and pebbles rolling. A simple linear relationship like this works for coarser material, but when fine grain sizes are involved things are more complicated.

The *Hjulström diagram* (Fig. 4.5) shows the relationship between water flow velocity and grain size (Hjulström 1939). There are two main lines on the graph. The lower line displays the relationship between flow velocity and particles which are already in motion. This shows that a pebble will come to rest at around 20–30 cm s^{-1}, a medium sand grain at 2–3 cm s^{-1}, and a clay particle when the flow velocity is effectively zero. The grain size of the particles in a flow can therefore be used as an indicator of the velocity at the time of deposition of the sediment if deposited as isolated particles. The upper, curved line shows the flow velocity required to move a particle from rest. On the right half of the graph this line parallels the first but at any given grain size the velocity required to initiate motion is higher than that to keep a particle moving. On the left-hand side of the diagram there is a sharp divergence of the lines: counter-intuitively, the smaller silt and clay particles require a higher velocity to move them than sand. This can be explained by the properties of clay minerals which will dominate the fine fraction in a sediment. Clay minerals are cohesive *(2.5.5)* and once they are deposited they tend to stick together, making them more difficult to entrain in a flow than sand grains. Note that there are two lines for cohesive material. 'Unconsolidated' mud has settled but remains a sticky, plastic material. 'Consolidated' mud has had much more water expelled from it and is rigid. In practice, many deposits of muddy material lie between these two lines.

The behaviour of fine particles in a flow, as indicated by the Hjulström diagram, has important consequences for deposition in natural depositional environments. Were it not for this behaviour, clay would be eroded in all conditions except standing water, but mud can accumulate in any setting where the flow stops for long enough for the clay particles to be deposited: resumption of flow does not re-entrain the deposited clay unless the velocity is relatively high. Alternations of mud and sand deposition are seen in environments where flow is intermittent, such as tidal settings *(11.2.4)*.

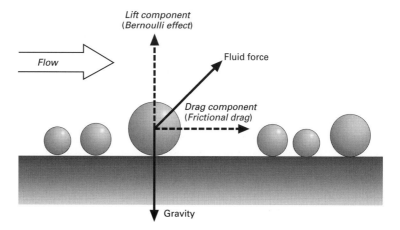

Fig. 4.4 Forces acting on a grain in a flow. (After Middleton & Southard 1978; Collinson & Thompson 1982.)

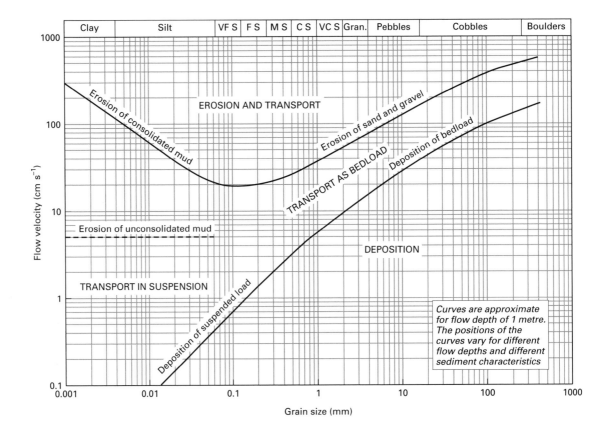

Fig. 4.5 The Hjulström diagram, showing the relationship between the velocity of a water flow and the transport of loose grains. Once a grain has settled it requires more energy to start it moving than to keep it moving when it is already in motion. The cohesive properties of clay particles mean that fine-grained sediments require relatively high velocities to re-erode them once they are deposited, especially once they are compacted. (From *Earth*, 2nd edn by Frank Press and Raymond Siever. ©1974, 1978, 1982 and 1986 by W.H. Freeman and Company. Used with permission.)

4.2.5 Clast size variations: graded bedding

If the velocity changes during a period of flow the sizes of the clasts which are deposited will reflect the changes in flow strength. A flow decreasing from $20 \, cm \, s^{-1}$ to $1 \, cm \, s^{-1}$ will initially deposit coarse sand but will progressively deposit medium and fine sand as the velocity falls. The bed of sand which is formed from a decelerating flow will show a reduction in grain size from coarse at the bottom to fine at the top. This pattern of clast size change in a single bed is referred to as *normal grading*. Conversely, an increase in flow velocity

through time may result in an increase in grain size up through a bed, known as *reverse grading*. Normal grading is more common because many natural flows commence with a strong surge followed by a gradual waning of the flow velocity. Flows which gradually increase in strength through time to produce reverse grading are less frequent. Material deposited from static water also exhibits grading on account of the relationship between grain size and settling velocity described by Stokes' law. Larger particles achieve a higher terminal velocity and settle out of suspension faster than smaller grains (see Leeder 1982).

Grading can occur in a wide variety of depositional settings: normal grading is an important characteristic of many turbidity current deposits *(4.6.2)* but may also result from storms on continental shelves *(14.3)*, overbank flooding in fluvial environments *(9.3)* and delta top settings *(12.1.1)*.

It is useful to draw a distinction between grading which is a trend in grain size within a single bed and trends in grain size which occur through a number of beds. A pattern of several beds which starts with a coarse clast size in the lowest bed and finer material in the highest is considered to be *fining-upward*. The reverse pattern with the coarsest bed at the top is a *coarsening-upward* succession (Fig. 4.6). Note that there can be circumstances where individual beds are normally graded but are in a coarsening-upward succession of beds. The recognition and interpretation of fining- and coarsening-upward patterns are important in the analysis of sedimentary environments.

4.2.6 Fluid density and particle size

The forces acting on a particle are a function of the viscosity and density of the fluid medium as well as the mass of the particle. Higher-viscosity fluids exert greater drag and lift forces for a given flow velocity. The two

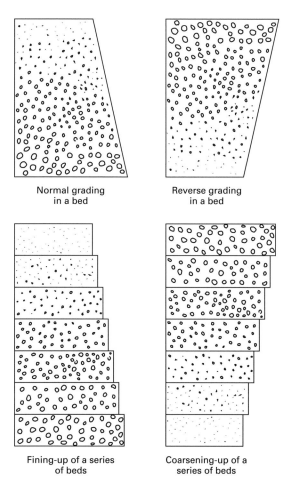

Normal grading
in a bed

Reverse grading
in a bed

Fining-up of a series
of beds

Coarsening-up of a
series of beds

Fig. 4.6 Normal and reverse grading in single beds; fining-upward and coarsening-upward patterns in a series of beds.

most important fluids on the Earth's surface are water and air. Water flows are able to transport clasts as large as boulders at the velocities recorded in rivers, but even at the very high wind strengths of storms the largest rock and mineral particles carried are likely to be around a millimetre in size. This limitation to the particle size carried by air is one of the criteria which may be used to distinguish material deposited by water from that transported and deposited by wind *(8.2)*. Higher-viscosity fluids such as ice and debris flows *(4.6.1)* can transport boulders metres or tens of metres across. The large clasts may be carried on the top of a laminar flow.

4.3 Flows, sediment and bedforms

A *bedform* is a morphological feature formed by the interaction between a flow and sediment on a bed. Ripples in sand in a flowing stream and sand dunes in deserts are both examples of bedforms, the former resulting from flow in water, the latter from air flow. To explain how bedforms are generated and why different types of bedform occur a further brief foray into fluid dynamics is required.

The presence of frictional forces within a flow was noted when considering the Bernoulli equation *(4.2.3)*. Friction is greatest at the edges of the flow—for instance, at the base of flow in a channel where eddies in turbulent flow interact with the solid boundary. A number of layers within the fluid can be recognized (Fig. 4.7). At the margin there is an *adsorbed layer* where the fluid particles are attached to the solid surface; this is only a few molecules thick. There is then a *boundary layer*, a zone which shows a flow velocity gradient from zero at the adsorbed layer to the mean flow velocity in the *free stream*, which is the portion of the flow unaffected by boundary effects. Within the boundary layer lies a *viscous sub-layer*, a region typically fractions of a millimetre thick in which viscous forces are important at low velocities.

The relationship between the thickness of the viscous sub-layer and the size of grains on the bed of flow defines a further property of the flow. If all the particles lie within the viscous sub-layer the surface is hydraulically *smooth*. If there are particles which project up through this layer then the flow surface is *rough*. In aqueous flows which exceed the critical velocity required to move sediment, the flow surface is always rough if the grain diameter exceeds 0.6 mm. The significance of this will become apparent when the

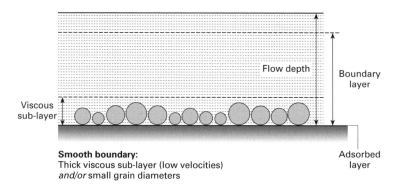

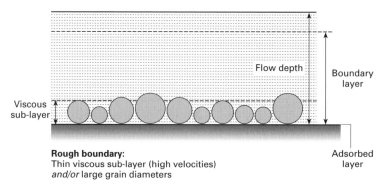

Fig. 4.7 Layers within a flow and flow surface roughness: a thin adsorbed layer where there is no fluid movement, the viscous sub-layer and the boundary layer within the flow.

relationship between grain size and bedform type is discussed below.

Bedforms in flows of both air and water are considered together in the rest of this section. There are many similarities of form and process between the behaviour of sand in flows of water and in air currents, but there are also some features which are unique to aeolian bedforms. The depositional processes and sedimentary structures of aeolian bedforms are considered further in Chapter 8.

4.3.1 Current ripples

Once the critical flow velocity for movement of sand grains has been reached saltation starts to occur. If a flow over an even sand bed is observed it is apparent that the grains start to arrange themselves in clusters. These clusters are only a few grains high, but once they have formed they create steps which influence the flow in the boundary layer. Flow can be visualized in terms of *streamlines* in the fluid, imaginary lines which indicate the direction of flow (Fig. 4.8). Streamlines lie parallel to a flat bed or the sides of a cylindrical pipe, but where

there is an irregularity, such as a step in the bed caused by an accumulation of grains, the streamlines converge and there is an increased transport rate. At the top of the step, a streamline separates from the bed surface and a region of *boundary layer separation* forms between the *flow separation point* and the *flow attachment point* downstream (Fig. 4.8). Beneath this streamline lies a region called the *separation bubble* or *separation zone*. Expansion of flow over the step results in an increase in pressure (the Bernoulli effect, *4.2.3*) and the sediment transport rate is reduced, resulting in deposition on the lee side of the step.

Current ripples (Figs 4.9 & 4.10) are small bedforms formed by the effects of boundary layer separation on a bed of sand. The small cluster of grains eventually forms the *crest* of a ripple and separation occurs near this point (Allen 1968). Sand grains roll or saltate up to the crest on the upstream or *stoss side* of the ripple. Avalanching of grains occurs down the downstream or *lee side* of the ripple as accumulated grains become unstable at the crest. Within the separation bubble there is a weak vortex (a *roller vortex*: Fig. 4.8). Grains which avalanche on the lee slope tend to come to rest at an angle close to the

1. Erosion in the trough of a bedform

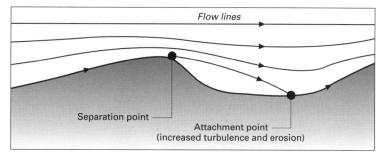

2. Development of counter-currents in lee of bedform

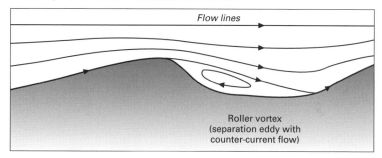

Fig. 4.8 Flow over a bedform: imaginary streamlines within the flow illustrate the separation of the flow at the brink of the bedform and the attachment point at which the streamline meets the bed surface where there is increased turbulence and erosion. A separation eddy may form in the lee of the bedform and produce a minor counter-current (reverse) flow.

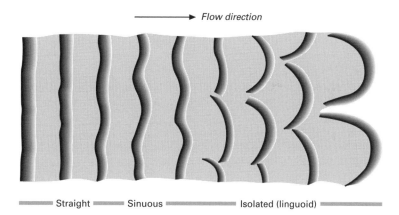

Fig. 4.9 Current ripples in plan view with straight, sinuous and isolated crests.

maximum critical slope angle for sand at around 30°. At the flow attachment point there are increased stresses on the bed which result in erosion and the formation of a small scour, the *trough* of the ripple.

CURRENT RIPPLES AND CROSS LAMINATION

A ripple migrates downstream as sand is added to the crest and accretes on the lee slope. This moves the crest and hence the separation point downstream. The effect of this is to move the attachment point and trough downstream as well. Scour in the trough and on the base of the stoss side supplies the sand which moves up the gentle slope of the stoss side of the next ripple and so a whole train of ripple troughs and crests advances downstream. The sand which avalanches on the lee slope during this migration forms a series of layers at the angle of the slope. These thin, inclined layers of sand are called

Fig. 4.10 Current ripples formed in sand in an estuary: field of view 1 m.

cross laminae; they build up to form the sedimentary structure referred to as *cross lamination* (Fig. 4.11).

When viewed from above, current ripples show a variety of forms (Fig. 4.9). They may have straight to sinuous crests (*straight* or *sinuous ripples*) which are

relatively continuous or form a pattern of unconnected arcuate forms called *linguoid ripples*. Eddies and irregularities in the roller vortices appear to be responsible for the more complex linguoid ripples. Straight and linguoid ripple crests give rise to different patterns of cross lamination in three dimensions. A perfectly straight ripple would generate cross laminae all dipping in the same direction and lying in the same plane: this is *planar cross lamination*. Sinuous and linguoid ripples have lee slope surfaces which are curved, generating laminae which dip at an angle to the flow as well as downstream. As linguoid ripples migrate curved cross laminae are formed mainly in the trough-shaped low areas between adjacent ripple forms, resulting in a pattern of *trough cross lamination* (Fig. 4.11).

CREATING AND PRESERVING
CROSS LAMINATION

Current ripples migrate by the removal of sand from the

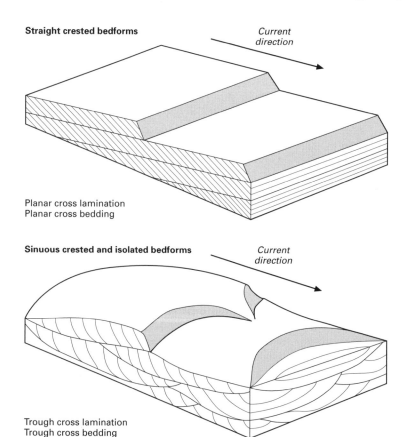

Fig. 4.11 Migrating straight crested ripple and dune bedforms form planar cross lamination and planar cross bedding. Sinuous or isolated (linguoid or lunate) ripple and dune bedforms produce trough cross lamination and trough cross bedding. (After Tucker 1991.)

stoss side of the ripple and deposition on the lee side. If there is a fixed amount of sand available the ripple will migrate over the surface as a simple ripple form, with erosion in the troughs matching addition to the crests. These *starved ripple* forms are preserved if blanketed by mud. In circumstances where there is a net addition of sand and the current is carrying and depositing sand particles, the amount of sand deposited on the lee slope will be greater than that removed from the stoss side. There will be a net addition of sand to the ripple and it will grow in altitude as it migrates. Most importantly, the depth of scour in the trough is reduced, leaving cross laminae created by the earlier migrating ripple preserved. In this way a layer of cross laminated sand is generated (Fig. 4.11).

When the rate of addition of sand is high there will be no net removal of sand from the stoss side and each ripple will migrate up the stoss side of the ripple form in front. These are *climbing ripples* (Allen 1972) (Fig. 4.12). When the addition of sediment from the current exceeds the forward movement of the ripple, deposition will occur on the stoss side as well as on the lee side. Climbing ripples are therefore indicators of rapid sedimentation as their formation depends upon the addition of sand to the flow at a rate equal to or greater than the rate of downstream migration of the ripples.

CONSTRAINTS ON CURRENT RIPPLE FORMATION

The formation of current ripples requires moderate flow velocities over a hydrodynamically smooth bed (see

Lee-side laminae   Stoss-side laminae

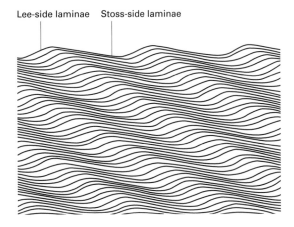

Fig. 4.12 Climbing ripple cross lamination produced by rapid deposition from a flow carrying a high proportion of sand. (After Collinson & Thompson 1982.)

above). They only form in sands in which the dominant grain size is less than 0.7 mm (coarse sand grade) because bed roughness created by coarser sand inhibits the small-scale boundary layer separation required for ripple formation. Because ripple formation is controlled by processes within the boundary layer there are no water depth constraints and current ripples may form in waters ranging in depth from a few centimetres to kilometres. This is in contrast to most other subaqueous bedforms (subaqueous dunes, sand waves, wave ripples) which are dependent on water depth.

Current ripples vary in height from 5 to 30 mm and the wavelengths (crest to crest or trough to trough) range from 50 to 400 mm (Allen 1968). The ripple wavelength is approximately 1000 times the grain size, although this relationship is subject to a good deal of variation. It is important to note the upper limit to the dimensions of current ripples and to emphasize that ripples do not 'grow' into larger bedforms.

4.3.2 Dunes

Beds of sand in rivers, estuaries, beaches and marine environments also have bedforms on them which are distinctly larger than ripples. These large bedforms are called *dunes*, although other terms such as 'megaripples', 'sand waves' (see below) and 'bars' have also been used (see Leeder 1982; Collinson & Thompson 1982; J.R.L. Allen 1994; P.A. Allen 1997). Evidence that these larger bedforms are not simply large ripples comes from measurement of the heights and wavelengths of all bedforms (Fig. 4.13). The data fall into clusters which do not overlap, indicating that they form by distinct processes which are not part of a continuum. The morphology of a subaqueous dune is similar to that of a ripple: they have a stoss side leading up to a crest and sand avalanches down a lee slope towards a trough. Flow separation is once again important, with a roller vortex developing over the lee slope and scouring occurring at the reattachment point in the trough. Beyond that, the similarities with ripples are less apparent as there is more variability of form and process in subaqueous dunes.

DUNES AND CROSS BEDDING

Migration of a subaqueous dune results in the construction of a succession of sloping layers formed by the avalanching on the lee slope, which are referred to as *cross beds*. At low flow velocities the roller vortex is

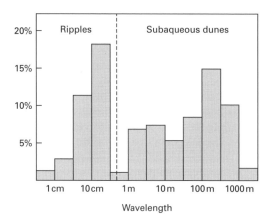 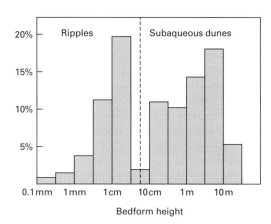

Fig. 4.13 Graphs of subaqueous ripple and subaqueous dune bedform wavelengths and heights. (After Collinson & Thompson 1982.)

Fig. 4.14 Planar cross bedding in Eocene shallow marine sandstone beds, Bighorn Basin, Wyoming, USA. The scale is in inches (1 inch = 2.54 cm).

weakly developed and there is little scouring at the reattachment point. The cross beds formed simply lie at the angle of rest of the sand, and as they build out into the trough the basal contact is angular. Bedforms which develop at these velocities usually have low-sinuosity crests, so the three-dimensional form of the structure is similar to planar cross lamination. This is *planar cross bedding*, and the surface at the bottom of the cross beds is flat and close to horizontal because of the absence of scouring in the trough. Cross beds bound by horizontal surfaces are sometimes referred to as *tabular cross bedding* (Figs 4.11 & 4.14). Cross beds may form a sharp angle at the base of the avalanche slope or may be asymptotic (tangential) to the horizontal (Figs 4.15 & 4.16). At high flow velocities the roller vortex is a strong feature creating a counter-current at the base of the slip face which may be strong enough to generate ripples (*counter-flow ripples*) which migrate a short distance up the toe of the lee slope (Fig. 4.15).

A further effect of the stronger flow is the creation of a marked scour pit at the reattachment point. The avalanche lee slope advances into this scoured trough so the bases of the cross beds are marked by an undulating erosion surface. The crest of a subaqueous dune formed under these conditions will be highly sinuous or will have broken up into a series of linguoid dune forms. *Trough cross bedding* formed by the migration of sinuous subaqueous dunes typically has asymptotic bottom contacts and an undulating lower boundary.

CONSTRAINTS ON THE FORMATION OF DUNES

Dunes range in wavelength from 60 cm to hundreds of metres and in height from 5 cm to over 10 m (Leeder 1982). The smallest are larger than the biggest ripples.

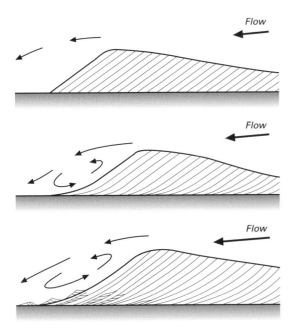

Fig. 4.15 Tangential toe at the bottom of a set of cross beds. Counter-current ripples at the toe of a subaqueous dune bedform formed by localized flow in the separation 'bubble'.

Fig. 4.16 Cross bedding in Cretaceous shallow marine sandstone beds, Morondava basin, western Madagascar.

They form in fine to very coarse sand and gravel but are not found in very fine sands. There is a relationship between the boundary layer thickness and the wavelength and height of dunes; in water flows in rivers, etc., the boundary layer is approximately the depth of flow. With increased flow depth these dimensions generally become larger but it is difficult to establish a clear depth–size relationship (Allen 1970a). As a consequence

of this depth dependence, subaqueous dunes are mainly found in the channels of rivers, deltas, estuaries and shelves with strong tidal currents (see Chapters 9, 11, 12 and 14).

SAND WAVES

Surveys of continental shelf seas have revealed the presence of large, linear bedforms in sandy areas of the sea floor. These features have wavelengths of tens to hundreds of metres and may be over 10 m high. The crests are straight to moderately sinuous and the troughs do not have well-developed scour pits. The presence of subaqueous dunes on the backs of some of these sand waves indicates that they may be distinct, but there is so much overlap between the sizes and shapes of sand waves and subaqueous dunes that clear separation of the two is not easy. They are typically 1–8 m high with wavelengths of 50–300 m and occur in tidally influenced estuaries and shelves. The characteristics of bedforms formed in tidally influenced environments are discussed further in Chapter 11.

SUPERIMPOSED BEDFORMS

Figure 4.17 shows current ripples and subaqueous dunes coexisting in a river estuary. Ripples form in the current on the stoss side of the dunes and in the trough, where the complex eddies can give rise to complex ripple patterns. In the case of bedforms in tidal environments the superimposed bedforms may be a consequence of changing flow strengths and flow depths.

4.3.3 Cross stratification, cross bedding and cross lamination

It is worth summarizing the terms used in this context to ensure consistency of terminology (Collinson & Thompson 1982). *Cross stratification* is any layering in a sediment or sedimentary rock which is orientated at an angle to the depositional horizontal. These inclined strata most commonly form in sand and gravel by the migration of bedforms. As a bedform migrates sand is deposited on its lee slope at an angle of up to 30° to the horizontal, forming thin layers at this angle which may be preserved if there is net accumulation. If the bedform is a ripple the resulting structure is referred to as *cross lamination*. Ripples are limited in crest height to about 3 cm so cross laminated beds do not exceed this

Fig. 4.17 Ripple bedforms on the upstream side of dune bedforms exposed in an estuary (Barmouth, Wales).

thickness. Migration of larger bedforms such as dunes and sand waves forms *cross bedding* which may be tens of centimetres to tens of metres in thickness. Cross stratification is the more general term and is used for inclined stratification generated by processes other than the migration of bedforms—for example, the inclined surfaces formed on the inner bank of a river by point bar migration *(9.2.2)*. Other terms which have been used are 'current bedding', 'festoon bedding' and 'false bedding', but these are now virtually obsolete. A single unit of cross bedded material is referred to as a *set*, and a stack of similar sets is called a *co-set* (Fig. 4.18).

4.3.4 Plane bedding and planar lamination

Plane bedding is the simplest of all sedimentary structures. It is simply layers of sand deposited from the flow to produce *planar lamination*. A bedform stability diagram (Fig. 4.19) has two regions where plane beds are stable. *Lower-stage plane beds* form in sands of coarse grain size and above (over 0.7 mm) when the critical velocity is reached and the grains start to move along the bed surface. Ripples do not form at these coarse grain sizes because the bed surface is rough *(4.3)* and prevents flow separation occurring. The horizontal planar lamination produced under these circumstances tends to be rather poorly defined.

At high flow velocities *upper-stage plane beds* occur in all sand grain sizes producing well-defined planar lamination with laminae typically 5–20 grains thick (Fig. 4.20). The bed surface is also marked by elongate ridges a few grain diameters high, separated by furrows orientated parallel to the flow direction (Allen 1964a). This feature is referred to as *primary current lineation* (often abbreviated to *pcl*) and it is characteristic of upper-stage plane bedding. Primary current lineation forms on beds as a result of a characteristic of flow within the viscous sub-layer *(4.3)*, the formation of

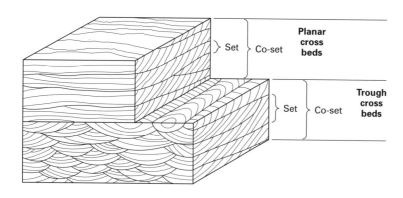

Fig. 4.18 Sets and co-sets of cross stratification. (After Collinson & Thompson 1982.)

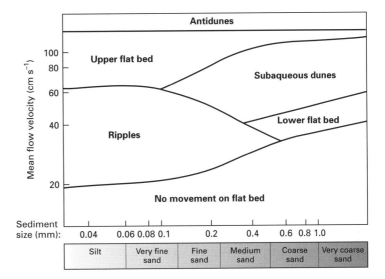

Fig. 4.19 A bedform stability diagram showing the stability fields of different bedforms formed in sediment of different grain sizes at different flow velocities. (After Harms *et al.* 1975; Walker 1992b.)

Fig. 4.20 Parallel laminated sandstone deposited in an overbank environment (Cretaceous, Alexander Island, Antarctica).

'bursts' and 'sweeps'. When turbulent flow over a smooth surface is examined in detail it is apparent that there is a 'streaking' parallel to the direction of flow. The flow is made up of regions where fluid is 'bursting' up from the viscous sub-layer into the main boundary layer and of parallel zones of 'sweeps' of fluid down into the viscous sub-layer. This effect dies out rapidly up through the flow but at the bed boundary it creates the ridges and furrows seen as primary current lineation. The effect is subdued when the bed surface is rough and is therefore less well defined in coarser sands.

4.3.5 Rapid (supercritical) flow

Flow may be considered to be *tranquil*, often with a smooth water surface, or *rapid*, with an uneven surface of wave crests and troughs in some circumstances. These flow states can be expressed in terms of a parameter, the *Froude number*, which relates the velocity of the water to the speed at which a wave can be transmitted through the water. In its simplest form the Froude number can be considered to be a ratio of the flow velocity to the velocity of a wave in the flow (Leeder 1982). When the value is less than one, a wave (formed, for example, by a pebble dropped into the water or by wind on the surface: *4.4*) can propagate upstream because it is travelling faster than the flow. This is the tranquil or *subcritical flow* state. A Froude number greater than one indicates that the flow is too fast for a wave to propagate upstream and the flow is rapid or *supercritical*. An analogy can be made between subcritical and supercritical flow in water and subsonic and supersonic motion through air: the latter refers to a sound wave which is different in form from a water wave, but in both cases there is a threshold between motion slower than the wave and motion faster than a wave can propagate. In water this threshold is associated with a change at the flow surface called a *hydraulic jump* which may sometimes be seen in a stream as a distinct breaking wave between regions of rapid and tranquil flow.

In circumstances where the Froude number is just below or above one, for flow in water over a bed of sand, standing waves may temporarily form on the surface of the water before steepening and sometimes breaking in an upstream direction. Sand on the bed forms a ridge referred to as an *antidune* (or *in-phase wave*) and as the

---

wave breaks accretion of sand occurs on the upstream side of the antidune. Were it to be preserved, antidune cross bedding would be seen as cross stratification dipping upstream. However, such preservation is rarely seen simply because as the flow velocity drops the sediment is reworked into upper-stage plane beds by subcritical flow. Well-documented occurrences of antidune cross stratification are known from pyroclastic surge deposits *(16.3.4)* where high-velocity flow is accompanied by very high rates of sedimentation (Schminke *et al.* 1975).

4.3.6 Bedform stability diagrams and flow regimes

The relationship between the grain size of the sediment and the flow velocity is summarized in Fig. 4.19. This *bedform stability diagram* indicates the most likely bedform to occur for a given grain size and velocity and has been constructed from experimental data (modified from Harms *et al.* 1975 and Walker 1992b). It should be noted that the boundaries between the fields are not sharp and there is a lot of overlap where either of two bedforms may be stable. Note also that the scales are logarithmic on both axes. In addition to the fields of stability of bedforms, two general flow regimes are recognized: a *lower flow regime* in which ripples, sand waves, dunes and lower plane beds are stable; and an *upper flow regime* where plane beds and antidunes form. Flow in the lower flow regime is always subcritical and

the change to supercritical flow lies within the antidune field.

4.4 Waves

Waves are produced in bodies of water by wind acting on the surface or by an input of energy from an earthquake, landslide or similar phenomenon. Any body of water, ranging from a pond to an ocean, is subject to the formation of wind-generated waves on the surface. The height and energy of waves are determined by the strength of the wind and the *fetch*, the expanse of water across which the wave-generating wind blows. Waves generated in open oceans can travel well beyond the areas in which they were generated. A simple wave form involves an *oscillatory motion* of the surface of the water; there is no net horizontal water movement. The wave form moves across the water surface in the manner seen when a pebble is dropped into still water. When a wave enters very shallow water the amplitude increases and then the wave *breaks*, creating the horizontal movement of waves seen on the beaches of lakes and seas.

4.4.1 Formation of wave ripples

The oscillatory motion of the top surface of a water body produced by waves generates a circular pathway for water molecules in the top layer (Fig. 4.21). This circular

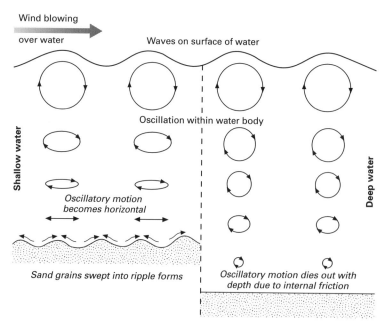

Fig. 4.21 The formation of wave ripples in sediment produced by oscillatory motion in the water column due to wave ripples on the surface of the water. Note that there is no overall lateral movement of the water, or of the sediment.

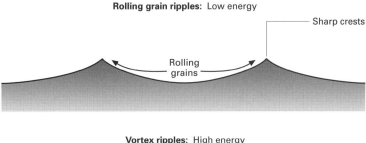

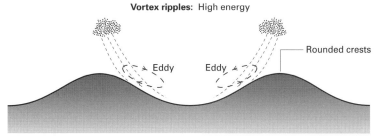

Fig. 4.22 Forms of wave ripple: rolling grain ripples produced when the oscillatory motion is capable only of moving the grains on the bed surface; and vortex ripples formed by higher-energy waves relative to the grain size of the sediment.

motion sets up a series of circular cells in the water below. With increasing depth internal friction reduces the motion and the effects of the surface waves die out. The depth to which surface waves affect a water body is referred to as the *wave base (11.3)*. In shallow water, the base of the water body interacts with the waves. Friction causes the circular motion at the surface to become transformed into an elliptical pathway which is flattened at the base into a horizontal oscillation. This horizontal oscillation may generate wave ripples in sediment.

At low energies *rolling grain ripples* form (Fig. 4.22) (Bagnold 1946). The peak velocity of grain motion is at the mid-point of each oscillation, reducing to zero at the edges. This sweeps grains away from the middle where a trough forms to the edges where ripple crests build up. Rolling grain ripples are characterized by broad troughs and sharp crests. At higher energies grains can be kept temporarily in suspension during each oscillation. Small clouds of grains are swept from the troughs on to the crests where they fall out of suspension. These *vortex ripples* (Fig. 4.22) (Bagnold 1946) have more rounded crests but are otherwise symmetrical. Where waves move into shallow water the forward and backward movement becomes uneven and asymmetric wave ripples may form.

4.4.2 Characteristics of wave ripples

In cross-section wave ripples are generally symmetrical in profile. Laminae within each ripple dip in both directions and are overlapping. These characteristics are seen in cross lamination generated by the accumulation of sediment influenced by waves (Fig. 4.23). In plan view wave ripples have long, straight to gently sinuous crests which may split or *bifurcate* (Fig. 4.24). These characteristics may be seen on bedding planes. Wave ripples can form in any non-cohesive sediments and are principally seen in coarse silts and sand of all grades. If the wave energy is high enough wave ripples can form in gravel grade material including granule and pebble deposits. These gravel ripples have wavelengths of several metres and heights of tens of centimetres.

4.4.3 Distinguishing wave and current ripples

It can be quite critical to a palaeoenvironmental

Fig. 4.23 Wave ripple cross lamination in fine-grained sandstone (Carboniferous, County Clare, Ireland).

Fig. 4.24 Wave ripples in sand exposed on a beach. These were produced by wind blowing over shallow standing water.

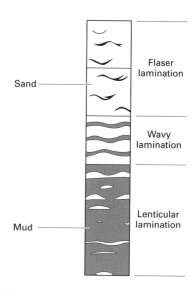

Fig. 4.25 Mixtures of sand and mud in different proportions produce different forms of lenticular and wavy bedding. (After Reineck & Singh 1973.)

interpretation to know whether the ripples preserved on a bedding surface or the cross lamination within a bed were formed by wave action or current flow. They can be distinguished in the field on the basis of their respective shapes. In plan view wave ripples have the characteristics described in Section 4.4.2 whereas current ripples are commonly very sinuous and broken up into short, curved crests. When viewed from the side, wave ripples are symmetrical with cross laminae dipping in both directions either side of the crests. In contrast, current ripples are asymmetrical with cross laminae dipping only in one direction, the only exception being climbing ripples which have distinctly asymmetric dipping laminae.

4.5 Sedimentary structures in sand–mud mixtures

Both sand and mud may be deposited in environments which experience variations in current or wave activity or sediment supply due to changing current strength or wave power. For example, tidal settings *(11.2)* display regular changes in energy in different parts of the tidal cycle, allowing sand to be transported and deposited at some stages and mud to be deposited from suspension at others. This may lead to simple alternations of layers of sand and mud, but if ripples form in the sands due to either current or wave activity then an array of sedimentary structures (Fig. 4.25) may result depending on the proportions of mud and sand. *Flaser bedding* is characterized by isolated thin drapes of mud amongst the cross laminae of a sand. *Lenticular bedding* is composed of isolated ripples of sand completely surrounded by

mud. Intermediate forms made up of approximately equal proportions of sand and mud are called *wavy bedding* (Reineck & Singh 1973, 1980).

4.6 Mass flows

Mixtures of detritus and fluid move under gravity by several different physical mechanisms which may act individually or in combination. These types of flow are known collectively as *mass flows* or *gravity flows* (Middleton & Hampton 1973). All require a slope to provide the potential energy to drive them, but once a flow is initiated it may continue under its own momentum.

4.6.1 Debris flows

These are dense, viscous mixtures of sediment and water in which the volume and mass of sediment present exceed those of water (Leeder 1982). The water may constitute less than 10% of the flow. A dense, viscous mixture of this sort will typically have a very low Reynolds number so the flow is likely to be laminar *(4.2.1)*. In the absence of turbulence no dynamic sorting of material into different sizes occurs during flow and the resulting deposit is very poorly sorted. Some sorting may develop by slow settling and there may be locally reverse grading produced by shear at the bed boundary. Material of any size from clay to large boulders may be present.

Debris flows occur on land, principally in arid environments where water supply is sparse, and in the submarine environment where they transport material down continental slopes. Once a debris flow has started the slope required to overcome friction need only be about 1°. Deposition occurs when internal friction becomes too great and the flow 'freezes'. There is not necessarily any change in the thickness of the deposit in a proximal to distal direction and the clast size distribution may be the same throughout the deposit. The deposits of debris flows on land are typically matrix-supported conglomerates, although clast-supported deposits also occur if the relative proportion of large clasts is high in the sediment mixture. They are poorly sorted and show a chaotic fabric — that is, there is usually no preferred orientation to the clasts — except within zones of shearing which may form at the base of the flow. Large clasts carried by the flow may remain proud of the top of the flow unit and protrude from the layer when deposited. This gives debris flow deposits an irregular top surface.

When a debris flow travels through water it may partly mix with it and the top part of the flow may become dilute. The tops of subaqueous debris flows are therefore characterized by a gradation up into better-sorted, graded sediment which may have the characteristics of a turbidity current *(4.6.2)*. Depositional environments in which debris flows occur are principally alluvial fans *(8.4.2)* and ephemeral stream flows *(8.3.1)* in con-

tinental environments. In marine environments they occur on continental slopes *(15.2.3)* and adjacent parts of the basin plain and around volcanic seamounts and islands *(16.4.4)*.

4.6.2 Turbidity currents

Turbidity currents are mixtures of sediment and water less dense than debris flows and hence have a higher Reynolds number. They are mixtures of sediment and water which move under gravity due to the density contrast with the less dense medium of sea water or fresh water. Most turbidity currents initially move down a slope which provides potential energy, but movement on a horizontal surface over long distances is also possible provided that the density contrast is maintained. A turbidity current may lose density by deposition of sediment if the flow is overloaded with sediment, as is the case for all but the most dilute turbidity currents (Allen 1997). The limit of flow of a turbidity current is reached when the density contrast is no longer sufficient to maintain momentum and it decelerates to zero at the end point of the flow. Sorting occurs in the turbulent flow, separating the coarser material which is deposited first from the finer which can be kept in turbulent suspension for longer. *Turbidites*, the deposits of turbidity currents (Fig. 4.26), are therefore mostly normally graded (Middleton 1966).

In detail, the internal characteristics of a turbidite show more than just a simple grading: a pattern of textures and sedimentary structures in these deposits was

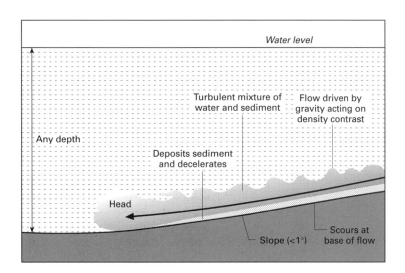

Fig. 4.26 Features of a turbidity current.

first noted by Bouma (1962) after whom the *Bouma sequence* of internal characteristics is named. An ideal turbidite deposit contains five divisions ('a'–'e') in the Bouma scheme (Fig. 4.27), although most turbidites do not contain all five divisions.

BOUMA DIVISION 'a' (T_a)

The lowest part consists of poorly sorted, structureless sand. This is attributed to deposition under a decelerating flow where the zone close to the bed is hyperconcentrated and turbulence is reduced. There is little sorting in this basal layer and no sedimentary structures form.

BOUMA DIVISION 'b' (T_b)

Laminated sand characterizes this layer; the grain size is normally finer than in the 'a' layer and the material is better sorted. The parallel laminae are generated by the separation of grains in upper flow regime transport *(4.3.6)*.

BOUMA DIVISION 'c' (T_c)

Cross laminated medium to fine sand, sometimes with climbing ripple lamination, forms the middle division of the Bouma sequence. Ripples form in fine- to medium-grained sand at moderate flow velocities (Fig. 4.19) and hence represent a reduction in flow velocity compared to the underlying 'b' division with its plane bedding. Climbing ripples form where the rate of sedimentation is

comparable to the rate of migration of the ripple, a condition which is commonly reached in sediment-laden turbidity currents.

BOUMA DIVISION 'd' (T_d)

Fine sand and silt in this layer are the products of waning flow in the turbidity current. Horizontal laminae may occur due to separations of fine grain sizes but the lamination is commonly less well defined than in the 'b' layer.

BOUMA DIVISION 'e' (T_e)

The top part of the turbidite consists of fine-grained sediment of silt and clay grade. This material is deposited from suspension as the turbidity current comes to rest. It is often indistinguishable from 'background' sedimentation from suspension in the ambient water body.

PROXIMAL TO DISTAL CHANGES IN
TURBIDITE DEPOSITS

When a turbidity current flows through a water body it may become less dense due to the deposition of sediment at the base, the dissipation of dense fluid in vortices at the head of the flow (Fig. 4.26) and some entrainment of the ambient fluid in the flow. A reduction in density causes the flow to decelerate, and at lower velocities the capacity of the turbidity current for carrying dense,

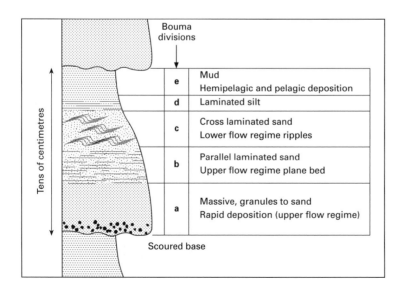

Fig. 4.27 The vertical pattern of grain size variation and sedimentary structures formed in a typical medium-grained turbidite. This is the Bouma sequence, consisting of five divisions: a, b, c, d and e. (After Bouma 1962.)

coarse sediment is reduced. As most turbidity currents are waning flows of this type (Middleton & Hampton 1976), with increasing distance the deposits become finer as the coarser material is progressively deposited from the flow (Lowe 1982; Stow 1994). The lower parts of the Bouma sequence are only present in the more proximal parts of the flow. Distally the lower divisions are progressively lost as the flow carries only finer sediment (Fig. 4.28) and only the 'c' to 'e' or perhaps just 'd' and 'e' parts of the Bouma sequence are deposited. The thickness of an individual turbidity current deposit may be anything from tens of metres to a few millimetres.

EROSION WITHIN SUCCESSIONS OF TURBIDITES

Sedimentary structures are common on the bases of turbidites. Strong turbulent flows scour into the underlying sediment as they pass over it to produce flute marks and grooves and other erosive features *(4.8)*. These features are useful palaeocurrent indicators in turbidite deposits. The scouring may be strong enough to remove completely the upper parts of a previously deposited bed, especially in the more proximal parts of the flow where the turbulent energy is highest. The 'd' and 'e' divisions may therefore be absent due to this erosion. The eroded

Fig. 4.28 Proximal to distal changes in the deposits formed by a turbidity current.

sediment may be incorporated into the overlying deposit as mud clasts.

HIGH-CONCENTRATION TURBIDITES

The Bouma sequence characterizes some turbidites, although many deposits do not fit into the scheme. These are beds of rather poorly sorted structureless sands which have a thin layer of silt and mud at the top. Applying the Bouma sequence terminology, the 'b', 'c' and sometimes 'd' divisions are missing. These beds are interpreted as the deposits of turbulent flows which contain a higher proportion of sediment in the mixture than 'normal' turbidity currents. A division is drawn at a bulk density of $1.1\,\mathrm{g\,cm^{-3}}$ between low-concentration and high-concentration turbidites, although there is a gradation between the two (Pickering *et al.* 1989). The effect of a higher concentration of sediment is that the turbulence is less effective at separating the grain sizes. Most of the sediment carried is deposited simultaneously as a poorly sorted mixture, with only the finer suspended material separating at the top of the flow (Lowe 1982).

OCCURRENCE AND COMPOSITION OF TURBIDITES

Turbidity currents may occur in any subaqueous environment from an inland lake to the deepest ocean.

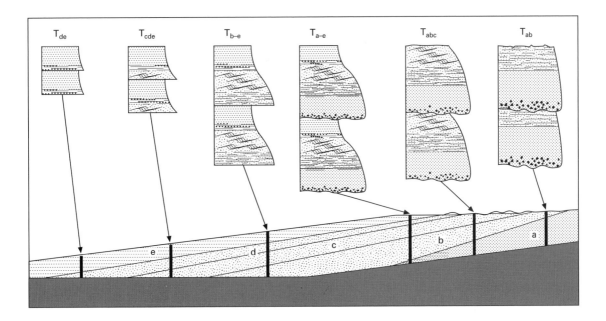

They are most commonly seen in the deposits of deep lakes *(10.3.2)* and in deep marine environments *(15.2)*. Terrigenous clastic turbidites with the texture of lithic wackes (greywackes) are perhaps the most commonly seen but turbidite deposits may have a wide range of textures and composition, including carbonate turbidites in basins flanked by carbonate shelves *(14.5)*. Turbidity current processes are particularly important in volcanic settings *(16.4.3)*.

TIME AND TURBIDITY CURRENTS

Turbidity currents are individual flow events. They occur over a geologically very short period of time, with most of the deposition occurring in a few hours to a few days. In fact, in the context of geological time turbidites deposit 'instantaneously'. The time taken for the thin layer of suspended sediment to be deposited on the top of the turbidite is many orders of magnitude longer (months to hundreds of years).

4.6.3 Grain flows

The mechanism of mass transport in an avalanche of material down a steep slope is that of a *grain flow* (Leeder 1982). Particles are kept apart in the fluid medium by repeated collisions. Grain flows rapidly 'freeze' as soon as the kinetic energy of the particles falls below a critical value. This mechanism is most effective in well-sorted material falling under gravity down a steep slope such as the slip face of an aeolian dune or subaqueous bedform. Grain flows are typically reverse-graded. They may occur in coarser sediment in combination with other mass flow processes in a steep subaqueous setting such as the foreset of a fan delta *(12.3)*.

4.6.4 Liquefied flows

When a mixture of sediment and water is subject to a high-energy vibration such as a seismic shock from an earthquake, liquefaction occurs. In this state of *liquefied flow* any density contrast within layers of the fluid–sediment mixture will result in the upward movement of the lighter material (Leeder 1982). Vertical fluid escape pipes form 'pillars' which disrupt the layering in the sediment into 'dishes', and sediment may reach the surface to erupt as a sand volcano *(17.1.1)*.

4.7 Mudcracks

Clay-rich sediment is cohesive *(2.5.5)* and the individual grains tend to stick to each other as the sediment dries out. As water is lost the volume reduces and clusters of clay minerals pull apart, developing cracks in the surface. Under subaerial conditions a polygonal pattern of cracks develops when muddy sediment dries out completely: these are *desiccation cracks* (Fig. 4.29). The spacing of desiccation cracks depends upon the thickness of the layer of wet mud, with a broader spacing occurring in thicker deposits. In cross-section desiccation cracks taper downwards and the upper edges may roll up if all of the moisture in the mud is driven off. The edges of desiccation cracks are easily removed by later currents and may be preserved as mud-chips or *mud-flakes* in the overlying sediment. Desiccation cracks are most clearly preserved in sedimentary rocks when the cracks are filled with silt or sand washed in by water or blown in by the wind. The presence of desiccation cracks is a very reliable indicator of the exposure of the sediment to subaerial conditions.

Synaeresis cracks are shrinkage cracks in clayey sediments which form under water. As the clay layer settles and compacts it shrinks to form single cracks in the surface of the mud. In contrast to desiccation cracks, synaeresis cracks are not polygonal but are simple, straight or slightly curved, tapering cracks. These sub-aqueous shrinkage cracks have been formed experimentally and have been reported in sedimentary rocks, although some of these occurrences have been reinterpreted as desiccation cracks (Astin 1991). Neither desiccation cracks nor synaeresis cracks form in silt or sand because these coarser materials are not cohesive.

Fig. 4.29 Desiccation cracks formed in mud deposited in a small pond which has dried out.

4.8 Erosional sedimentary structures

The sedimentary structures described in the preceding sections are formed as a product of the transport and deposition of material. Fluid flow over sediment which has already been deposited can result in the partial and localized removal of sediment from the bed surface. The features which remain on the bed surface are referred to as *sole marks* (Fig. 4.30). They are preserved in the rock record when another layer of sediment is deposited on top, leaving the feature on the bedding plane. Sole marks may be divided into those which form as a result of turbulence in the water causing erosion (*scour marks*) and impressions formed by objects carried in the water flow (*tool marks*). They may be found in a very wide range of depositional environments but are particularly common in successions of turbidites *(4.6.2)* where the sole mark is preserved as a cast at the base of the overlying turbidite.

4.8.1 Scour marks

Turbulent water flow over a bed surface results in

Fig. 4.30 Sole marks at the base of a flow: scours produced by flow eddies (flute marks) and turbulence around obstacles (obstacle scours); and tool marks formed by movement of an object along the bed surface (grooves) or saltating on the surface (prod, skip and bounce marks).

localized eddies even when the bed surface is smooth and flat. These turbulent eddies erode into the bed and create a distinctive erosional scour called a *flute cast*. Flute casts are asymmetric in cross-section, with one steep edge opposite a tapered edge (Fig. 4.30). In plan view they are narrower at one end widening out on to the tapered edge. The steep, narrow end of the flute marks the point where the eddy initially eroded into the bed and the tapered, wider edge marks the passage of the eddy as it is swept away by the current. Flute marks can therefore be used as palaeocurrent indicators *(5.4.1)*. They vary in size from 5 to 50 cm long and from 1 to 20 cm wide (Collinson & Thompson 1982). As with many sole marks, it is as common to find the cast of the feature formed by the infilling of the depression as it is to find the depression itself (Fig. 4.31).

An obstacle on the bed surface such as a pebble or shell can produce eddies which scour into the bed (*obstacle scours*). Linear features on the bed surface caused by turbulence are elongate *ridges and furrows* if on the scale of millimetres or *gutter casts* if the troughs are a matter of centimetres wide and deep, extending for several metres along the bed surface.

4.8.2 Tool marks

An object being carried in a flow over a bed can create marks on the bed surface. *Grooves* are sharply defined elongate marks created by an object (tool) being dragged

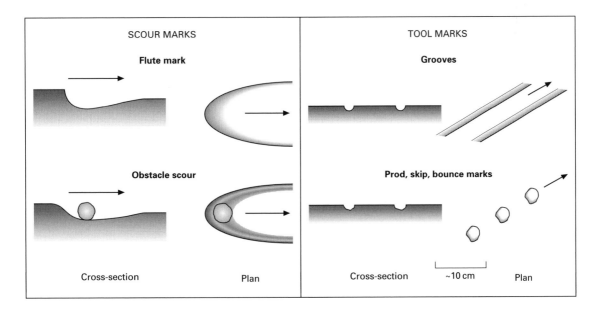

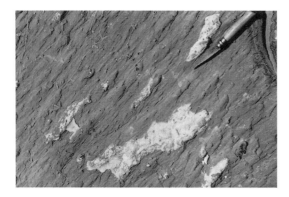

Fig. 4.31 Flute marks on the base of a sandstone bed produced by scour into an underlying mudstone bed which has since been removed; the knife blade indicates the direction of flow.

along the bed. Grooves are sharply defined features, in contrast to *chevrons* which form when the sediment is still very soft. An object saltating *(4.2.2)* in the flow may produce marks known variously as *prod*, *skip* or *bounce marks* at the points where it lands. These marks are often seen in lines along the bedding plane. The shape and size of all tool marks are determined by the form of the object which created them and irregular-shaped fragments, such as fossils, may produce distinctive marks. The nature of the tool is often not known unless it is preserved at the end of the trail, as may sometimes happen.

4.8.3 Channels and slump scars

A distinction may be drawn between scours, which are small-scale features caused by turbulence in a flow, and the much larger-scale features of channels and slump scars. A *channel* may be considered to be a depression on the land or sub-sea surface which wholly or partly confines flow. Channels are a fundamental component of

fluvial environments, deltas, estuaries and submarine fans. Channels in all these settings are distinctly larger than scours formed on the bed surface beneath either confined (channelized) flow or unconfined flow (e.g. sheetfloods, overbank flows, turbidites).

Slump scars (Fig. 4.32) form as a result of gravitational instabilities in sediment piles. When a mass of sediment is deposited on a slope it is to some extent unstable even if the slope is only a matter of a degree or so. If subjected to a shock from an earthquake or sudden addition of sediment load on part of the pile, failure may occur on surfaces within the sediment body. This leads to slumping of material. The surface left as the slumped material is removed is a slump scar which is preserved when later sedimentation subsequently fills in the scar. Slump scars can be recognized in the stratigraphic record as smooth-profiled spoon-shaped surfaces in three dimensions, and they range from a few metres to hundreds of metres across. They are common in deltaic sequences but may also occur within any material deposited on a slope.

4.9 Sedimentary structures and sedimentary environments

Bernoulli's equation and the Reynolds and Froude numbers may seem far removed from sedimentary rocks exposed in a cliff but if we are to interpret those rocks in terms of the processes which formed them a little fluid dynamics is useful. Understanding what sedimentary structures mean in terms of physical processes is one of the starting points for the analysis of sedimentary rocks in terms of environment of deposition. Most of the sedimentary structures described are familiar from terrigenous clastic rocks, but it is important to remember that any particulate matter interacts with the fluid medium in which it is transported and many of these

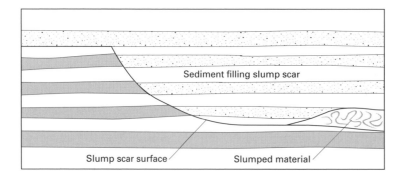

Fig. 4.32 Slump scars produced by movement of a mass of material on a failure surface.

Sediment filling slump scar

Slump scar surface   Slumped material

features also occur commonly in calcareous sediments made up of bioclastic debris and volcaniclastic rocks. The next chapter introduces the concepts used in palaeo-environmental analysis and is followed by chapters which consider the processes and products of different environments in more detail.

Further reading

Allen, J.R.L. (1970) *Physical Processes of Sedimentation*, 268 pp. Allen & Unwin, London.

Allen, J.R.L. (1982) *Sedimentary Structures: Their Character and Physical Basis,* Vol. 1. *Developments in Sedimentology*, 593 pp. Elsevier, Amsterdam.

Allen, J.R.L. (1985) *Principles of Physical Sedimentology*, 272 pp. Unwin Hyman, London.

Allen, P.A. (1997) *Earth Surface Processes*, 404 pp. Blackwell Science, Oxford.

Collinson, J.D. & Thompson, D.B. (1982) *Sedimentary Structures*, 194 pp. George Allen & Unwin, London.

Leeder, M.R. (1982) *Sedimentology, Process and Product*, 344 pp. Unwin Hyman, London.

Pye, K. (1994) *Sediment Transport and Depositional Processes*, 397 pp. Blackwell Scientific Publications, Oxford.

5 Environments and facies

The nature of the material deposited anywhere will be determined by the physical, chemical and biological processes which have occurred during the formation, transport and deposition of the sediment. Those processes also define what we consider to be the environment of deposition. In the following chapters the range of depositional environments which occur around the surface of the Earth are considered in terms of the processes which occur in each and the character of the sediment deposited. By way of introduction to these chapters, the concepts of depositional environments and sedimentary facies are considered in this chapter. The methodology of analysing sedimentary rocks, recording data and interpreting them in terms of processes and environment is considered here in a general sense. Examples are quoted which relate to processes and products in environments considered in more detail in subsequent chapters.

5.1 Interpreting past depositional environments

The settings in which sediment accumulates are familiar as geomorphological entities such as rivers, lakes, coasts, shallow seas, and so on. One of the goals of sedimentary geology is to determine the environment in which any given succession of sedimentary rocks accumulated. In pursuing this objective the sedimentologist is trying to establish the conditions on the surface of the Earth at different times and in different places, and hence to build up a picture of the history of the surface of the planet. The first stage is the investigation of sedimentary rocks by means of a rigorous, scientific methodology which is known as facies analysis (Walker 1992a; Reading & Levell 1996).

5.2 The concept of 'facies'

A fundamental tool in the description and interpretation of sedimentary rocks is the concept of *sedimentary facies*. The word 'facies' is defined in slightly differing ways by different authors, but the consensus is that it refers to the sum of the characteristics of a sedimentary unit (Middleton 1973). These characteristics include the dimensions, sedimentary structures, grain sizes and types, colour and biogenic content of the sedimentary rock. It is a means of classifying sedimentary rocks in a way which is infinitely adaptable to individual circumstances. An example would be 'cross bedded medium sandstone': this would be a rock consisting mainly of sand grains of medium grade, exhibiting cross bedding as the primary sedimentary structure. Not all aspects of the rock are necessarily indicated in the facies name and in other instances it may be important to emphasize different characteristics. The fact that the rock is red rather than grey may be important and form part of the facies or the occurrence of mica flakes may be significant. In other situations the facies name for a very similar rock might be 'red micaceous sandstone' if the colour and grain types were considered to be more important than the grain size and sedimentary structures. The full range of the characteristics of a rock would be given in the facies description which would form part of any study of sedimentary rocks.

Different terms are used where some aspect of the facies is the focus of interest: a *lithofacies* description is one confined to the characteristics of a rock which are the product of only physical and chemical processes; a *biofacies* description is one in which the observations are concerned with the fauna and flora present; and an *ichnofacies* description focuses on the trace fossils in the rock. As an example, a single rock unit may be described in terms of its lithofacies as a grey bioclastic packstone, as having a biofacies of echinoid and crinoids and a *Cruziana* ichnofacies: the sum of these and other characteristics would constitute the sedimentary facies.

5.2.1 Facies analysis

The facies concept is not just a convenient means of describing rocks and grouping sedimentary rocks seen in the field, it also forms the basis for the interpretation of strata. The lithofacies characteristics result from the physical and chemical processes which were active at the time of deposition of the sediments, and the biofacies and ichnofacies provide information about the palaeo-ecology during and after deposition. With a knowledge of the physical, chemical and ecological conditions it becomes possible to reconstruct the environment at the time of deposition. This process of *facies analysis*, the interpretation of strata in terms of depositional environments, can be considered central to the main objective of sedimentology and stratigraphy which is the recon-

struction of the past (Fig. 5.1) (Anderton 1985; Reading & Levell 1996).

The interpretation of sedimentary environments from facies can be a very simple exercise or require complex consideration of many factors before a tentative deduction can be made. In some cases there is a characteristic of the rock which is unique to a particular environment. As far as we know, hermatypic corals have only ever grown in shallow, clear and fairly warm sea water: the presence of these fossil corals in life position in a sedimentary rock may therefore be used to indicate that the sediments were deposited in shallow, clear and warm sea water. Where there are such direct indicators of conditions, the interpretation of sedimentary rocks in terms of past environments is straightforward. In contrast, cross bedded sandstone can form during deposition in deserts, rivers, deltas, lakes, beaches and shallow seas: a 'cross bedded sandstone' lithofacies would therefore not provide us with an indicator of a specific environment.

Interpretation of facies should be objective and based only on a recognition of the processes likely to have formed the beds. From the presence of symmetrical ripple structures in a fine sandstone it can be deduced that the bed was formed under shallow water, with wind over the surface of the water creating waves which stirred the sand to form symmetrical wave ripples. The 'shallow water' interpretation is made because wave ripples do

Fig. 5.1 Facies analysis flow chart.

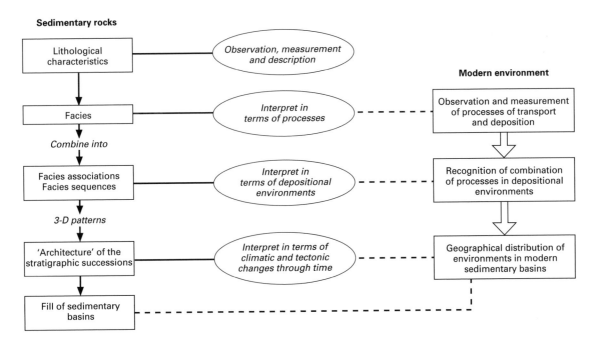

not form in deep water *(4.4.1)*, but it cannot be shown on the basis of the ripples alone whether the water was in a lake, lagoon or open shelf environment. The facies should therefore be referred to as 'symmetrically rippled sandstone' or perhaps 'wave rippled sandstone', but not 'lacustrine sandstone' because further information is required before that interpretation can be made.

In most cases it is the combination of different lithofacies, biofacies and ichnofacies which provides us with the information needed to deduce the environment of deposition from sedimentary strata. Observations which suggest deposition in a channel (a channel-fill facies) with deposits which show evidence of deposition by sheets of water which dry out (an overbank facies) would allow the interpretation of the rocks as the deposits of a river channel and floodplain (fluvial) environment *(9.4.1)*. The recognition of *lithofacies associations* is therefore an important part of facies analysis as it is most commonly the associations which provide the clues to the environment of deposition (Collinson 1969; Reading & Levell 1996).

5.2.2 Facies associations

Once all the beds in a succession have been assigned to facies, patterns in the distribution of these facies can then be investigated. For example (Fig. 5.2), do beds of the 'bioturbated mudstone' occur more commonly with (above or below) the 'shelly fine sandstone' or the 'medium sandstone with rootlets'? Which of these three occurs with the 'coal' facies? When attempting to establish associations of facies it is useful to bear in mind the processes of formation of each. Of the four examples of facies just mentioned, the 'bioturbated mudstone' and the 'shelly fine sandstone' both probably represent deposition in a subaqueous, possibly marine, environment whilst the 'medium sandstone with rootlets' and the 'coal' would both have formed in a subaerial setting. Two facies associations may therefore be established if, as would be expected, the pair of subaqueously deposited facies tend to occur together, as do the pair of subaerially formed facies.

Distinct facies can be interpreted in terms of the processes which led to the formation of the sediment. As noted above, many of these processes are not unique to a particular environment but one way of viewing depositional environments is to consider them in terms of the combinations of processes which occur in them. A tidal estuary, for example *(12.7)*, is a distinct physiographic

setting in which there is a channel supplied with fresh water entering a marine environment which experiences tidal currents and mudflats which are periodically flooded by the sea: this represents a very distinct combination of physical, chemical and biological processes. The products of these processes are seen as the sedimentary facies deposited in the channel and on the mudflats. The association of facies therefore reflects the combination of processes which occur in the depositional environment.

The procedure of facies analysis can therefore be thought of as a two-stage process: the recognition of facies which can be interpreted in terms of processes; and establishing facies associations which reflect combinations of processes and therefore environments of deposition (Fig. 5.1). The temporal and spatial relationships between depositional facies as observed in the present day and recorded in sedimentary rocks was recognized by Walther (1894). Walther's law can be simply summarized as stating that if one facies is found superimposed on another without a break in a stratigraphic succession then those two facies would have been deposited adjacent to each other at any one time.

Not all lithofacies have to be grouped into associations. A single facies may have been formed by such distinctly different processes that it is not appropriate to include it in another facies association. As an example, a succession of deposits formed in an arid region *(8.1)* may include different gravelly facies which may be grouped into an association as the deposits of an alluvial fan and a playa lake facies association consisting of evaporite and mudstone facies: a cross bedded, well-sorted medium sandstone facies does not fit into either the alluvial fan or playa lake facies association and must therefore be considered as a separate entity (a product of aeolian dune deposition: *8.2.3*).

5.2.3 Facies sequences

A facies sequence is simply a facies association with the facies occurring in a particular order (Reading & Levell 1996). They occur when there is a repetition of a series of processes as a response to regular changes in conditions. If, for example, a bioclastic wackestone facies is always overlain by a bioclastic packstone facies which is in turn always overlain by a bioclastic grainstone (Fig. 5.2), these three facies may be considered to be a facies sequence. Such a pattern may result from repeated shallowing upward due to deposition on shoals of bioclastic

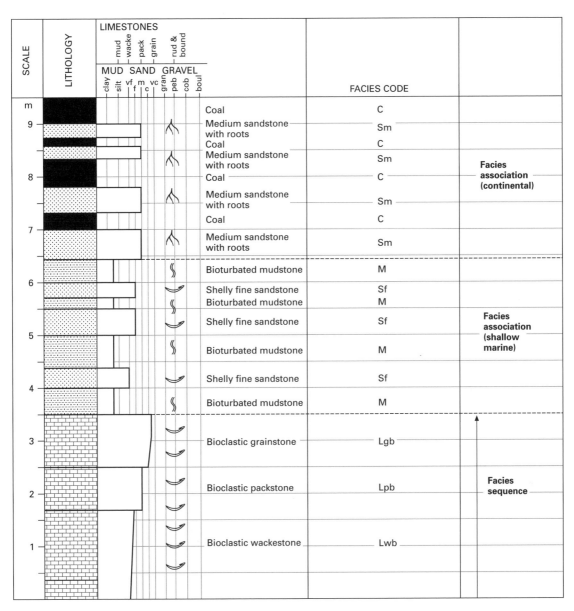

Fig. 5.2 Facies associations, facies sequences and facies codes.

sands and muds in a shallow marine environment *(14.6.2)*. Recognition of facies sequences can be on the basis of visual inspection of graphic sedimentary logs or by using a statistical approach to determining the order in which facies occur in a succession, such as a *Markov analysis* (Till 1974; Swan & Sandilands 1995). This technique simply requires a transition grid to be set up with all the facies along both the horizontal and the vertical axes of a table: each time a transition occurs from one facies to another (e.g. from bioclastic wackestone to bioclastic packstone facies) in a vertical succession this is entered on to the grid. Facies sequences show up as higher than average transitions from one facies to another.

5.2.4 Facies names and facies codes

In the process of carrying out a facies analysis of a

succession of sedimentary rocks the question of naming facies and facies associations arises. One option is to refer to facies simply by number or letter in an alphanumeric order. The drawback of this approach is that 'facies 1', 'facies 2', 'facies association A' and so on convey no descriptive information and provide no clues to the character of the sediments. A better way of doing things is to give a short, descriptive name to each facies—for example, 'laminated grey siltstone facies', 'foraminiferal wackestone facies' or 'cross bedded pebbly conglomerate facies'. A compromise has to be reached so that a given name adequately describes the facies but is not too cumbersome. A general rule would be to provide sufficient adjectives to distinguish the facies from each other but no more. For example, 'mudstone facies' is perfectly adequate if only one mudrock facies is recognized in the succession. On the other hand, the distinction between 'trough cross bedded coarse sandstone facies' and 'planar cross bedded medium sandstone facies' may be important in the analysis of successions of shallow marine sandstone.

The names for facies should normally be purely descriptive but it is quite acceptable to refer to facies associations in terms of the interpreted environment of deposition. An association of facies such as 'symmetrically rippled fine sandstone', 'black laminated mudstone' and 'grey graded siltstone' may have been interpreted as having been deposited in a lake on the basis of the facies characteristics, and perhaps some biofacies information indicating fresh water fauna. This association of facies may therefore be referred to as a 'lacustrine facies association' and may have been distinguished from other continental facies associations deposited in river channels ('fluvial channel facies association') and as overbank deposits ('floodplain facies association').

To make long facies names easier to deal with, a system of coded abbreviations is often employed when summarizing large amounts of facies information (Fig. 5.2). It helps if the codes are easy to interpret and related to the facies name. One convention used in the description of facies in terrigenous clastic sediments is a system which has the basic grain size indicated by the first letter followed by suffixes which describe sedimentary structures (Miall 1978). By this scheme conglomerates have the principal letter 'G' (for gravel), with 'S' for sands and 'F' for fine-grained mudrocks; suffixes may provide further information about the grain size (e.g. 'Sc' to indicate 'sand, coarse'), sedimentary structures

('Gx' for cross stratified conglomerates, the letter 'x' being a common abbreviation for 'cross'), colour or other distinctive characteristics. There are no rules for the code letters used, and there are many variants on this theme (some workers use the letter 'Z' for silts, for example) including similar schemes for carbonate rocks based on the Dunham classification (3.1.4). As a general guideline, it is best to develop a system which is consistent in pattern (e.g. all sandstone facies starting with the letter 'S') and which uses abbreviations which can be readily interpreted.

5.3 Distribution of palaeoenvironments in time and space

Once the likely palaeoenvironment for a suite of sedimentary rocks has been established by facies analysis, the relationship to rocks which were deposited at the same time in different places can be considered as well as the changes in palaeoenvironment through time in any one place. This can only be carried out once a temporal framework has been established using stratigraphic correlation techniques outlined in Chapters 18–21. Palaeoenvironmental analysis is then combined with stratigraphy into the field of study known as basin analysis, which is considered briefly in Chapter 23.

One of the elements of palaeoenvironmental studies which is important in basin analysis is a determination of the direction in which rivers flowed, deltas built out, coastlines prograded, submarine fans spread out, and so on. For this some sort of directional information is useful and this can be obtained from sedimentary rocks using palaeocurrent indicators.

5.4 Palaeocurrents

A *palaeocurrent indicator* is evidence for the direction of flow at the time a sediment was deposited. The value of knowing the direction of flow is that it makes it possible to start making palaeogeographic reconstructions. Facies and facies associations deposited in different depositional environments can be linked on the basis of connections indicated by palaeocurrent data (Potter & Pettijohn 1977). For instance, a knowledge of the direction of flow in the channels of a fluvial deposit makes it possible to relate these deposits to deltaic or estuarine sediments, as these will be expected to occur downstream. This sort of interpretation is extremely useful in making predictions about the characteristics of rocks which cannot be seen because they are covered by younger strata. Palaeocurrent analysis is therefore carried out as part of a facies analysis to learn more about palaeoenvironments.

5.4.1 Palaeocurrent indicators

Certain sedimentary structures formed by the flow of water or air can be used as indicators of palaeocurrent or *palaeoflow*. Two groups of palaeocurrent indicators can be distinguished.

Unidirectional indicators are features which give the direction of flow.

1 Cross lamination *(4.3.1)* is produced by ripples migrating in the direction of the flow of the current. The dip direction of the cross laminae is measured in sedimentary rocks.

2 Cross bedding *(4.3.2)* is formed by the migration of aeolian and subaqueous dunes, and the direction of dip of the lee slope is approximately the direction of flow. In sedimentary rocks the direction of dip of the cross strata in cross bedding is measured.

3 Large-scale cross bedding and cross stratification formed by large bars in river channels *(9.2.1)* and shallow marine settings *(14.4)*, or the progradation of foresets of Gilbert-type deltas *(12.3)*, are an indicator of flow direction. The direction of dip of the cross strata is measured. An exception is epsilon cross stratification produced by point bar accumulation which lies perpendicular to flow direction *(9.2.2)*.

4 Clast imbrication *(2.3.3)* is formed when discoid gravel clasts become orientated in strong flows into a stable position, with one of the two longer axes dipping upstream when viewed from the side. Note that this is opposite to the measured direction in cross stratification.

5 Flute casts *(4.8.1)* are local scours in the substrata generated by vortices within a flow. As the turbulent vortex forms it is carried along by the flow and lifted up away from the basal surface, to leave an asymmetric mark on the floor of the flow with the steep edge on the upstream side. The direction along the axis of the scour away from the steep edge is measured.

Flow axis indicators are structures which provide information about the axis of the current but do not differentiate between upstream and downstream directions. They are nevertheless useful in combination with unidirectional indicators—for example, grooves and flutes may be associated with turbidites *(4.6.2)*.

1 Primary current lineations *(4.3.4)* on bedding planes are measured by determining the orientation of the lines of grains.

2 Groove casts *(4.8.1)* are elongate scours caused by the indentation of a particle carried within a flow which give the flow axis.

3 Elongate clast orientation may provide information if needle-like minerals, elongate fossils such as belemnites, or pieces of wood show a parallel alignment in the flow.

4 Channel and scour margins can be used as indicators because the cut bank of a channel lies parallel to the direction of flow.

5.4.2 Measuring palaeocurrents from cross stratification

The measurement of the direction of dip of an inclined surface is not always straightforward especially if the surface is curved in three dimensions, as is the case with trough cross stratification. Normally an exposure of cross bedding which has two faces at right angles is needed (Fig. 5.3). Where a horizontal surface cuts through trough cross bedding, determination of the palaeoflow direction is easier and only the horizontal surface is required (Fig. 5.4). The determination of palaeoflow direction from planar cross stratification is straightforward because the plane only dips in one direction. In all cases a single vertical cut through the cross stratification is unsatisfactory because this only gives an apparent dip which is not necessarily the direction of flow.

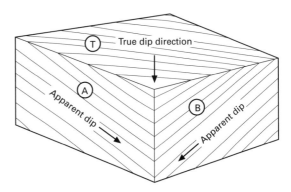

Fig. 5.3 The true direction of dip of planes (e.g. planar cross beds) cannot be determined from a single vertical face (face A or B): a true dip can be calculated from two different apparent dip measurements or measured directly from the horizontal surface (T).

Fig. 5.4 Trough cross bedding exposed on the top surface of a bed of sandstone, Cambrian, Sinai Peninsula, Egypt. The concave trace of the troughs on the bed surface indicates a flow direction away from the viewpoint.

5.4.3 Presentation and analysis of directional data

Directional data are commonly collected and used in geology. Palaeocurrents are most frequently encountered in sedimentology but similar data are collected in structural analyses and in palaeoecological studies. Once a set of data has been collected it is useful to be able to determine parameters such as the mean direction and the spread about the mean (or standard deviation). Calculating the mean of a set of directional data is not as straightforward as, for example, determining the average of a set of bed thickness measurements. Palaeocurrents are measured as a bearing on a circle of 360 degrees. Determining the average of a set of bearings by adding them together and dividing by the number of readings does not give a meaningful result: to illustrate why, two

bearings of 010° and 350° obviously have a mean of 000°/360°, but simple addition and division by two gives an answer of 180°, the opposite direction. Calculation of the circular mean and circular variance of sets of palaeocurrent data can be carried out with a calculator or by using a computer program. The mathematical basis for the calculation (Till 1974; Swan & Sandilands 1995) is outlined below.

In order to handle directional data mathematically it is first necessary to translate the bearings into rectangular coordinates and express all the values in terms of x and y axes. For each bearing β, determine the x and y values, given by:

$$x = \cos \beta$$

$$y = \sin \beta$$

Then add all the x values together and determine their mean $\bar{x}$. Finally, add all the y values together and determine their mean $\bar{y}$. The result will be a mean value for the average direction expressed in rectangular coordinates, with the values of $\bar{x}$ and $\bar{y}$ each between −1 and +1. To determine the bearing that this represents, calculate:

$$\bar{\beta} = \tan^{-1}(\bar{y}/\bar{x})$$

This value of $\bar{\beta}$ will be between +90° and −90°. To correct this to a true bearing, it is necessary to determine which quadrant the mean will lie in. This can be established by taking the sine and cosine of $\bar{\beta}$: if both are positive, the bearing is 000–090°, cosine negative 090–180°, both negative 180–270° and sine negative 270–360°.

The spread of the data around the calculated mean is proportional to the length of the line, R. If the end lies very close to the perimeter of the circle, as happens when all the data are very close together, R will have a value close to 1. If the line R is very short it is because the data have a wide spread: as an extreme example, the mean of 000°, 090°, 180° and 270° would result in a line of length 0 as the mean values of x and y for this group would lie at the centre of the circle. The length of the line R is calculated using Pythagoras's theorem:

$$R = \sqrt{(\bar{x}^2 + \bar{y}^2)}$$

Palaeocurrent data are normally plotted on a *rose diagram* (Fig. 5.5). This is a circular histogram on which directional data are plotted. The calculated mean can also be added. The base used is a circle divided up with radii at 10° or 20° intervals and containing a series of

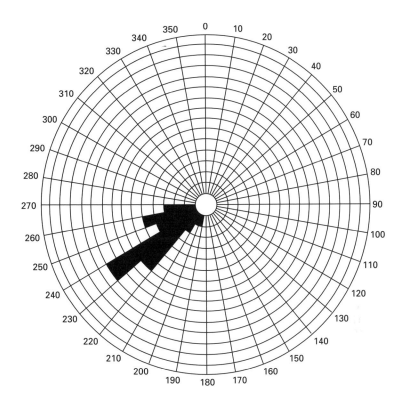

Fig. 5.5 A rose diagram is used as a way of representing palaeocurrent data (*N*=33; scale from centre is one division for each reading).

concentric circles. The data are first grouped into blocks of 10° or 20° (000–019°, 020–039°, etc.) and the number which fall within each range marked by gradations out from the centre of the circular histogram. In this example (Fig. 5.5) three readings are between 260° and 269°, five between 250° and 259°, and so on. The scale from the centre to the perimeter of the circle should be indicated, and the total number, *N*, in the data set indicated.

Palaeocurrent data collected from strata which have been tectonically deformed and tilted must be reorientated back to the depositional horizontal. This manipulation of directional data requires the use of stereonet techniques used commonly in structural geology.

5.5 Provenance

Palaeocurrent data provide an indication of the direction of transport of sediment, which in turn gives clues about where the clastic detritus originated. Further information about the source of the sediment, or *provenance* of the material, may be obtained from an examination of the clast types present (Pettijohn 1975). If a clast present in a sediment can be recognized as being characteristic of a particular source area by its petrology or chemistry then

its provenance can be established. In some circumstances this makes it possible to establish the palaeogeographical location of a source area and provides information about the timing and processes of erosion in uplifted areas (Dickinson & Suczek 1979).

Provenance studies are generally relatively easy to carry out in coarser clastic sediments because a pebble or cobble may be readily recognized as having been eroded from a particular bedrock lithology. Many rock types may have characteristic textures and compositions which allow them to be identified with confidence. It is more difficult to determine the provenance where all the clasts are sand-sized because many of the grains may be individual minerals which could have come from a variety of sources. Quartz grains in sandstones may have been derived from granite bedrock, a range of different metamorphic rocks or reworked from older sandstone lithologies, so although very common, quartz is often of little value in determining provenance. It has been found that certain heavy minerals *(2.4.2)* are very good indicators of the origin of the sand (Table 5.1). Provenance studies in sandstones are therefore often carried out by separating the heavy minerals from the bulk of the grains and identifying them individually (Mange & Maurer

Table 5.1 Heavy minerals are used as indicators of the source (provenance) of detrital sediment.

Source rock	Common characteristic minerals
Acid igneous	Apatite
	Zircon
	Biotite
	Magnetite
	Hornblende
Basic igneous	Rutile
	Augite
	Ilmenite
	Hypersthene
High-rank metamorphic	Garnet
	Kyanite
	Sillimanite
	Staurolite
	Epidote
Low-rank metamorphic	Tourmaline
	Biotite
Pegmatite	Fluorite
	Garnet
	Tourmaline
Sedimentary	Zircon (rounded)
	Tourmaline (rounded)
	Leucoxene

1992). This procedure is called *heavy mineral analysis*, and it can be an effective way of determining the source of the sediment. Clay mineral analysis is also sometimes used in provenance studies because certain clay minerals are characteristically formed by the weathering of particular bedrock types (Blatt 1985): for example, weathering of basaltic rocks produces the clay minerals in the smectite group *(2.5.3)*.

5.6 Graphic sedimentary logs

A sedimentary log is a graphical method for representing a series of beds of sediments or sedimentary rocks. It is also an efficient method of collecting data systematically. There are many different schemes in use, but they are all variants on a theme. The format presented here (Fig. 5.6) closely follows that of Tucker (1982, 1996); other commonly used formats are illustrated in Collinson and Thompson (1982). The objective of any graphic sedimentary log should be to present the data in a way which is easy to recognize and interpret using simple

symbols and abbreviations which should be understandable without reference to a key (although a key should always be included to avoid ambiguity). Facies analysis and ultimately complete palaeoenvironmental analyses are made on the basis of information presented in graphic sedimentary logs.

5.6.1 Drawing a graphic sedimentary log

The vertical scale used is determined by the amount of detail required. If information on beds only a centimetre thick is needed then a scale of 1 : 10 is appropriate. A log drawn through tens or hundreds of metres may be drawn at 1 : 100 if beds less than 10 cm thick need not be recorded individually. Summary logs which provide only an outline of a succession of strata may be drawn at a scale of 1 : 1000. Intermediate scales are also used, with multiples of 2 or 5 used to make scale conversion easier.

Most of the symbols for lithologies in common use are more or less standardized: dots are used for sands and sandstone, bricks for limestone, and so on. The scheme can be modified to suit the succession under description, for example, by the superimposition of the letter 'G' to indicate a glauconitic sandstone, by adding dots to the brickwork to represent a sandy limestone, and so on. In most schemes the lithology is shown in a single column. Alongside the lithology column (to the right) there is space for additional information about the sediment type and for the recording of sedimentary structures (see below). A horizontal scale is used to indicate the grain size in clastic sediments. The Dunham classification for limestones *(3.1.4)* can also be represented using this type of scale. This scheme gives a quick visual impression of any trends in grain size in normally or reverse-graded beds, and fining-upward or coarsening-upward successions of beds.

By convention, the symbols used to represent sedimentary structures bear a close resemblance to the appearance of the feature in the field or in core (Fig. 5.7). This representation is somewhat stylized for the sake of simplicity and to clarify the interpretation of the structure. Again, symbols can be adapted to suit individual circumstances. Where space allows, symbols are placed within the bed but may also be drawn alongside. Bed boundaries may be sharp, erosional, where the upper bed cuts down into the lower one, or transitional/gradational, in which there is a gradual change from one lithology to another.

Any other details about the succession of beds can

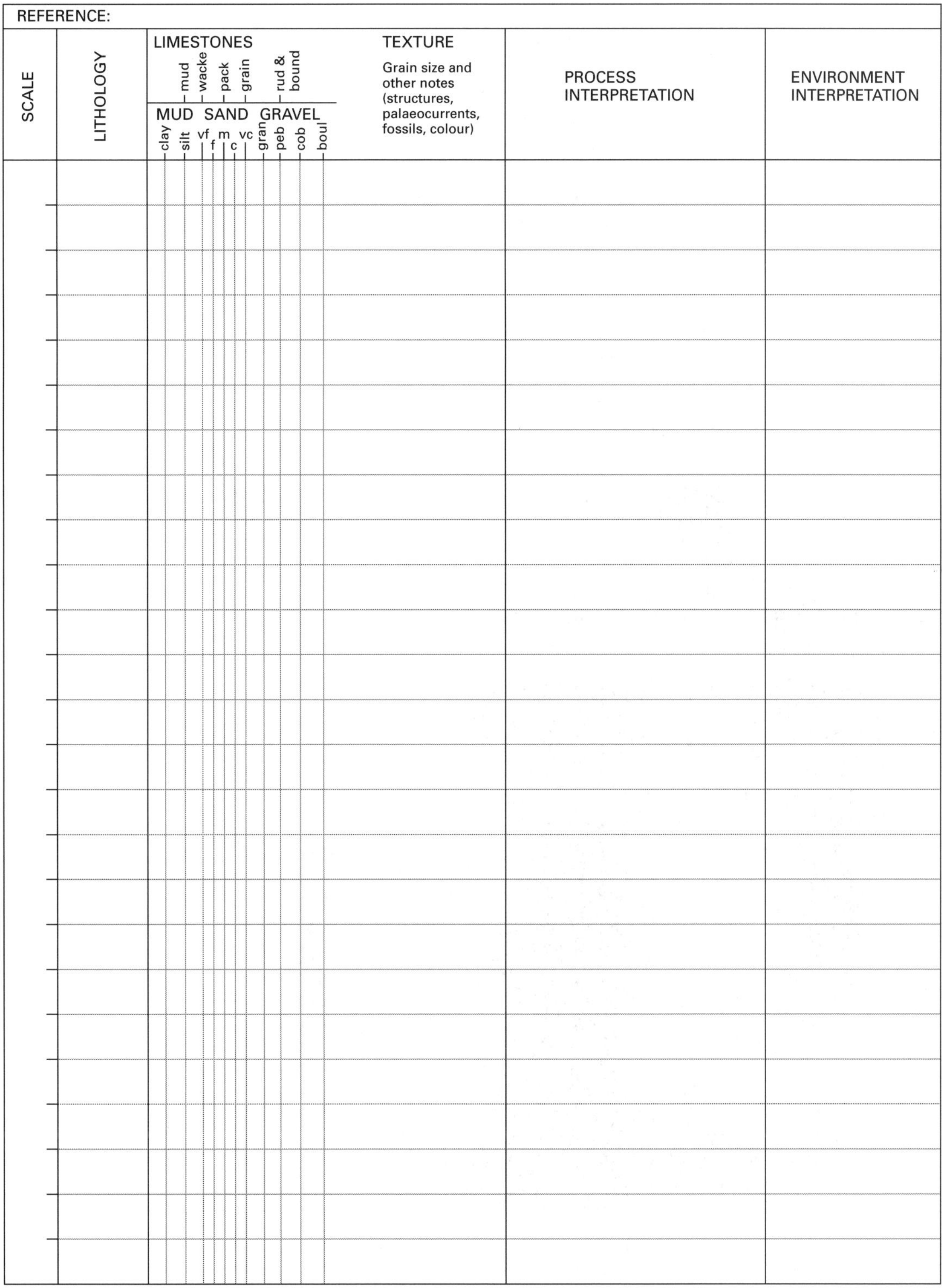

Fig. 5.6 A template for a graphic sedimentary log.

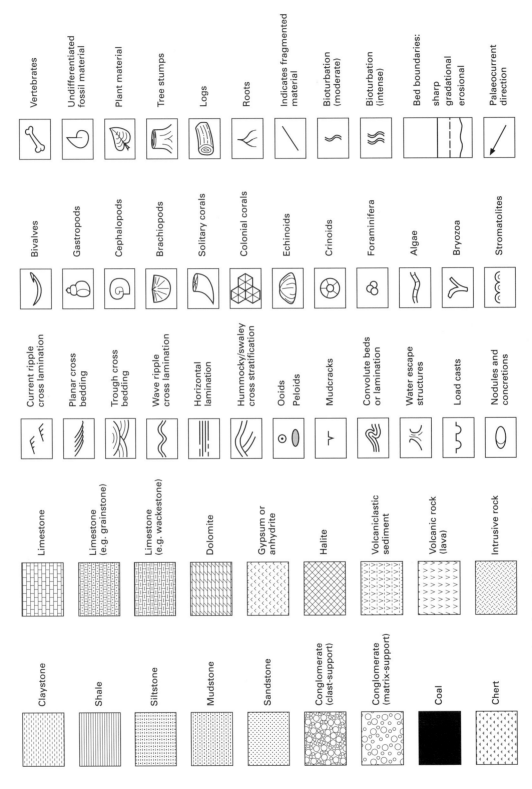

Fig. 5.7 Symbols commonly used on graphic sedimentary logs.

also be recorded on the graphic log (Fig. 5.8). Palaeo-current data may be presented as a series of arrows orientated in the direction of palaeoflow measured or summarized for a unit as a rose diagram *(5.4.3)* along-side the log. Colour is normally recorded in words or abbreviations, and any further remarks or observations may simply be written alongside the graphic log in the appropriate place.

Interpretation of the information in terms of processes and environment is normally carried out back in the laboratory. If a full facies analysis is to be done, litho-facies must be identified and all intervals on the graphic logs assigned to one of these facies. The relationships between facies can be more readily seen on a graphic log than in any other form of presentation of the data.

Computer-aided graphic log presentation has become popular in recent years. The widespread use of computer drawing packages has resulted in a tendency for the

symbols used on logs to become more standardized and stylized. Dedicated log drawing packages exist but more commonly an ordinary drawing package is used to generate the log. A drawback of this trend is that these computer-generated graphic logs often do not contain as much information as a more individualistic one drawn by hand. Subtle variations in the form of sedimentary struc-tures can be included on a hand-drawn log but would be lost by the use of standard symbols (Anderton 1985). There is still a place for the graphic log drawn by pen and pencil, and the log drawn in the field must still be considered as the fundamental raw data.

5.6.2 Other graphical presentations: sketches and photographs

A graphic log is a one-dimensional representation of beds of sedimentary rock which is the only presentation possible with drill-core and is perfectly adequate for 'layer-cake' strata (beds which do not vary in thickness

Fig. 5.8 Example of a graphic sedimentary log.

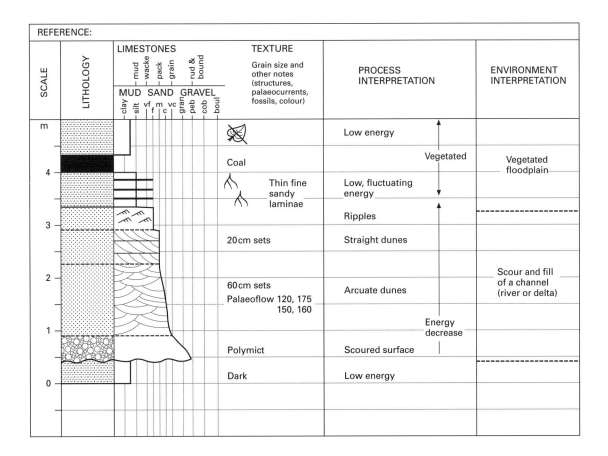

or characteristics laterally). Where an exposure of beds reveals that there is significant lateral variation—for example, river channel and overbank deposits in a fluvial environment—a single, vertical log does not adequately represent the nature of the deposits. A two-dimensional representation is required in the form of a section drawn of a natural or artificial exposure in a cliff or cutting (Fig. 5.9).

A carefully drawn sketch section showing all the main sedimentological features (bedding, cross stratification, etc.) is normally satisfactory and may be supplemented by a photograph if possible. In an ideal case, a photograph taken on a previous visit can be used as a template for a field sketch. A photograph should never be considered as a substitute for a field sketch: sedimentological features are never as clear on a photograph as they are in the field and a lot of information can be lost if important features and relationships are not drawn at the time. A good geological sketch need not be a work of art. Geological features should be clearly and prominently represented whilst incidental objects such as trees and bushes can often be ignored. All sketches and photographs must include a scale of some form and the orientation of the view must be recorded.

Further information on the field description of sedimentary rocks is provided in Tucker (1996).

Fig. 5.9 Example of a field sketch. The complex lateral variations present in some facies, such as the fluvial deposits sketched here (Chapter 9), cannot be adequately represented by a single vertical graphic.

5.7 Facies and environments: summary

An objective, scientific approach is essential for successful facies analysis. A succession of sedimentary strata should be described first in terms of the lithofacies (and sometimes biofacies and ichnofacies) present, at which stage interpretations of the processes of deposition can be made. The facies can then grouped into lithofacies associations which can be interpreted in terms of depositional environments on the basis of the combinations of physical, chemical and biological processes which have been identified from analysis of the facies. There are facies associations and sequences which commonly occur in particular environments, and these are illustrated in the following chapters as 'typical' of these environments. However, there is a danger of making mistakes by 'pigeon-holing', that is, trying to match a succession of rocks to a particular 'facies model'. Whilst general characteristics usually give a good clue as to the depositional environment, small details can be vital and must not be overlooked. The analysis of palaeocurrent data is a very useful adjunct to facies interpretation, and together these form the basis for determining environments of deposition in the past. To achieve all of these analyses, an effective method of presenting data from sedimentary rocks is required: this is provided by graphic sedimentary logs.

The objectives of facies analysis are to determine the environment of deposition of successions of rocks in the sedimentary record. A general assumption is made that the sedimentary environments which exist today (Fig.

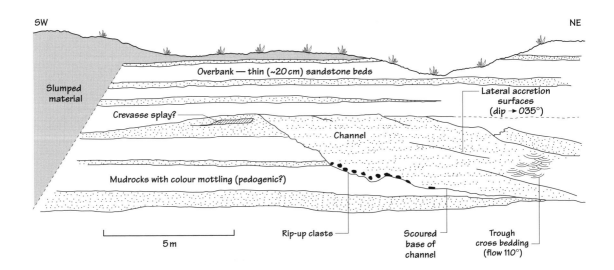

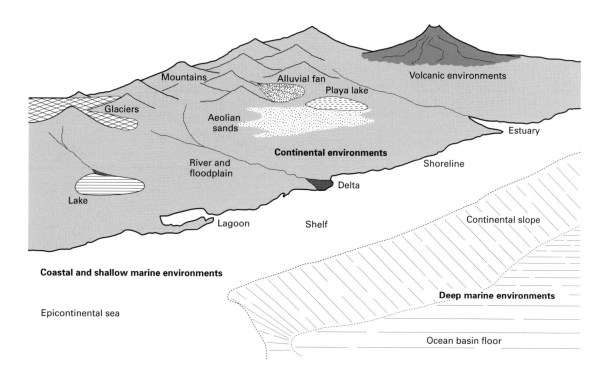

Fig. 5.10 Depositional sedimentary environments.

5.10) have existed in the past. In broad outline this is the case, but there is evidence from the stratigraphic record of conditions which existed during periods of Earth history which have no modern counterparts. This 'non-uniformitarian' aspect of stratigraphy is considered in the context of changes in global climate and vegetation patterns in Chapter 24. The characteristics of continental depositional environments are considered in Chapters 6–10, marine environments in Chapters 11–15, and volcanic settings in Chapter 16.

Further reading

Anderton, R. (1985) Clastic facies models and facies analysis. In: *Sedimentology, Recent Developments and Applied Aspects* (eds P.J. Brenchley & B.P.J. Williams), pp. 31–47. Blackwell Scientific Publications, Oxford.

Pettijohn, F.J. (1975) *Sedimentary Rocks*, 3rd edn, 718 pp. Harper and Row, New York.

Reading, H.G. (ed.) (1996) *Sedimentary Environments: Processes, Facies and Stratigraphy*, 688 pp. Blackwell Scientific Publications, Oxford.

Tucker, M.E. (1996) *Sedimentary Rocks in the Field*, 2nd edn, 153 pp. Wiley, Chichester.

Walker, R.G. & James, N.P. (eds) (1992) *Facies Models: Response to Sea Level Change*, 409 pp. Geological Association of Canada, St Johns, Newfoundland.

6 Continents: sources of sediment and environments of deposition

A little under one-third of the surface of the Earth is land. As man inhabits continental environments it is relatively easy to study land surface processes and their products in the form of sediments and sedimentary rocks. Our understanding of these processes (geomorphology) is rather more complete than our knowledge of the oceans (oceanography) although from a geological point of view marine environments and the rocks formed in them are more important. Strata deposited in seas are volumetrically greater than continental deposits in the stratigraphic record, they contain a more complete record of Earth history and they are economically more significant, containing most of the known reserves of oil and gas. On the other hand, the ultimate source of the clastic and chemical deposits in the oceans is the continental realm, where weathering and erosion generate the sediment which is carried as bedload, in suspension or as dissolved salts to the marine realm. The tectonic and climatic controls on continental environments not only determine the patterns and nature of sedimentation in rivers, lakes and deserts but also fundamentally determine the supply of sediment to the world's oceans. Tectonic forces create regions of uplift and subsidence which, respectively, act as sources and sinks for sediment. The composition and texture of the sediment is determined by climate-dependent weathering processes; the rates at which erosion occurs and detritus is transported to depositional environments are governed by the amounts and periodicity of rainfall.

6.1 From source of sediment to formation of strata

In considering the formation of sedimentary rocks and the creation of a stratigraphic sequence, the starting point is pre-existing rocks of igneous, metamorphic or sedimentary origin. The processes involved in the pathway to the creation of sedimentary strata are illustrated in Fig. 6.1. The ultimate source of clastic detritus is bedrock, which must be uplifted by tectonic processes. Once

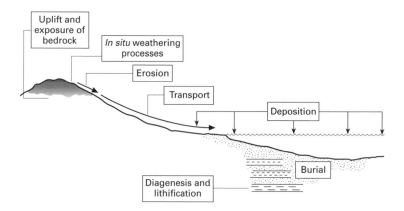

Fig. 6.1 The pathway of processes involved in the formation of a succession of clastic sedimentary rocks.

elevated, this bedrock undergoes weathering at the land surface to create clastic particles and release ions into solution in surface and near-surface waters. Erosion follows, the process of removal of the weathered material from the bedrock surface, allowing the transport of material as dissolved or particulate matter by a variety of mechanisms. Eventually the sediment will be deposited by physical, chemical and biogenic processes in a sedimentary environment on land or in the sea. The final stage is the lithification of the sediment to form sedimentary rocks which may then be exposed at the surface by tectonic processes. The first stages almost always occur on continents and in some cases the whole cycle takes place in the continental realm.

6.2 The formation of mountains and hills

In the past few decades the theory of plate tectonics has been developed out of the older concepts of continental drift. Plate tectonics is an elegant and unifying theory which links the formation of igneous and metamorphic rocks and all the superficial uplift and subsidence to the movement of separate plates of Earth's crust. New crust is created by upwelling of magmas from the upper mantle in the ocean basins as plates move apart; oceanic crust is destroyed when it goes down into the mantle at subduction zones (destructive plate margins). Continental crust is less dense than oceanic crust and where two plates of continental crust collide, the crust is thickened to form mountain chains; beneath the mountains some of the crust is forced down into the mantle where it is assimilated. Volcanism is associated with

areas where the crust is formed and with destructive plate margins. For further details of plate tectonic processes the reader is referred to texts such as Kearey and Vine (1996).

The plate movements and associated igneous activity create the contours of the surface of the Earth which are then modified by erosion and deposition. Areas of high ground on the surface of the globe today can be related to plate boundaries (Fig. 6.2). For example, the Himalayas have formed as a result of the collision of the continental plates of India and Asia, and the Andes have a core of igneous rocks related to the subduction of oceanic crust of the east Pacific beneath South America.

6.3 Continental climatic regimes

The climate belts around the world are controlled principally by latitude (Fig. 6.3). The amount of energy from the Sun per unit area is less in polar regions than in the equatorial zones so there is a temperature gradient from each pole to the equator. Latitude and temperature determine the air pressure belts. There are regions of high pressure at the poles and in sub-tropical latitudes (around 30° north and south). Around the equator there is a zone of low pressure and also a low pressure belt lies in the sub-polar region around 60° north and south. These differences in pressure give rise to winds which move air masses between areas of high pressure in the sub-tropical and polar zones to regions of low pressure in between them. The Coriolis force imparted by the rotation of the globe influences these air movements to produce a basic pattern of winds around the Earth.

Fig. 6.2 Mountain ranges are the sites of weathering and erosion of bedrock, the origin of terrigenous clastic sediments (Pindos Mountains east of Arta, western Greece).

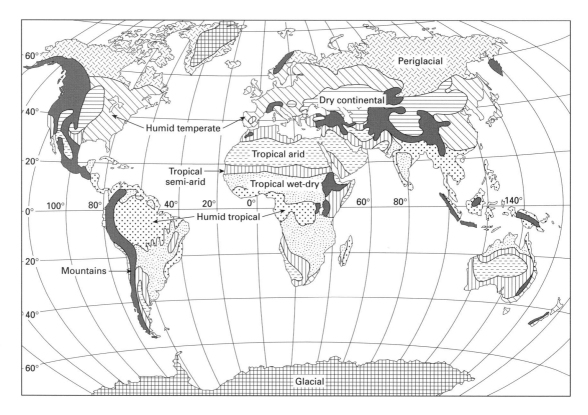

Fig. 6.3 Present-day world climate belts.

The combination of temperature distribution and wind belts gives rise to four main climate zones. *Polar* regions lie mainly north and south of the Arctic and Antarctic circles. They are regions of high pressure and low temperatures with conditions above freezing only part of the year, if at all. Between about 60° and 30° either side of the equator lie the *temperate*, moist mid-latitude climate belts which have strongly seasonal climates and moderate levels of precipitation. The *dry sub-tropical* belts are variable in width, depending on the configuration of land masses in the areas of the tropics of Cancer and Capricorn. Over large continental areas these dry areas are regions of high pressure, high temperatures and low precipitation. In the middle lies the *wet equatorial* zone of high rainfall and high temperatures.

These climate zones are not uniform in width around the world and have different local climatic characteristics which are determined by the extent of continental land masses and the elevation of the land. As both the positions and heights of continents vary through geological time due to plate movements, palaeoclimate belts can only be related to the modern belts in a relatively simplistic way unless complex climate modelling is carried out *(24.7.2)*.

6.4 Surface processes

Areas of high ground are found on all the continents and larger islands on the Earth. These mountains and hills are the regions where most of the clastic detritus is generated. Bedrock exposed on high ground is open to the elements: wind, rain, snow, and extremes of temperature all attack the rock, breaking it down both physically and chemically in a series of processes known collectively as *weathering*. The products of weathering are carried away from the high ground by gravity, water, ice or wind, bringing about the removal of material from the area, the process of *erosion*. The difference in height between these regions and the sea (or, indeed, the sea bed) provides the gravitational potential to move weathered material to the sites where it ultimately forms a sedimentary deposit. To a limited extent, gravity works alone, but more importantly, water and ice moving downslope carry the bulk of the weathered detritus. The wind can be an important medium for the transport of fine particulate matter.

Climate and geomorphology are the most important controlling forces which govern the processes which operate in mountainous regions. Weathering processes are particularly sensitive to temperature (both the average and the range) and the amount of water available. Transport processes are governed by the shape of the mountain slopes (the geomorphology), by the supply of water to rivers and streams, and by temperature (creation of the cold conditions under which glaciers form).

Some localized deposition of sediment may occur in upland regions, but these pockets of wind-blown sand, rock falls, river bed or lake deposits have a very low *preservation potential* and so they are rarely recorded in the stratigraphic record. The products of transport and deposition by glaciers also have a low preservation potential in the long term but are common features in present mountain areas because of the retreat of glaciers following the most recent ice age.

6.5 Weathering processes

As a body of rock comes close to the surface of the Earth, it becomes exposed to weathering processes, the first stage in the transformation of the material into sedimentary rocks. *Weathering* is the breakdown and alteration of original rock into material which is then available for transport away from the site (Fig. 6.4).

6.5.1 Physical weathering

These are processes which break the solid rock into pieces and may separate the different minerals without involving any chemical reactions. The most important agents in this process are as follows.

FREEZE–THAW ACTION

Water entering cracks in rock expands upon freezing, forcing the cracks to widen; this process is also known as *frost shattering*, and it is extremely effective in areas which regularly fluctuate around 0°C such as high mountains in temperate climates and in polar regions (Fig. 6.5).

SALT GROWTH

Sea water or other water containing dissolved salts may also penetrate into cracks, especially in coastal areas. Upon evaporation of the water, salt crystals form and their growth generates localized, but significant, forces which can further open cracks in the rock.

TEMPERATURE CHANGES

Changes in temperature alone have often been considered to play a role in the physical breakdown of rock. Rapid changes in temperature occur in some desert areas where the temperature can fluctuate by several tens of degrees Celsius between day and night; if different minerals expand and contract at different rates, the internal forces created could cause the rock to split. However, it has been found that most common minerals have similar coefficients of thermal expansion and it seems that temperature changes in the absence of water are not significant (Leeder 1982).

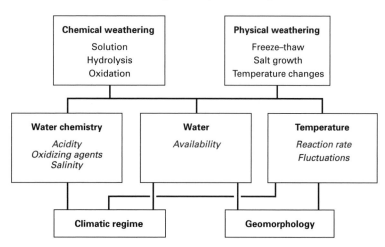

Fig. 6.4 Weathering processes and their controls.

Fig. 6.5 Broken-up granite affected by frost shattering in a mountainous area of the Sierra Nevada west of Tahoe, California, USA.

6.5.2 Chemical weathering

These processes involve changes to the minerals which make up a rock. The reactions which can take place are as follows.

SOLUTION

Most rock-forming silicate minerals have very low solubility in pure water at the temperatures at the Earth's surface and so most rock types are not susceptible to rapid solution. However, evaporite minerals such as halite (sodium chloride), which may be important in some sedimentary strata, will dissolve rapidly in both surface waters and *groundwater* (water passing through bedrock close to the surface). The solubility of many minerals is affected by the acidity of the water. Carbonate minerals which form limestone become more soluble as the acidity increases (Fig. 6.6).

HYDROLYSIS

Some minerals, including many silicates, react with water to form new minerals with different chemical structures; this may involve the addition of OH^- (hydroxide) ions. An example of this type of reaction is the formation of kaolinite (a clay mineral) from ortho-clase (a feldspar) by reaction with water. Hydrolysis reactions depend upon the dissociation of H_2O into H^+ and OH^- ions which occurs when there is an acidifying agent present. Natural acids which are important in

Fig. 6.6 Limestone weathers principally by solution of the rock to form karstic landforms. This pinnacle karst in Jurassic limestone beds in western Madagascar, north of Morondava, is a spectacular example of karst scenery.

promoting hydrolysis include carbonic acid (formed by the solution of carbon dioxide in water) and humic acids, a range of acids formed by the bacterial breakdown of organic matter in soils.

OXIDATION

Most sulphides and some silicates (e.g. glauconite) react with oxygen in the air or in water to form new compounds. The most widespread evidence of oxidation is the formation of iron oxides and hydroxides from minerals containing iron. The distinctive red-orange rust colour of iron (III) oxide may be seen in many rocks exposed at the surface, even though the amount of iron present may be very small.

6.5.3 Controls on weathering processes

All the chemical weathering processes are affected by factors which control the rate and the pathway of the reactions. First of all, water is essential to all chemical weathering processes and hence these reactions take place on a very small scale where water is scarce (e.g. in deserts). Temperature is also important, because most chemical reactions are more vigorous at higher temperatures; hot climates therefore favour chemical weathering, although there is an important exception in the case of calcium carbonate minerals which are more soluble in cold water. Finally, water chemistry affects the reactions: the presence of acids enhances hydrolysis and dissolved oxidizing agents facilitate oxidation reactions (Einsele 1992).

From the above it is clear that climate plays a very important role in both physical and chemical weathering processes (Fig. 6.7). In hot, wet, tropical areas, chemical weathering is enhanced because of the higher temperatures and abundance of water. Bedrock in these areas is typically deeply weathered and highly altered at the surface: seemingly resistant lithologies such as granite are reduced to quartz grains and clay as the feldspars and other silicate minerals are altered by surface weathering processes. Deep lateritic soils develop: *laterites* are red soils composed mainly of iron oxides and aluminium oxides which, along with quartz, are the products of the most extreme weathering of silicate bedrock. In general, chemical weathering results in fine-grained detritus and partial solution of the bedrock.

In contrast, chemical weathering is much less significant in cold, dry regions where chemical reactions are

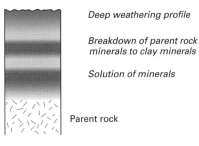

HOT AND HUMID

Deep weathering profile

Breakdown of parent rock minerals to clay minerals

Solution of minerals

Parent rock

Weathering products
Abundant clay
Quartz sand

COLD AND ARID

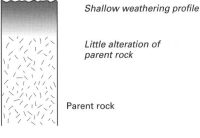

Shallow weathering profile

Little alteration of parent rock

Parent rock

Weathering products
Lithic fragments
'Unstable' minerals
(e.g. feldspars)
Little clay

Fig. 6.7 The contrast between processes and products of weathering in hot and wet conditions and cold and dry conditions.

slower. In these areas physical weathering processes are more effective, although these too rely on the presence of water. The products of weathering in high mountains are typically debris of the bedrock, broken up but with little or no change in the mineral composition. A granite breaks down into gravel clasts, as well as grains of quartz, feldspar and other rock-forming minerals. Most of the products of physical weathering are hence coarse material with little clay generated or solution of the rock.

In temperate climates both physical and chemical weathering processes tend to be subdued. Deep weathering profiles take much longer to form in the cooler climate, and colonization of soils by plants produces a vegetation cover which serves to protect the bedrock.

6.5.4 The products of weathering

The terms used for the products of weathering and of erosion of material exposed on continental land masses are *terrigenous* (meaning derived from land) and *epiclastic* (indicating that breakup occurred at the surface, although this term is often just used for the surface breakup of volcanic rocks). Weathered material on the surface is an important component of the layer of unconsolidated detritus, or *regolith*, which occurs on top of the bedrock in most places. Terrigenous clastic detritus comprises minerals weathered out of bedrock unaltered, lithic fragments and new minerals formed by weathering and soil-forming processes.

Rock-forming minerals can be categorized in terms of their stability in the surface environment. Stable minerals such as quartz are relatively unaffected by chemical weathering processes and physical weathering simply separates the quartz crystals from each other and from other minerals in the rock. Micas and orthoclase feldspars are relatively resistant to these processes, whilst plagioclase feldspars, amphiboles, pyroxenes and olivines all react very readily under surface conditions and are only rarely carried away from the site of weathering in an unaltered state. Depending on the extent of the weathering, many minerals may be partially altered when they are transported away.

The most important products of the chemical weathering of silicates are clay minerals *(2.5.3)*. A wide range of clay minerals form as a result of the breakdown of different bedrock minerals under different chemical conditions; the most common are kaolinite, illite, chlorite and montmorillonite. Oxides of aluminium (bauxite) and iron (mainly haematite) also form under conditions of extreme chemical weathering.

In places where chemical weathering is subdued, lithic fragments may form an important component of the detritus generated by physical processes. The nature of these fragments will directly reflect the bedrock type and can include any lithology found at the Earth's surface. Some lithologies do not last very long as fragments: rocks made of evaporite minerals are readily dissolved and other lithologies are very fragile, making them susceptible to breakup. Detritus composed of basaltic lithic fragments can form around volcanoes and broken-up limestone can make up an important clastic component of some shallow marine environments *(16.3.6)*.

6.5.5 Formation of soils

Soil formation is an important stage in the transformation of bedrock and regolith into detritus available for transport and deposition. *In situ* (in place) physical and chemical weathering of bedrock creates a soil which may be further modified by biogenic processes (Fig. 6.8). The roots of plants penetrating into bedrock can enhance breakup of the underlying rock and the accumulation of vegetation (humus) leads to a change in the chemistry of the surface waters as humic acids form. Soil profiles become thicker through time as bedrock is broken up and organic matter accumulates, but an unconsolidated soil is also subject to erosion. Movement under gravity and by the action of flowing water may remove part or all of a soil profile. These erosion processes may be acute on slopes and important on flatter-lying ground where gullying may occur. The soil becomes disaggregated and contributes detritus to rivers. In temperate and humid tropical environments most of the sediment carried in rivers is likely to have been part of a soil profile at some stage.

Continental depositional environments are also sites of soil formation, especially the floodplains of rivers. These soils may become buried by overlying layers of sediment and are preserved in the stratigraphic record as fossil soils (palaeosols: *9.7*).

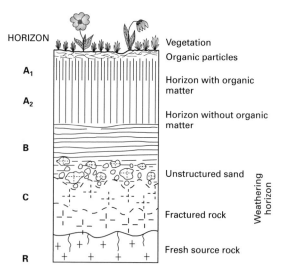

Fig. 6.8 An *in situ* soil profile with a division into different horizons according to presence of organic matter and degree of breakdown of the regolith.

6.6 Erosion and transport

Weathering is the *in situ* breakdown of bedrock. The weathered material lying on the surface is referred to as *regolith* and erosion is the removal of regolith material. In addition to the products of bedrock weathering, unconsolidated sediment and soil are also subject to erosion. Loose material on the land surface may then be transported downslope under gravity, washed by water, blown away by wind, scoured by ice or moved by a combination of these processes.

6.6.1 Erosion and transport under gravity

Transport of material by water and ice is gravity-driven in the sense that it provides the potential energy for the flow. On steep slopes in mountainous areas and along cliffs, processes which do not involve a flow of water or ice but are driven by gravity alone are important. These movements downslope under gravity are commonly the first stages in the erosion and transport of weathered material.

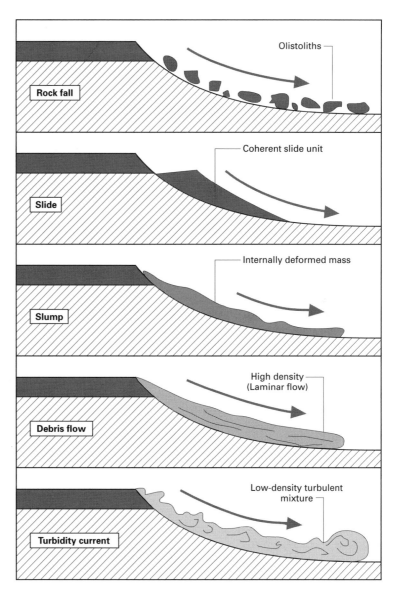

Fig. 6.9 Mechanisms of gravity-driven transport on slopes.

DOWNSLOPE MOVEMENT

There is a spectrum of processes of movement of material downslope (Fig. 6.9). A *landslide* is a coherent mass of bedrock which has moved downslope without significantly breaking up in the process. Many thousands of cubic metres of rock can be translated downhill, retaining the internal structure and stratigraphy of the unit. In other circumstances the rock breaks up during its movement as a *rock fall* to accumulate as a chaotic mass of material at the base of the slope. Both of these are the movement of material under gravity alone. They may be triggered by a shock caused by an earthquake, or by other mechanisms, such as waterlogging of a potentially unstable slope by a heavy rainfall or undercutting at the base of the slope.

Movement downslope may also occur when the regolith is lubricated by water and there is *soil creep*. This is a much slower process than falls and slides and may not be perceptible unless a hillside is monitored over a number of years. A process which may be considered to be intermediate between creep movement and slides is *slumping*. Slumps are instantaneous events like slides but the material is plastic due to saturation by water and it deforms during movement downslope. With sufficient water a slump may break up into a debris flow *(4.6.1)*.

SCREE AND TALUS CONES

In mountain areas weathered detritus falls as grains, pebbles and boulders down mountainsides to accumulate on or at the bottom of the slope. These accumulations of *scree* are often reworked by water, ice and wind but sometimes remain preserved as *talus cones*, concentrations of debris at the base of gullies (Fig. 6.10) (Tanner & Hubert 1991). These deposits are characteristically made up of angular to very angular clasts because transport distances are very short, typically only a few hundred metres, so there is little opportunity for the edges of the clasts to become abraded. They are typically poorly sorted as the short transport distance allows little time for the larger clasts to become separated from the smaller, and there is very little difference in the angle of rest of different clast sizes. The only sorting effect is for the smaller clasts to fall between the larger ones, giving a crude inverse grading in places. Water percolating through talus deposits will also tend to flush the smaller particles down through the pile of sediment. Screes tend to display only a crude stratification, such that the layers

Fig. 6.10 A talus cone at the side of a glacial valley in the Rocky Mountains, Lake Louise, Alberta, Canada. The scree cone is supplied by debris weathered from higher up the mountainside falling by gravity down a gully at the apex of the cone.

within the deposit can be picked out only by slight changes in the concentrations of larger and smaller clasts. Bedding is therefore difficult to see in talus deposits. Where bedding can be seen the layers are close to the angle of rest (around 30°) of loose aggregate material. Talus deposits are distinct from alluvial fans *(8.4)* because water does not play a role in the transport and deposition.

6.6.2 Erosion and transport by water

Falls, slides and slumps are responsible for moving vast quantities of material downslope in mountain areas but they do not move detritus very far, only down to the floor of the valleys. The transport of detritus over greater distances normally involves water, although ice and wind also play a role in some environments *(6.6.3, 6.6.4)*.

Erosion by water on hillsides is initially as a *sheet wash*, unconfined surface run-off down a slope following rain. This overground flow may pick up loose debris from the surface and erode into the regolith. The quantity of water involved and its carrying capacity depend not only on the amount of rainfall but also on the characteristics of the surface: water runs faster down a steep slope, vegetation tends to reduce flow and trap debris and a porous substrate results in infiltration of the surface water. Surface run-off is therefore most effective at carrying detritus during flash flood events on steep, impermeable slopes in sparsely vegetated arid regions. Vegetation cover and thicker, permeable soils in temperate and tropical climates tend to reduce the transport capacity of surface run-off.

In places where the sheet wash starts to become concentrated into runnels and gullies, the confined flow is more effective at eroding superficial material. As these gullies coalesce into channels the headwaters of streams and rivers are established. Rivers can be very effective at eroding into regolith and bedrock. Rapid, turbulent flow scours at the floor and margins of the channel, weakening them until pieces large and small fall off into the stream. Continued flow over soluble bedrock such as limestone also gradually removes material in solution. Some of the eroded material may be small enough to be carried away in the stream flow as bedload or in suspension; larger fragments may remain in the floor of the channel until either they break up *in situ* or there is a high-magnitude flow event capable of transporting the clasts. The confluence of streams forms larger rivers which may feed alluvial fans, fluvial environments of deposition, lakes or seas.

Climate and topography both affect the way in which erosion by water occurs. Steep slopes give rise to high-velocity water flows which are more able to erode material than flow down gentle gradients. The amount of water available to cause erosion is determined by the rainfall, but the total amount of rainfall is not the only important climatic factor; in arid regions very infrequent but violent rainstorms can result in far more erosion than much greater amounts of water falling steadily throughout the year.

6.6.3 Erosion and transport by wind

Winds are the result of pressure differences between two parts of the atmosphere at the same level. Masses of air move between regions of high pressure and regions of low air pressure. On a global scale regions of high pressure occur at the poles where cold air sinks, and low pressure at the equator where the air is heated up, expands and rises. Local variations in pressure result from the temperature of water masses which move with ocean currents, heat absorbed by land masses and cold air over high glacial regions. A complex and shifting pattern of high pressure (anticyclones) and low pressure (depressions) regions generates winds all over the surface of the Earth.

Winds are capable of picking up loose debris from the land surface. This *winnowing* effect is most important in mountain areas where winds can be strong and effective at picking up silt- and sand-sized debris and can carry it great distances away. In lowland areas wind action is also an important force if loose material on the surface is not bound in soil and covered by plants. Dry floodplains of rivers and beaches may be particularly susceptible to wind erosion.

Eroded fine material (up to sand grade) can be carried over distances of hundreds or thousands of kilometres by the wind (Schutz 1980; Pye 1987). The size of material carried is related to the strength (velocity) of the air current. This is in turn dependent upon the difference in air pressure and distance between areas of high and low pressure. Winds experienced in the present day vary in strength up to around $200\,\mathrm{km\,h^{-1}}$. Although they may have a devastating effect on the areas through which they pass, hurricanes and tornadoes are relatively localized and of short duration. Wind-transported detritus is most conspicuous in areas of sparse vegetation (including beaches) and low rainfall (such as hot deserts and periglacial regions). The processes of transport and deposition in these arid regions are considered in Chapter 8.

6.6.4 Erosion and transport by ice

Glaciers in mountain regions can exert very great forces on bedrock. A moving body of ice can cut deeply into unweathered bedrock as well as removing any loose detritus on the surface (Sugden & John 1976; Drewry 1986). Erosion occurs by two processes, abrasion and plucking.

Glacial abrasion occurs by the frictional action of blocks of material embedded in the ice ('tools') on the bedrock. These tools cut grooves, *glacial striae*, in the bedrock a few millimetres deep and elongate parallel to the direction of ice movement (Fig. 6.11): striae can hence be used to determine the pathways of ice flow long after the ice has melted. The scouring process creates *rock flour*, fine-grained debris which is incorporated into the ice.

Glacial plucking is most common where a glacier flows over an obstacle. On the upflow side of the obstacle abrasion occurs but on the downflow side the ice dislodges blocks which range from centimetres to metres across. The blocks plucked by the ice and subsequently incorporated into the glacier are often loosened by subglacial freeze–thaw action *(6.5.1)*. The landforms created by this combination of glacial abrasion and plucking are called *roches moutonées*, apparently because they resemble sheep from a (very) great distance. Meltwater

associated with glaciers and ice sheets also plays an important role in erosion in periglacial areas. Glacially related sedimentation occurs mainly around the edges of glaciers where melting releases entrained material which is deposited at the ice margin or transported further away by meltwater. Glacial processes and products are considered in Chapter 7.

6.7 Factors which influence erosion rates

The interrelated factors of relief, climate and vegetation control rates of erosion in an area, along with the nature of the bedrock (Ibbeken & Schleyer 1991; Einsele 1992).

6.7.1 Topography and relief

Slope is an important factor in determining rates of erosion, and all the physical processes of removal of material are in some way gravity-driven. Rock falls and landslides are clearly more frequent on steep slopes than in areas of subdued topography: stream flow and overland water flow are faster across steeper slopes and hence have more erosive power.

A distinction needs to be drawn between the altitude of the terrain and the *relief*. A plateau region may be thousands of metres above sea level but if it is flat there will be little difference between the rates of erosion across the plateau and a lowland region: only at the edges of the plateau will there be steep slopes to facilitate erosion. The plateau may have a significant altitude but it has little relief, that is, there is little change in the height of the ground over the area. A deeply incised topography consisting of steep-sided valleys separated by narrow ridges provides the greatest area of steep slopes for bedrock and regolith to be eroded.

Relief tends to be greatest in areas which are undergoing uplift due to tectonic activity and thermal doming due to hot-spots in the mantle (Kearey & Vine 1996). Rejuvenation of the landscape by uplift occurs mainly around plate boundaries, particularly convergent margins such as orogenic belts. In tectonically stable areas the relief is subdued due to weathering and erosion resulting in a low, gentle topography. The cratonic centres of continental plates are typically regions of low relief and hence rates of erosion are low. There is therefore a correlation between tectonic activity and rates of erosion. Uplift of continental crust is necessary for relief to be created and maintained on the Earth's surface.

Fig. 6.11 Granite bedrock in the Sierra Nevada, California, USA, scoured by ice during the Pleistocene leaving glacial striae on the surface.

6.7.2 Climate

Climate plays an important role in determining both the amount of erosion which occurs and the processes which cause erosion. Availability of water is a key factor. Erosion by wind is most effective in the absence of water, but other processes require a water supply, ultimately from rain and snow.

In glacial areas there must be sufficient precipitation in upland areas to feed valley glaciers; if there is insufficient snow fall the glacier may stagnate or retreat as the snout melts *(7.1)*. *Periglacial* regions (areas which border glaciers) have a seasonal cover of snow which melts in the summer months. However, the ground may remain frozen at depths of a few metres all year round in a *permafrost* state. Water accumulating in the layer above the permafrost layer may eventually saturate the regolith and promote slumping on slopes. Repeated freezing and thawing of the regolith may also lead to creep downslope.

In temperate and arid climates it is not just the total amount of water which flows as sheet wash or in rivers but the regularity with which rain falls which is important in erosion. A steady rainfall all year will tend to result in a vegetation cover which protects the regolith. Highly *ephemeral* (irregular) flow due to infrequent but intense rainstorms tends to remove more material in each event than much greater volumes of water flowing continuously. Rainfall events in desert regions may be catastrophic in terms of the amount of erosion that occurs.

6.7.3 Vegetation

Plant cover is an important factor in determining the rate of erosion of weathered material. The types of plant present and the coverage they have over the land surface are determined by the climate regime, which is in turn influenced by the latitude, altitude and proximity of marine environments (continental versus maritime vegetation types). The relative importance of different plant groups has varied through geological time and hence vegetation is not a uniformitarian factor in rates of erosion *(24.4.2)*.

A dense vegetation cover is very effective at protecting the bedrock and its overlying regolith from erosion by rain impact and overland flow of water. Only where the vegetation cover is pierced is the regolith likely to be removed by erosion. In tropical regions destruction of the vegetation cover by natural events such as fires or by anthropogenic activity such as logging can have a catastrophic effect on erosion: the bedrock beneath the plant roots may be very deeply weathered and the regolith susceptible to being washed away by rainfall or floods. The effects of wind action on the regolith are also reduced where a vegetation cover binds fine detritus into soil.

A sparse plant cover in cold or arid regions leaves the regolith exposed to erosion by water and wind. In deserts overland flow following storms may be very infrequent, but in the absence of much plant life a lot of loose debris may be washed away in a single flash flood.

6.7.4 Bedrock

In a given topographic and climatic regime the type of bedrock is a fundamental control on the rates and styles of erosion. The main factor is the rate at which weathering processes break down the rock to make material available for erosion. The greatest variability is seen in humid climates where chemical weathering processes are dominant because different lithologies are broken down, and hence eroded, at widely different rates. The proportion of the rock-forming silicate minerals (Fig. 6.12) is the main factor: quartz-rich rocks are least susceptible to breakdown, whereas mafic rocks such as basalts are rapidly weathered and eroded. Where physical weathering processes are more important the fracture characteristics of the bedrock largely control the rate of breakdown. Lithologies such as shales and slates contain may lines of weakness and rapidly break up; homogeneous limestone and well-cemented quartz sandstone are highly resistant in dry regions.

6.8 Continental environments of deposition

The two principal controlling factors which govern the nature of the terrestrial depositional environments are *geomorphology*, the shape of the land, and *climate*, the temperature and amount and distribution of precipitation.

The underlying structure of the landscape is ultimately determined by tectonic activity which governs the areas of uplift and subsidence. Areas of uplift are regions of erosion (see above) and sediment accumulation occurs on the land surface where there are depressions to trap material. Surface depressions on the continents are created by subsidence driven by tectonic and thermal forces;

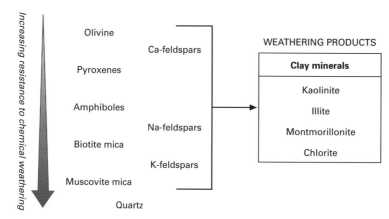

ROCK-FORMING SILICATE MINERALS

Increasing resistance to chemical weathering

Olivine

Ca-feldspars

WEATHERING PRODUCTS

Clay minerals

Kaolinite

Illite

Montmorillonite

Chlorite

Pyroxenes

Amphiboles

Na-feldspars

Biotite mica

K-feldspars

Muscovite mica

Quartz

Fig. 6.12 Chemical weathering of silicate minerals.

they form sedimentary basins (Chapter 23) which may be wholly continental or partly marine. The area of the sedimentary basin and the amount of subsidence in it determine the volume of material which can accumulate within it.

Terrestrial environments are strongly affected by climatic conditions. In cold climates with sufficient precipitation there is a build-up of snow and ice to form glaciers. These occur at high latitudes near the poles and at high altitudes. Temperate to tropical climates are regions with moderate to high rainfall and which are cool to very hot. Hot, arid climates have low rainfall and temperatures up to hot or very hot.

Four main continental environments are recognized, each strongly controlled by the geomorphology and climate. In cold mountainous or polar regions ice dominates the processes of transport and deposition of material in local pockets or across lower-lying plains (Chapter 7). Arid depositional environments (Chapter 8) are characterized by low rainfall. They are often in depressions in the interiors of continents and along certain coasts. They include three main sub-environments: wind-blown sands, alluvial fans deposited by ephemeral water flow and temporary playa lakes. The upper stretches of rivers are too steep for deposition to occur but at lower gradients in the lower stretches of rivers (Chapter 9) deposition occurs within the channel and on the floodplain. Lakes (Chapter 10) form in depressions in continental areas and are depositional environments which are strongly controlled by climate. Environments at the interface of terrestrial and marine conditions such as deltas, estuaries, beaches and lagoons (Chapters 12 and 13) are also strongly influenced by the climate.

6.9 Continental environments: summary

The continents may be divided into areas which are undergoing erosion and regions where sedimentation is taking place. Physical and chemical weathering processes break down bedrock to produce detritus which may be carried directly away or may form part of a soil. Weathering and soil-forming processes are strongly influenced by climate and determine the nature of the material generated from the bedrock, although the ultimate control on the composition of the detritus is the bedrock lithology. Surface processes driven by gravity, as well as the action of water, wind and ice, contribute to the erosion which removes this material and starts the cycle of transport into depositional environments. Terrigenous clastic material formed by weathering and erosion of the continents is volumetrically the most important constituent of the material deposited in the sedimentary environments considered in the following chapters.

Further reading

Duff, P.McL.D. (ed.) (1993) *Holmes' Principles of Physical Geology*, 4th edn, 791 pp. Chapman & Hall, London.
Einsele, G. (1992) *Sedimentary Basins, Evolution, Facies and Sediment Budget*, 628 pp. Springer-Verlag, Berlin.
Ollier, C.D. (1984) *Weathering*, 270 pp. Longman, London.
Selby, M.J. (1985) *Earth's Changing Surface,* 607 pp. Clarendon Press, Oxford.
Selby, M.J. (1994) Hillslope sediment transport and deposition. In: *Sediment Transport and Depositional Processes* (ed. K. Pye), pp. 61–88. Blackwell Scientific Publications, Oxford.

7   Glaciers and ice caps

In polar regions and high mountain areas the air temperature is at or below 0°C for most of the time. Precipitation occurs as snow which accumulates into permanent or semi-permanent bodies of ice known as glaciers. In mountainous areas glaciers occupy the upper slopes of valleys. They may be hundreds of metres thick and stretch for tens of kilometres down the mountainside. In polar regions ice sheets form covering almost the entire landscape of a continental land mass with a layer of ice which may be thousands of metres thick. The size and extent of glaciers and ice sheets around the Earth have varied considerably through geological time in response to global climatic changes. The Pleistocene period was marked by repeated advances and retreats of ice sheets in the Northern and Southern hemispheres, and similar glacial periods have occurred throughout Earth history. Ice is a powerful agent of erosion of bedrock and can transport detritus for long distances before melting and depositing sediment around the margins of the glacial area. The record of glacial events is seen in the landscape as erosional landforms in upland regions and as deposits of clastic material on land and in the marine environment. Such deposits are most common in the Quaternary geological record but may also be recognized in older parts of the stratigraphic record.

7.1 Formation of glaciers

In areas near the poles and in high mountainous regions throughout the world where the air temperature is below 0°C precipitation occurs as snow. If the local climate is seasonally warm the snow melts, but in colder places the rate of precipitation as snow exceeds the rate at which it melts. Under these conditions a *glacier* forms, a permanent body of ice which may be growing if precipitation outstrips removal (*ablation*) of ice by melting, evaporation or wind deflation. If ablation exceeds supply the glacier will retreat, becoming smaller with time. Large ice sheets which reach the coast in polar regions may be retreating by breakup of the ice margin, forming icebergs.

7.1.1 Geomorphology of continental glaciers

The shape of glaciers is governed by the relief of the landscape on which they form and the volume of ice present (Duff 1993; Miller 1996). In upland areas small *cirque glaciers* form in protected hollows on mountainsides. These are found at high altitudes all over the world, even within a few degrees latitude of the equator in New Guinea. Major mountain ranges in moderate and high latitudes also contain *valley glaciers*, bodies of ice which are confined within the valley sides (Fig. 7.1). In high latitudes valley glaciers may be fed by larger bodies of ice at higher altitudes. The lower slopes of a mountain range may be the site of formation of a *piedmont glacier*, where valley glaciers may merge and spread out as a body of ice hundreds of metres thick. Alaska and Iceland are areas of piedmont glacier formation today. Mountain regions can become partially or wholly blanketed by an *ice cap* to form an upland plateau such as Vatnajökull in Iceland.

Most of the world's continental ice (around 95%) is found near the poles as huge *ice sheets*. Greenland is the largest island in the world and is almost completely covered by an ice sheet. However, the volume of ice involved is small compared to the sheet which covers the Antarctic continent to a depth of up to 4000 m: if it were to melt completely the global sea level would rise by as much as 150 m *(21.8.2)*.

Fig. 7.1 A valley glacier: the Saskatchewan glacier is fed from the Columbia Icefield area of the Rocky Mountains in western Canada. Forward of the snout of the glacier lies a glacial outwash area where fluvial processes distribute debris transported by the glacier.

7.1.2 Sea ice

At the freezing point of pure water (0°C) ice has a density of 0.92 g cm^{-3} and therefore floats on fresh water (density 1.0 g cm^{-3}). The buoyancy of ice is increased in sea water which, because of dissolved salts, has a density of 1.025 g cm^{-3}. As a glacier or ice sheet reaches the coast it may extend out to sea as an *ice shelf*, a floating body of ice attached to the continental ice body. Ice shelves can only form in polar regions as the temperature must be low enough for permanent ice to exist at sea level. An ice shelf may be hundreds of metres thick and reach down to the sea floor, resulting in erosion well below sea level. Around the Antarctic continent ice shelves extend hundreds of kilometres out to sea, forming areas of floating ice which cover several hundred thousand square kilometres (Drewry 1986). The outer edges of ice shelves break up to form *icebergs*, free-floating bodies of ice which drift with the wind and surface currents out to sea. A large iceberg may drift thousands of kilometres across oceans into more temperate regions before melting completely.

7.1.3 Behaviour of glaciers and ice

Ice is normally thought of as a solid, but under pressure it behaves as a fluid, moving away from the point of higher pressure. Pressure can be provided by the weight of ice above any particular point, and the ice will flow slowly as an extremely viscous fluid. At high viscosities the flow is entirely laminar *(4.2.1)*. Measurements on modern glaciers indicate that ice moves at rates which vary from as little as a few metres per year to over 500 m per year (Allen 1970a). Different parts of a body of ice move at different rates because of different pressure gradients. Typically the flow rate is greatest at the surface of the ice, decreasing downwards towards the base of the glacier. Valley-confined glaciers have greatest flow in the middle of the valley, decreasing towards the margins.

7.1.4 Temperate and polar glaciers

The behaviour of a glacier in relation to the surrounding bedrock depends on the temperature of the ice (Miller 1996). At low temperatures in high latitudes (*polar glaciers*) the body of ice is frozen to the bedrock below. The glacier moves mainly by flow within the ice itself. The stresses exerted on the bedrock at the bases of polar glaciers may be very great, resulting in fracturing and removal of material. Melting of polar glaciers only occurs on the surface where streams may form in the warmer summer months. In lower latitudes the ice in glaciers is closer to melting point. These *temperate glaciers* are characterized by having a lot more associated water on the surface, in glacial tunnels and at the base of the ice. Meltwater between the glacier and the bedrock forms a lubrication zone allowing the ice to move more freely and there is less erosion by the ice. Water flowing under a glacier may be under considerable pressure and is therefore erosive, forming sub-glacial valleys.

7.2 Erosion by glaciers

Ice moving across bedrock erodes in two ways: glacial abrasion and glacial plucking *(6.6.4)*. The former results in fine-grained rock flour being incorporated into the moving glacier, whilst plucking can quarry out fragments which range from a few centimetres to rafts of rock tens of metres across. In addition to the erosion directly associated with the ice movement, meltwater at the base of the ice in temperate glaciers and in front of the snout of glaciers may scour into bedrock (Boulton 1972).

7.2.1 Characteristics of glacially transported material

Glacial erosion processes result in a wide range of sizes of detrital particles. As the ice movement is a laminar flow there is no opportunity for different parts of the ice body to mix and hence no sorting of material carried by the glacier will take place. Glacially transported debris is therefore typically very poorly sorted. Fragments plucked by the ice will be angular and debris carried within ice will not undergo any further abrasion, and only material on the top of an ice body will be subject to weathering processes. In addition to the poor sorting, debris carried by glaciers is very angular and the overall texture is therefore very immature.

The fine-grained material formed by glacial abrasion is known as *rock flour*. This silt- and clay-sized detritus is different in composition from similar-grade sediment produced by other mechanisms of weathering and erosion. Rock flour consists of very small fragments of many different minerals. In contrast, material of the same size produced by chemical weathering typically consists of clay minerals and fine-grained quartz. Unlike clay minerals, the fine particles in rock flour do not flocculate *(2.5.5)* and tend to remain in suspension for much longer periods of time. This high proportion of suspended sediment gives the characteristic green to white colour to lakes fed by glacial meltwaters.

Material carried by a glacier is not necessarily all the result of glacial erosion. Valley sides in cold regions are subject to extensive freeze–thaw weathering *(6.5.1)* the products of which fall down the valley sides on to the top surface of the glacier. In more temperate regions detritus may also be washed down the valley sides by overland flow and streams which are active during the summer thaw. Streams may also form on the surface of a glacier

or ice sheet during warmer periods and their action may contribute to the transport of debris.

Debris transported by a valley glacier tends to be concentrated in the ice near the base, lateral edges and the top of the body of ice, resulting from erosion along the valley bottom and sides and detritus falling or washing on to the surface. A concentration of debris towards the centre of a valley glacier may occur when two valley glaciers merge and the detritus carried in the margin of each forms a strip in the centre of the combined body.

7.3 Glacial deposits

Deposition of the detritus carried by the glacier occurs when the ice melts. The direct deposits of a glacier are known as *moraine*. Several different types of moraine can be recognized (Fig. 7.2). *Terminal moraines* form at the snout of the glacier where the melting of the ice keeps pace with the downslope migration of the glacier; they are typically ridges which lie across the valley. If a change in climate causes a general warming of the glaciated area, the glacier retreats as the melting at the snout is occurring faster than the addition of snow at the head of the glacier. When this happens the detritus which has accumulated at the sides of the glacier by abrasion of the valley walls and rock falls on to the top of the glacier is deposited as a *lateral moraine* (Figs 7.2 & 7.3). Lateral moraines lie at the sides of glaciated valleys, parallel to the valley walls. Where two glaciers in tributary valleys converge, detritus from the sides of each is trapped in the centre of the amalgamated glacier and the resulting deposit upon ice retreat is a *medial moraine* along the centre of a large glaciated valley.

7.3.1 The nature of glacial deposits

The deposits of a moraine are typically poorly sorted, consisting of mixtures of rock detritus of all sizes. These characteristics have given these glacial deposits the name *boulder clay*, reflecting the size range of material present. A more general term for any material deposited directly by ice is *till*. Consolidated and lithified tills are referred to as *tillites*. An alternative term for unconsolidated glacial deposits is *diamict*, with the consolidated equivalent being *diamictite* (Eyles & Eyles 1992; Miller 1996). Most of the deposits of the glacial episodes in the Pleistocene ice ages are tills as they have not become lithified. There are glacial deposits of Carboniferous to Permian age which occur as fully lithified

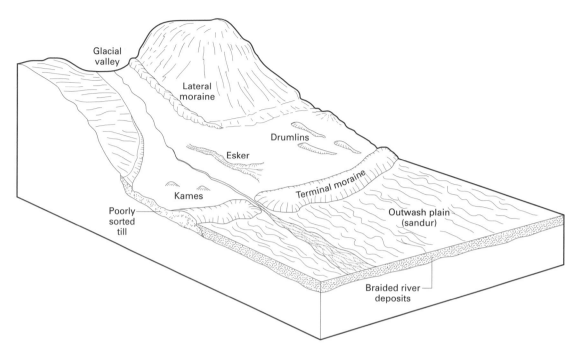

Fig. 7.2 Glacial deposits in continental environments.

Fig. 7.3 Lateral moraine forming a ridge at the edge of a retreating glacier, Athabasca Glacier, Rocky Mountains, Alberta, Canada.

tillites in the southern continents. The constituents of tills and tillites are the products of weathering in cold environments where physical weathering processes break up the rock but chemical weathering does not play an important role. For this reason, the mineral composition of the deposit may be very similar to that of the bedrock and unaltered lithic fragments are common. Clay

minerals are often rather uncommon even in the fine-grained fraction of a till because clays form principally by the chemical weathering of minerals, and in glacial environments this breakdown process is suppressed.

Tills formed in terminal moraines by the melting of the snout of a glacier are unstratified mounds of material, as are lateral and medial moraine deposits formed by glacial retreat. *Lodgement tills* formed along the base of a moving ice sheet may be better organized and show internal stratification formed during stages in the movement of the ice sheet. These tills may form sheets which can be tens of metres thick, irregular ridges or smoothed mounds known as *drumlins*. Drumlins are tens of metres high and may be several kilometres long, with the elongation in the direction of ice flow.

7.3.2 Deposits directly associated with glaciers

Moraines are sediment bodies deposited directly from the ice. Water associated with the melting of ice carries released sediment which is deposited beneath or in front of the glacier. In temperate glaciers partial melting of the ice results in rivers flowing in tunnels within or beneath the ice carrying with them any detritus held by the ice which melted. The deposits of these rivers form sinuous ridges of material known to glacial geomorphologists as *eskers* (Fig. 7.2). Eskers are typically a few metres to tens of metres high, tens to hundreds of metres wide and stretch for kilometres across the area formerly covered by an ice sheet (Warren & Ashley 1994). The processes of deposition of the material are similar to those of a river. Bars of gravel and sand form cross bedded and horizontally stratified lenses within the esker body. They may be distinguished from river deposits by the absence of associated overbank sediments *(9.3)* and by internal deformation (slump folds and faults) which forms when the sand and gravel layers collapse as the ice around the tunnel melts. *Kames* and *kame terraces* are mounds or ridges of sediment formed by the collapse of crevasse fills, sediment formed in lakes lying on the top of the glacier or the products of the collapse of the edge of a glacier.

7.3.3 Outwash plains

As the front of a glacier or ice sheet melts it releases large volumes of water along with any detritus being carried by the ice. Rivers flow away from the ice front over the broad area of the *outwash plain*, also known by their Icelandic name *sandur* (plural *sandar*) (Fig. 7.2). The rivers transport and deposit in the same manner as braided rivers *(9.2.1)* (Boothroyd & Ashley 1975; Boothroyd & Nummedal 1978; Maizels 1993). The large volumes of water and detritus associated with the melting of a glacier mean that the outwash plain is a very active region, with river channels depositing sediment

rapidly to form a thick, extensive braid plain deposit *(9.2.1)*. Outwash plain deposits can be distinguished from other braided river deposits by their association with other glacial features such as moraines.

Perhaps the most spectacular events associated with glacial sedimentation are sudden glacial outburst events known by their Icelandic name, *jökulhlaups*. Iceland lies across the spreading ridge of the North Atlantic and as such is an area of volcanic activity. The ice caps of central Iceland cover some of the volcanic centres and when an eruption occurs there is a catastrophic melting of the ice around the volcanic vent. This volume of meltwater usually bursts out of the edge of the ice cap, resulting in a dramatic surge of water and sediment on to the outwash plain. Deposits of glacial outbursts are thick beds of sand and gravel which are massive and poorly sorted or cross bedded and stratified (Maizels 1989).

Extreme catastrophic discharges need not always be associated with volcanic activity but may also result from sudden releases of meltwater from glacially dammed lakes and water trapped beneath glaciers. The deposits resulting from these high-magnitude events are characteristically poorly sorted and may include some very large blocks. Reworking of this material on the outwash plain may occur.

7.3.4 Wind erosion on outwash plains

Adjacent to glaciers the conditions are too cold for vegetation to flourish and the sediment deposited on the outwash plain is left exposed. Wind blowing across the plain picks up finer sand- and silt-sized material. Sands may be blown over the plain and accumulate into aeolian dunes. The silt is principally the rock flour formed by glacial abrasion and washed out of the glacier. This fine sediment can be carried further as dust clouds which can result in silt being deposited as loess *(2.5.2)* thousands of kilometres away from the source. Loess is typically made up of grains of quartz, feldspar, mica, carbonate and clay minerals 10–50 μm across. It occurs in deposits tens to hundreds of metres thick in China, central and eastern Europe and parts of North and South America which formed during glacial stages of the Pleistocene (Pye 1987). Wind-blown silt is also an important contributor to deep marine sediments *(15.5)*.

7.3.5 Marine glacial deposits

Ice shelves break up at the edges to form icebergs and

melt at the base in contact with sea water. Detritus released from the bottom of an ice shelf forms a *till sheet* (Fig. 7.4). Till sheets may be thick and extensive beneath a long-lived shelf (Eyles & Eyles 1992; Miller 1996). Reworking of the till by shallow marine currents results in stratification and sorting which is not encountered in sub-glacial lodgement tills formed on land. Icebergs may carry enclosed debris thousands of kilometres away from the shelf into deep waters. *Dropstones* are exotic clasts found in deep marine sediments carried by an iceberg and deposited as it melts. They can be pebble- to boulder-sized, in marked contrast to surrounding fine-grained, pelagic deposits *(15.5)*.

7.4 Distribution of glacial deposits

Quaternary valley and piedmont glaciers form distinctive moraines but are largely confined to upland areas which are presently undergoing erosion. In these upland areas glacial and periglacial deposits such as moraines, eskers, kames, and so on have a very poor preservation potential in the long term. Of more interest from the point of view of the stratigraphic record are the tills formed in lowland continental areas and in marine envi-

ronments as these are much more likely to lie in regions of net accumulation in a sedimentary basin. The volume of material deposited by ice sheets and ice shelves is also considerably greater than that associated with upland glaciation.

Extensive ice sheets are today confined to the polar regions within the Arctic and Antarctic circles. During the glacial episodes of the Quaternary the polar ice caps extended further into lower latitudes. The sea level was lower during glacial periods and many parts of the continental shelves were under ice. Upland glacial regions were also more extensive, with ice reaching beyond the immediate vicinity of the mountain glaciers. The growth of polar ice caps is known to be related to global changes in climate, with the ice at its most extensive when the globe was several degrees cooler. Other glacial episodes are known from the stratigraphic record to have occurred in the late Carboniferous and Permian (the Gondwana glaciation in the Southern hemisphere) and in the early Palaeozoic *(24.7.4)*.

7.5 Recognition of glacial deposits: summary

Glacial deposits are compositionally immature. Tills are typically composed of detritus which represents

Fig. 7.4 Glacial deposits in marine environments.

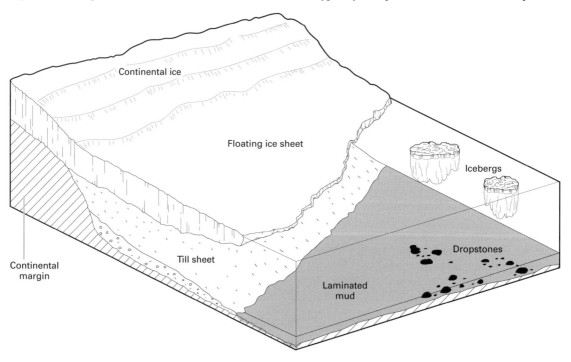

broken-up and powdered bedrock from beneath the glacier. Reworked glacial deposits on outwash plains may show a slightly higher compositional maturity as unstable clasts are broken down during fluvial transport. The deposits in glacial environments are also texturally immature, with angular fragments in coarser detritus deposited as tills. Moraines and lodgement tills show poor sorting, although fluvially and marine reworked sediment may show better sorting. There is a paucity of clay minerals in the fine-grained fraction because of the absence of chemical weathering processes in cold regions. Facies and facies associations are exclusively clastic: chemical and biochemical sediments do not form in cold conditions and the climate is not favourable for biogenic deposits. Fossils are rare or absent in glacial deposits.

Further reading

Drewry, D.D. (1986) *Glacial Geologic Processes*, 276 pp. Arnold, London.

Eyles, N. & Eyles, C.H. (1992) Glacial depositional systems. In: *Facies Models: Response to Sea Level Change* (eds R.G. Walker & N.P. James), pp. 73–100. Geological Association of Canada, St Johns, Newfoundland.

Hambrey, M.J. (1994) *Glacial Environments*, 296 pp. UCL Press, London.

Miller, J.M.G. (1996) Glacial sediments. In: *Sedimentary Environments: Processes, Facies and Stratigraphy* (ed H.G. Reading), pp. 454–484. Blackwell Science, Oxford.

Sugden, D.D. & John, B.S. (1976) *Glaciers and Landscape*, 376 pp. Edward Arnold, London.

8   Arid continental depositional environments

The scarcity of water in dry climates has a profound effect on the processes of transport and deposition of sediment. Vegetation is normally sparse, and in the absence of plants to trap dry sand- and silt-grade material on the land surface this detritus is picked up and transported by the wind to accumulate as aeolian sediments. Rain is sporadic, so rivers are ephemeral and may carry large quantities of detritus when flash floods occur. Debris is rapidly deposited in wadis or on alluvial fans which build up at basin margins. In hot deserts high rates of evaporation dry up lakes to deposit evaporite minerals in saline playa lakes. The deposits of arid continental environments are hence characterized by well-sorted sand deposited as aeolian dunes, poorly sorted conglomerate which accumulated on alluvial fans or in wadis and layers of evaporite minerals precipitated in playa lakes. These different facies may be found in association with each other in the stratigraphic record as the deposits of ancient desert environments.

8.1 Deserts

The term *desert* is normally used for a place where there is low rainfall and consequently little vegetation. Modern desert regions include the well-known deserts of the Sahara, Arabia and Namibia, but there are also large desert regions in central Asia, Australia and in North and South America. The popular image of deserts is that they are hot environments, but most are also very cold for part of the time either on a daily cycle or through the seasons. The deserts of central Asia are only hot during the summer months, with temperatures falling well below 0°C for much of the winter. The lack of rainfall (*aridity*) is more important than temperature in characterizing a desert environment. A region is defined as *arid* if it receives on average less than 250 mm of precipitation in a year. If the annual precipitation is between 250 and 500 mm the region is considered to be *semi-arid*.

Not all deserts are environments of accumulation of sediment. A *stony desert* is a non-depositional region where the removal of fine sediment by wind is greater than the supply from outside the area. As a consequence, only the larger clasts which cannot be moved by aeolian action remain on the land surface. With time these pebbles and boulders become abraded by the repeated collision of airborne particles. If the wind direction is reasonably constant, one, two or three sides may be worn by wind abrasion to produce *ventifacts*, or wind-abraded clasts. *Zweikanters* are ventifacts with two blown surfaces and *dreikanters* have three surfaces. The only evidence for a stony desert in the geological record would be scattered ventifacts on a surface of bedrock.

8.1.1 Arid zone environments and facies

Three principal depositional environments occur in arid regions (Fig. 8.1). Wind action in a desert transports and deposits sand: areas of accumulation of sand are known as *sand seas* or *ergs*, the latter being the Arabic name for a sandy desert area (Brookfield 1992). Prominent ergs of the present day are in parts of the Sahara, Namibia, Arabia, south-western North America and western central Australia. Coarser detritus which cannot be transported by wind tends to be restricted to ephemeral rivers and alluvial fans at the margins of sedimentary basins — for example, along the sides of Death Valley in California. Hot, arid environments are characterized by high rates of evaporation, leading to the precipitation of salts

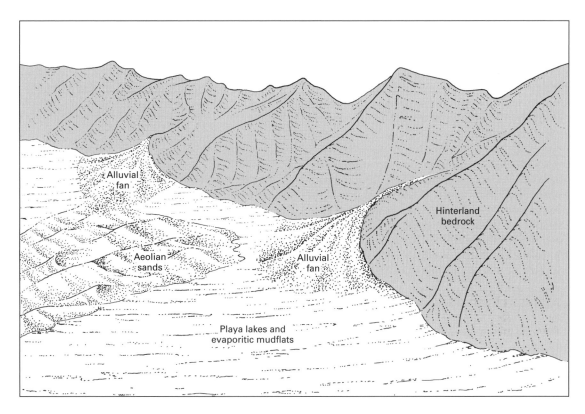

Fig. 8.1 Depositional environments in arid regions.

in temporary lakes and on mudflats adjacent to alluvial fans and areas of sand dunes.

8.2 Wind in deserts

One of the main characteristics of arid regions is the importance of air as a transporting medium. Movement of particles by the wind, or *aeolian transport*, can occur in any terrestrial environment, but in areas of frequent rainfall the effects of aeolian transport and deposition are masked by the action of water on the land surface. Aeolian processes are therefore most evident in arid environments where there is little surface water (Kocurek 1996).

8.2.1 Aeolian transport

Air has a very low viscosity compared to water, and consequently relatively high velocities are required to achieve sufficient lift force on a particle to bring it into

the main stream of air flow. Gusts of wind may exceed 200 km h^{-1} (55 m s^{-1}) but in order to transport large volumes of detritus a more regular air flow is required and these rarely exceed 30 m s^{-1}. At these velocities the largest particles that can be carried by the wind are around 0.5 mm in diameter (medium sand size), with coarser grains only carried during very strong storms (Pye 1987; Nickling 1994).

Transport in air can be by rolling, saltation and suspension *(4.2.2)*, but there are significant differences between the way saltation occurs in water and air. In water, the viscosity of the liquid limits the velocity of the grain on the downward part of its trajectory, and the water acts as a 'buffer' between the moving grain and other grains stationary on the surface. On impact, the saltating grain has little effect on the grains it lands on. A grain moving in air, which has a lower viscosity, rises higher and will impact on grains on the bed with sufficient force to dislodge them, sending them up into the free stream of flow. Hence once saltation has started the process will be enhanced by the 'splatter effect' of landing grains propelling stationary grains.

8.2.2 Texture and composition of aeolian sands

The absence of any significant buffering by a more viscous fluid and the higher jumps during saltation mean that grain impacts are more forceful in air than in water. Repeated impacts result in the gradual *attrition*, or wearing down, of airborne particles. Attrition results in the gradual smoothing of edges of grains as sharp corners are worn off. Particles which have undergone extensive aeolian transport are therefore characteristically well-rounded, and sometimes referred to as *millet-seed* grains. Another consequence of this violent interaction of the grains is that softer or more delicate minerals and lithic fragments are gradually destroyed by collision with harder grains such as quartz. Sands transported and deposited by wind are typically very mature texturally (the grains are well rounded and well sorted) and are also compositionally mature, consisting primarily of grains of quartz and other very stable grains, such as fragments of chert.

The colour of sands in modern deserts varies from pale yellow to bright red. The dry environment is oxidizing, breaking down any minerals containing iron into oxides and hydroxides. The yellow coloration is from the iron hydroxide *goethite*, whilst the deeper red colours are due to the presence of *haematite* which is an iron oxide. Fine particles of these iron minerals coat the sand grains, and although they may only represent a tiny fraction of the material present they give the desert sands their characteristic colours. Changes in these iron minerals may occur after deposition during diagenesis *(17.4.1)* such that a sand deposited with a grain coating of yellow goethite may become red as the hydroxide is altered to haematite.

8.2.3 Aeolian bedforms

Processes of transport and deposition by wind are principal features of sandy deserts, producing bedforms of a similar range in size and variety to the bedforms recognized in subaqueous environments. The sedimentary structures produced by the migration of these bedforms may therefore be considered to be characteristic of sandy desert environments. Although these aeolian features are only abundant in desert environments, localized wind-blown deposits can occur in association with other depositional settings. The most common of these smaller aeolian deposits occur in association with coastal facies *(13.2.2)* where wind-blown sand may accumulate along the coast above high water. These sands may be made up of terrigenous clastic detritus or carbonate grains, the latter forming white coastal sand dunes. Many of the features described below for aeolian sands formed in deserts may also be seen in these beach dune deposits.

AEOLIAN RIPPLE BEDFORMS

As wind blows across a bed of sand, grains will move by saltation *(4.2.2)* forming a thin carpet of moving sand grains (Bagnold 1954). The grains are only in temporary suspension, and as each grain lands, it has sufficient energy to knock impacted grains up into the free stream of air, continuing the process of saltation. Irregularities in the surface of the sand and the turbulence of the air flow will create patches where the grains are slightly more piled up. Grains in these piles will be more susceptible to being picked up by the flow and at a constant wind velocity all medium sand grains will move about the same distance each time they saltate. The result is a series of piles of grains aligned perpendicular to the wind and spaced equal distances apart. These are the crests of *aeolian ripples* (Fig. 8.2). The troughs in between are shadow zones where grains will not be picked up by the air flow and where few saltating grains land.

Aeolian ripples have extremely variable wavelengths (crest to crest distance) ranging from a few centimetres to several metres. Ripple heights (from the bottom of the trough to the top of the next crest) also show a big range, from less than a centimetre to tens of centimetres. Coarser grains tend to be concentrated at the crests, where the finer grains are winnowed away by the wind, and as aeolian ripples migrate they may form a layer of

Fig. 8.2 Aeolian ripples formed in sand in the Wahiba Sands desert area, Oman.

reverse-graded sand. In circumstances where a crest becomes well developed, grains may avalanche down into the adjacent trough, forming cross lamination (Sharp 1963).

AEOLIAN DUNE BEDFORMS

Sand dunes (Fig. 8.3) are the most obvious feature of any sandy desert. These *aeolian dunes* are bedforms 3–600 m in wavelength and between 10 cm and 100 m high. It should be noted that aeolian ripples do not grow into aeolian dunes. A plot of the range of sizes of aeolian ripples and dunes (Fig. 8.4) shows that they fall into two distinct fields (Wilson 1972). The distinction between ripple- and dune-scale bedforms seen in aeolian environments is analogous to the differences in processes and products between subaqueous ripples and subaqueous dunes.

Aeolian dunes migrate by the saltation of sand up the stoss side of the dune to the crest. This saltation may result in the formation of aeolian ripples which are commonly seen on the stoss sides of dunes (Fig. 8.5). Sand accumulating at the crest of the dune is unstable and will cascade down the lee slope as an avalanche or grain flow *(4.6.3)* to form an inclined layer of sand. Repeated avalanches build up a set of cross beds which may be preserved if there is a net accumulation of sand. There is also accretion to the lee face from the fallout of suspended sand: these *grain fall* deposits are reworked by avalanching at the top of the dune but may be preserved at the toe bedded with grain flow deposits.

The orientation and form (planar or trough) of the

Fig. 8.3 Aeolian dunes, Wahiba Sands desert area, Oman.

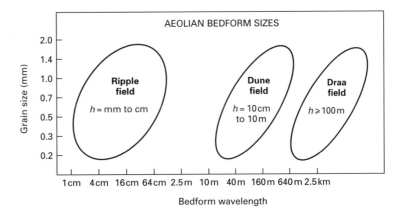

Fig. 8.4 Wavelengths of aeolian ripples, dunes and draas: the lack of overlap between the size ranges indicates that they are distinct bedforms. (After Wilson 1972; courtesy of the Society for Sedimentary Geology.)

Fig. 8.5 Aeolian ripples superimposed on the back of an aeolian dune (Wahiba Sands desert area, Oman). The sand migrates up the stoss of the dune by saltation, creating aeolian ripples; at the crest of the dune the sand avalanches down the lee slope to form cross bedding in the sand.

cross bedding will depend on the type of dune (Fig. 8.6) (McKee 1979). Planar cross beds will form by the migration of *transverse dunes* which are straight-crested forms aligned perpendicular to the prevailing wind direction. Transverse dunes form where there is an abundant supply of sand and as the sand supply decreases there is a transition to *barchan dunes*, which are lunate structures with arcuate slip faces forming trough cross bedding. Under circumstances where there are two prominent wind directions at approximately 90° to each other, *linear* or *seif dunes* form. The deposits of these linear dunes are characterized by cross bedding, reflecting avalanching down both sides of the dune and hence orientated in different directions. In areas of multiple wind directions *star dunes* have slip faces in many orientations and hence the cross bedding directions display a similar variability.

There are circumstances in which the whole dune bedform is preserved but more commonly the upper part of the dune is removed as more aeolian sand is deposited in an accumulating succession. The size of the set of cross beds formed by the migration of an aeolian dune can vary from around a metre to 10–20 m. Such large-scale cross bedding is common in aeolian deposits but is less frequently seen in subaqueous sands which are typically cross bedded in sets a few decimetres to metres thick. The sorting is also normally better in aeolian deposits when compared to subaqueous sands and clay is typically absent from sandy aeolian deposits.

DRAA BEDFORMS

When a sand sea is viewed from high altitudes in aerial

photographs or satellite images, a pattern of structures which are an order of magnitude larger than dunes becomes evident. The surface of the sand sea shows an undulation on a scale of hundreds of metres to kilometres in wavelength and tens to hundreds of metres in amplitude. These structures are known as *draas*, and there is again evidence that they are a distinct, larger bedform separate from the dunes which may be superimposed on them (Wilson 1972). Draas are usually made up of dunes on the stoss and lee sides, but a single slip face may develop on some lee slopes. They show a similar variability of shape to dunes with star, linear and transverse forms.

8.2.4 Palaeowind directions

The slip faces of aeolian dunes generally face downwind so by measuring the direction of dip of cross beds formed by the migration of aeolian dunes it is possible to determine the direction of the prevailing wind at the time of deposition (Fig. 8.6). The variability of the readings obtained from cross beds will depend upon the type of dune (McKee 1979). Transverse dunes generate cross beds with little variability in orientation, whereas the curved faces of barchan dunes produce cross beds which may vary by as much as 45° from the true downwind direction. Multiple directions of cross bedding result from the numerous slip faces of a star dune. In all cases the confidence with which the palaeowind direction can be inferred from cross bedding orientations is improved as more readings are taken.

Wind directions are normally expressed in terms of the

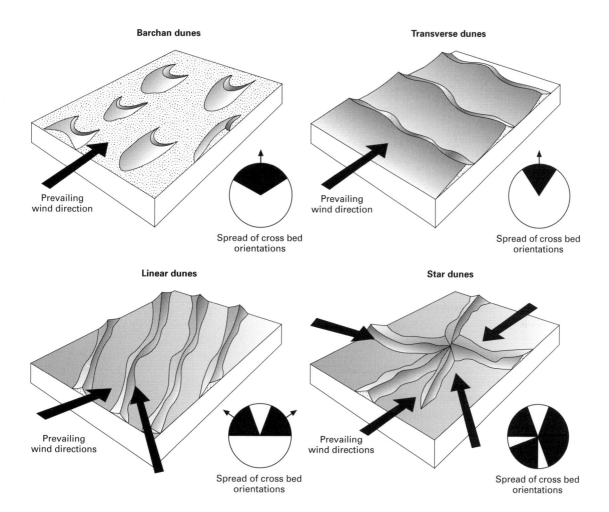

Fig. 8.6 Four principal types of aeolian dune formed in response to wind direction and abundance of sand. (After McKee 1979.)

direction from which the wind blows; that is, a south-westerly wind is one which is blowing from the south-west towards the north-east, and will generate dune cross bedding which dips towards the north-east. Note that this form of expression of direction is different from that of water currents which are normally presented in terms of the direction the flow is towards.

8.2.5 Other wind-blown deposits

Thick deposits of wind-blown silt (*loess*) are known from the Pleistocene of Asia and central Europe (Pye 1987) *(7.3.4)*. They are not necessarily associated with other desert sediments, having formed during dry periods of the glacial cycles when strong winds carried large amounts of dust thousands of kilometres across the continents. Other long-travelled wind-blown deposits found throughout the stratigraphic record are of pyroclastic origin *(16.3.1)*.

8.3 Water in a desert

When it rains in a desert, it often pours. The mean annual rainfall for an arid region is less than 250 mm, but this figure is a little misleading. Precipitation may be negligible for many months or years, then many centimetres of rain may fall in a single storm lasting a few hours. The large volumes of water associated with a desert rainstorm can be very effective at transporting and depositing detritus in a flash flood. Although they result from

Fig. 8.7 Active flow in a wadi following a rainstorm in the Sinai desert, Egypt. The slurry of water, sand and mud flowed for a few hours, leaving a thin bed of sandy mud. Much larger flows are required to produce beds decimetres to metres thick of coarser, gravelly material.

infrequent events, flash flood deposits can constitute a significant proportion of sediments laid down in a desert environment.

8.3.1 Wadi gravels

A *wadi* is a desert river or stream with ephemeral flow. In upland areas detritus formed by weathering and erosion falls by gravity or is washed down valley sides on to the wadi floor, where it accumulates for months or years. Following a rainstorm, water flow in a wadi may be sufficiently powerful to move the detritus (Fig. 8.7) either as bedload of the stream flow or as a debris flow *(4.6.1)*. Material may be carried in the channel of a wadi for many kilometres out into a sedimentary basin to form a channel deposit associated with aeolian and playa lake deposits. Wadi deposits are characteristically poorly sorted, consisting of angular or subangular gravel clasts in a matrix of sand and mud (Fig. 8.8). Sedimentary structures are absent, except where the deposit has been winnowed by water flowing over the surface. This reworking may wash out the finer matrix and result in a horizontal stratification of the deposit and the development of imbrication. The deposits of an individual wadi are not laterally extensive as they are restricted by the width of the channel, but the channel may migrate laterally or there may be multiple channels on an alluvial plain which merge to form a more extensive deposit.

8.4 Alluvial fans

Alluvial fans are cones of detritus which form at a break

Fig. 8.8 Poorly sorted pebbly sandstone and conglomerate interpreted as wadi gravel deposits. The clasts are angular and beds are crudely stratified. Permian, Devon, England.

in slope at the edge of an alluvial plain. They are formed by deposition from a flow of water and sediment coming from an erosional realm adjacent to the basin. The term 'alluvial fan' has been used in geological and geographical literature to describe a wide variety of deposits with an approximately conical shape including deltas and the depositional tracts of distributary river systems. It is probably simplest to restrict usage of the term to deposits which are unchannelized and occur on relatively steep slopes (greater than 1°) forming cones up to a few kilometres in radius (Blair & McPherson 1994).

The 'classic' modern alluvial fans described from such places as Death Valley in California, USA (Bull 1972; Blair & McPherson 1994) (Fig. 8.9) occur in arid and semi-arid environments. Many ancient examples of

Fig. 8.9 The Badwater alluvial fan in Death Valley, California, USA. This fan is only a few hundred metres in radius and has a steep depositional slope.

alluvial fan deposits are also thought to have formed in similar settings. Alluvial fans also form in much more humid climates with similar morphologies and processes *(8.4.5)*.

8.4.1 Morphology of alluvial fans

The area of high ground being drained by the valley system which supplies the alluvial fan is the *drainage basin* of the fan. At the point where a valley opens out at the edge of a sedimentary basin there will be a distinct break in the topography. Flows of water and sediment mixtures from the *feeder canyon* spread out to form a sheet. The flow quickly loses energy and deposits the sediment load. Repeated depositional events will build up a deposit which has the form of a segment of a cone radiating from the point where the feeder canyon meets the basin margin.

On a typical alluvial fan, a number of morphological features can be recognized (Fig. 8.10). The *fan apex* is the highest, most proximal point adjacent to the feeder canyon from which the fan form radiates. A *fan-head trench* may be incised into the fan surface near the apex. The depositional slope will be steepest in the proximal area, as much as 15°, but the slope rapidly decreases to around 5° over most of the fan surface and 1–2° at the edges. A break in slope is seen at the *fan toe*, which is the limit of the deposition of coarse detritus and marks the edge of the alluvial fan. They are thickest at the apex and taper as a conical wedge towards the toe (Fig. 8.10).

8.4.2 Processes of deposition on alluvial fans

On present-day alluvial fans the flows can be seen to be of two main types. Debris flows occur where there is a dense mixture of water and sediment. These viscous flows do not travel far and a small, relatively steep alluvial fan cone is built up. More dilute mixtures flow as traction currents with a bedload of coarse sand and gravel and a suspended load of finer material. These flows spread out on the fan surface as *sheetfloods* (Bull 1972; Blair & McPherson 1994).

SUBAERIAL DEBRIS FLOWS

When an aggregate of weathered material is mixed with a small amount of water it becomes a dense slurry which may flow for a few hours as a debris flow. The detritus carried may include clasts of all sizes, ranging from individual clay particles to boulders. Such a mixture is mobile on a slope and will move as a very viscous mass until it runs out of momentum, usually when the gradient decreases or the mixture loses water content. A high concentration of clay particles increases the viscosity of the slurry, allowing it to transport large clasts, and flow is normally laminar.

Elongate clasts can become orientated parallel to the movement direction in laminar flows, although if the viscosity is very high clasts can adopt and maintain any orientation. Some cobbles and boulders may be carried along on the top of the flow, buoyed up by the viscosity

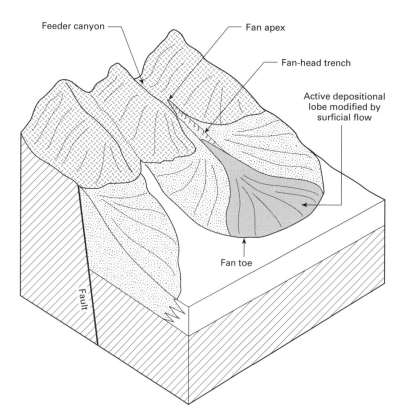

Fig. 8.10 Morphological elements of an alluvial fan.

of the slurry. It is possible for boulders metres across to be carried by debris flows. There is little opportunity for the separation of different grain sizes in a viscous, laminar flow. The only internal structure which forms in viscous debris flows is due to friction at the base of the flow. Drag along the underlying surface results in a zone of shearing in the basal few tens of centimetres of the debris flow. A crude horizontal stratification and parallel alignment of clasts may be seen in this sheared zone.

Debris flows continue to move until the momentum of flow can no longer overcome the internal friction: flows tend to stop abruptly, 'freezing' the flow in its tracks. Debris flows normally maintain a constant thickness and very little thinning down the flow is observed. Individual beds are typically matrix-supported conglomerates which lack internal sedimentary structures. Bed thicknesses range from a few tens of centimetres to a couple of metres, although individual beds can often be difficult to distinguish in a succession of debris flow deposits (Figs 8.11 & 8.12).

The characteristics of a bed deposited by a debris flow in an arid environment are:

1 the conglomerate normally has a matrix-supported fabric such that the clasts do not always touch each other and are almost entirely separated by the finer matrix;
2 sorting of the conglomerate into different clast sizes within or between beds is usually very poor;
3 the clasts may show a crude alignment parallel to flow in the basal sheared layer but otherwise the beds are structureless with clasts randomly orientated;
4 outsize clasts metres across may occur sometimes rafted on the top of the debris flow unit;
5 beds deposited by debris flows are rarely thinner than 30 cm and may be metres thick.

SHEETFLOOD DEPOSITION

If the volume of water mixed with the detritus is sufficiently large the clasts are moved by traction processes of rolling and saltation and in suspension *(4.2.2)*. *Sheetfloods* on alluvial fans follow heavy rainfall in the drainage basin. Accumulated detritus is washed out through the feeder canyon on to the fan surface where it spreads out as a sheet typically covering a third or half of

Debris flow deposits

~1–2 m

- Poorly sorted
- Random clast orientation
- Matrix-supported
- No sedimentary structures

Sheetflood deposits

~0.5–2 m

- Moderately sorted
- Imbricated clasts
- Clast-supported
- Normally graded

Fig. 8.11 The characteristics of beds formed by debris flow and sheetflood processes on alluvial fans.

Fig. 8.12 A debris flow deposit 2 m thick showing poor sorting and a chaotic arrangement of clasts.

the fan surface area. As the flow leaves the steep valley system and spreads out on to the fan it loses energy rapidly. Deposition occurs as a result of this loss of energy with the larger clasts being deposited first. If a large amount of the coarser material is deposited nearer the apex of the fan the sheetflood deposit may thin towards the toe. Once the flow has reached the toe of the fan virtually all the bedload has been deposited and only the suspended load is carried out beyond the fringe of the fan. The flow velocity decreases as a flash flood subsides and the sheetflood deposits successively finer and finer bedload as the flow wanes. The result is a normally graded bed, typically a few tens of centimetres thick (Bull 1972). These depositional events are short-lived, involving the rapid dumping of detritus. Some imbrication of clasts may form but other sedimentary structures resulting from the flow are rare.

The characteristics of a sheetflood deposit on an alluvial fan are as follows (Figs 8.11 & 8.13):

1 it has a sheet geometry with beds at least a hundred times wider than they are thick;

2 it is generally normally graded, with boulders at the base and mainly sand at the top, but the rate of deposition is normally too great to allow clear sorting of all the coarser fraction into the lower part of the bed and the finer at the top;

3 imbrication of discoid clasts and alignment of elongate clasts parallel to the flow are often present;

4 other sedimentary structures such as cross stratification may occur but are uncommon.

8.4.3 Modification of alluvial fan deposits

Deposition on an alluvial fan occurs very infrequently (in terms of the human time-scale). It has been estimated that significant depositional events on one of the alluvial fans in the south-west United States occur every 300 years or so on average (Beaty 1970). The sheetfloods and

Fig. 8.13 Bedded sandstone and conglomerate interpreted as sheetflood deposits on an Oligocene alluvial fan, Salto de Roldan, north of Huesca, northern Spain.

debris flows which deposit sediment normally last only a matter of hours, so there are very long periods between these events. Water flows from the drainage basin over the surface of the fan sporadically after rainfall events which were too small to transport any material on to the fan. These superficial water flows can locally winnow out sand and mud from between the gravel clasts, removing the matrix of the deposit (Blair & McPherson 1994). If the spaces are not filled in later an *open framework* or *matrix-free conglomerate* may be preserved. Matrix-free conglomerate beds (which are rarely preserved in the stratigraphic record) have also been considered to be primary *sieve deposits* which were deposited without any matrix in the first place (Hooke 1967).

A more significant modification of the alluvial fan surface is by streams which become established on the fan surface between depositional episodes. These rework debris flow and sheetflood deposits, form a channel and remove some material from the fan surface. On many modern alluvial fans this is an important process (Blair & McPherson 1994). These steep channels have the form of a braided river *(9.2.1)* with bars of gravel redeposited or left uneroded within the channel. These reworked sediments can be referred to as *stream channel deposits*, although other authors (Bull 1972) use this term for the deposits of true braided rivers which form on a fan surface. The angle of a fan surface (greater than 1–2°) would normally be considered too steep for a braided river to deposit material (about one-third of a degree: *9.1.2*), but it is possible that some parts of rivers have a steep fan morphology and are composed of stream channel deposits.

In arid regions aeolian processes also act to modify the fan surface by winnowing away sand and silt from the fan surface. The surfaces of gravel clasts too coarse to be moved by the wind can be polished by the abrasion of wind-borne particles, and a *desert varnish* of oxides may coat the surfaces of pebbles and boulders. These effects are rarely seen in ancient alluvial fan deposits.

8.4.4 Controls on alluvial fan deposition

CLIMATE

Arid-zone alluvial fans are examples of episodic sedimentation systems. Most of the time there is no deposition occurring on the fan. At intervals of years to hundreds or thousands of years a depositional event occurs, adding to the body of sediment on the fan. In order to deposit a lobe of sediment on the surface of an arid-zone alluvial fan, there must be a supply of detritus in the catchment area and water to carry the detritus out on to the fan. Weathering processes in the catchment area gradually produce fragments of bedrock which accumulate on the lower slopes and in the valley bottoms. This will build up until there is a sufficient supply of water to wash it out on to the fan. Rainfall in arid areas is infrequent, generally occurring in short-duration rainstorms which happen every few years, tens or hundreds of years. For a depositional event to occur a rainstorm of sufficient intensity to wash detritus out of the hinterland valleys and into the feeder canyon is required. In such a storm many centimetres of rainfall will occur over a matter of hours or days. The longer the period between these rainstorms, the more time available for bedrock to weather and debris to accumulate in the hinterland valleys. Very infrequent depositional events hence involve a greater volume of detritus. Alluvial fan deposition is therefore strongly controlled by climatic factors which influence the periodicity and intensity of rainstorms.

BEDROCK

The characteristics of an alluvial fan succession will also be determined by the nature of the bedrock in the hinterland (Hooke 1968). Debris flow deposition is favoured if the detritus in the flow contains a high proportion of mudrock, as abundant clay minerals will increase the density and viscosity of the flow. Fans built up adjacent to a hinterland rich in fine-grained lithologies tend to be steeper, built up largely of debris flows with short transport histories.

TECTONIC ACTIVITY

An alluvial fan cannot continue to build upwards as it would start to block its own feeder canyon. To create a thick succession of alluvial fan deposits, continued subsidence is required. In the absence of adequate subsidence, a fan will build outwards rather than upwards (Heward 1978). Alluvial fans develop where there is a distinct break of slope at the edge of a sedimentary basin. This topographic break is most distinct if there is continued tectonic activity to maintain uplift in the hinterland and subsidence in the adjacent basin. This situation occurs at the edges of rift valleys and along the mountain fronts at the margins of orogenic belts (Nichols 1987a).

Although alluvial fan deposits are not by any means the most significant volumetrically in a sedimentary basin, they are important because they occur at the basin margins where tectonic activity is most easily recognized. Changes in the amount or style of faulting at a basin margin will be reflected in the nature of the deposits in the alluvial fans—for example, if there is a strike-slip fault at the edge of the basin (Haughton 1989).

8.4.5 Alluvial fans in humid environments

Although alluvial fans have been considered here under the heading of arid region deposits, they also occur in much wetter depositional settings. The first description of an alluvial fan was of a deposit formed after a flood in the humid climate of northern England (Smith 1754 in

Blair & McPherson 1994). Alluvial fans can occur in any continental setting where there is a drainage basin to supply coarse detritus and a sharp break in slope to act as a site for deposition. Steep edges of sedimentary basins are therefore the most likely sites for the formation of alluvial fans.

In circumstances where a lot of water is involved in the transport, the deposits are typically from sheetfloods rather than debris flows. Modification by soil processes and reworking by streams flowing over the surface are common. In humid environments the fan deposits will interfinger with floodplain deposits of rivers (Chapter 9) and lakes (Chapter 10).

In recent years there has been some debate about the use of the term 'alluvial fan'. 'Fan' is one of the most used and abused terms in geomorphology and sedimentology. Used in a purely morphological sense it simply means a deposit which is approximately semi-circular or at least forms a segment of a circle in plan view. The processes of deposition within that fan shape, the products of those processes and the area that the fan may cover vary considerably. The term 'alluvial fan' is broadly applied to any cone of detritus built up in a continental environment, although some argue that this definition is too broad and would prefer a more restricted use of the term (Blair & McPherson 1994). A scree cone formed primarily of rock fall and rock avalanche is perhaps best called a 'talus cone' *(6.6.1)* to distinguish it from other bodies formed by the build-up of water-lain sediment. Fans formed by the unconfined flow of sheetfloods and debris flows are the 'original' alluvial fans and may be typified by fans in Death Valley, California (Fig. 8.9). In cases where the flow is confined to river channels which migrate laterally to form a cone of detritus, the term 'alluvial fan' or *humid fan* (Collinson 1986, 1996) is sometimes also applied because of the morphology of the deposit, although the processes and products of deposition are clearly quite different. The Kosi Fan in India is often quoted as the type example of this type of fan: it is 150 km from apex to toe, and has been formed by the lateral migration of a tract of a tributary to the River Ganges (Wells & Dorr 1987). River systems which are distributary in character also form a cone of detritus which may be tens of kilometres across. Distributary river systems are commonly terminal (they do not feed into a lake or sea), so these are referred to as *terminal fans* (Kelly & Olsen 1993) or simply *fluvial distributary systems* (Nichols 1987b) *(9.1.4)*.

8.5 Playa lakes

After a rainstorm, water accumulates in low-lying areas which range in size from ponds between sand dunes to vast inland lakes such as Lake Eyre in central Australia. In hot arid regions the rate of evaporation in these ponds and lakes is greater than the supply of water from direct rainfall or ephemeral streams, so the bodies of water dry up. These ephemeral water bodies in deserts are called *playa lakes* (Figs 8.14 & 8.15) (Ward 1988). The temporary streams flowing into the lake carry with them a suspended load of clay and silt which is deposited in the still water. A layer of mud is laid down and as the lake shrinks by evaporation the mud is exposed and dries up to form desiccation cracks *(4.7)* in the mud around the lake margins.

In addition to the suspended load, the lake waters also contain ions in solution dissolved from the bedrock over which the ephemeral streams flowed. As the lake dries up, these ions become concentrated in the water, which becomes saline. The water becomes saturated in certain ions which are precipitated as salts (evaporite minerals: *3.3*). The minerals formed in this way will vary according to the chemistry of the lake water and this will vary

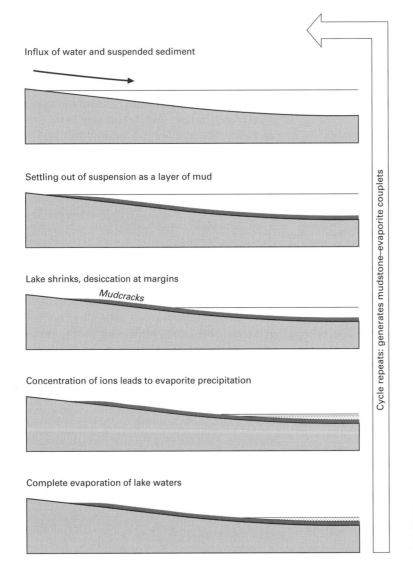

Influx of water and suspended sediment

Settling out of suspension as a layer of mud

Lake shrinks, desiccation at margins

Mudcracks

Concentration of ions leads to evaporite precipitation

Complete evaporation of lake waters

Cycle repeats: generates mudstone–evaporite couplets

Fig. 8.14 Sequence of events which generates couplets of mud and evaporite minerals a few centimetres thick in playa lakes.

Fig. 8.15 A playa lake in a desert basin, Badwater, Death Valley, California, USA. Around the edges of the very saline waters of the lake evaporite minerals precipitate on the mudflats.

according to the ions supplied by the feeder streams. The water chemistry will be unique to each lake so the types and proportions of evaporite minerals will be similarly variable; this is in contrast to marine evaporites *(13.5)* which are all similar, precipitated from marine waters of constant composition. The commonest evaporite minerals in saline lakes are calcium sulphate (gypsum) and sodium chloride (halite). Most of the other salts which occur are carbonates, sulphates and chlorides of magnesium, sodium, calcium and potassium. A crust of evaporite minerals is left by the retreating lake waters, which may dry up completely to form a *salt pan* in the lake bed (Ward 1988). These areas are also referred to as *saline pans* or *salinas*.

The deposition of mud out of suspension, followed by precipitation of evaporite minerals as the lake dries up, leads to the formation of a *depositional couplet* consisting of a mud layer a few millimetres thick overlain by evaporite minerals in a bed a few centimetres thick. When the lake bed receives another influx of water the most soluble evaporite minerals may be wholly or partly dissolved but a layer of evaporite is likely to remain. This layer will be overlain by mud falling out of suspension. Repeated influxes of water into the lake build up a succession of mud–evaporite couplets which are characteristic of playa lake sedimentation (Fig. 8.14).

Evaporite minerals also form within the sediments surrounding playa lakes. In these areas the sediment is saturated with saline groundwater which evaporates at the ground surface. This concentrates the dissolved minerals and leads to crystallization of evaporite minerals.

These regions are sometimes referred to as *inland sabkhas* (cf. coastal sabkhas: *13.5.1*). The most common to be formed is gypsum which grows within the sediment in an interconnected mass of bladed crystals known as *desert rose*.

8.6 Life in the desert

The irregular supply of water to desert regions makes them harsh environments for most organisms. Only drought-tolerant plants survive, and animals which live on these plants or each other are present but not abundant. The preservation potential of these sparse flora and fauna is very low, so the deposits of arid environments are typically devoid of fossils. Plant and soft animal tissue is either destroyed in the oxidizing environment of a desert or is removed by scavengers. Only the bones of large mammals and reptiles survive to be preserved as fossils.

Trace fossils *(11.7.2)* are more common than body fossils in the deposits of arid environments. Tracks and trails left in mud as it dries out around the edges of playa lakes or in interdune ponds have a good chance of being preserved as the impressions are filled by overlying deposits.

8.7 Characteristics of the deposits of arid continental environments: summary

Aeolian sandstone is characteristically well-sorted and composed of well-rounded grains of stable minerals such as quartz. A grain coating of iron oxides and hydroxides typically gives the sandstone a yellow to red colour. Small-scale sedimentary structures include planar lamination and rarer cross lamination. Cross bedding formed by migration of dunes can occur in sets which are typically metres thick, with orientations depending on the wind direction and the form of the dune. Wadi and alluvial fan gravels are texturally similar deposits which form poorly sorted conglomerates which are matrix-supported if deposition was by a debris flow or clast-supported and graded if deposition was by a sheetflood. These deposits are confined to channels in wadis but spread out as sheets on alluvial fans to build up cones of detritus at the margins of a basin. Playa lake mudstone and evaporite beds are characterized by depositional couplets made up of mudstone beds deposited from suspension which may show desiccation cracks and layers of evaporite minerals formed by the ephemeral water body drying out. The association of these facies can be considered characteristic of arid environments but each can occur in association with other facies. Aeolian deposits may also form beach ridges *(13.2.2)*, alluvial fans also occur in humid continental settings and evaporite deposits are found associated with arid coastal environments *(13.5)*.

Further reading

Blair, T.C. & McPherson, J.G. (1994) Alluvial fans and their natural distinction from rivers based on morphology, hydraulic processes, sedimentary processes and facies assemblages. *Journal of Sedimentary Research* **A64**, 450–589.

Brookfield, M.E. (1992) Eolian systems. In: *Facies Models: Response to Sea Level Change* (eds R.G. Walker & N.P. James), pp. 143–156. Geological Association of Canada, St Johns, Newfoundland.

Frostick, L.E. & Reid, I. (eds) (1987) *Desert Sediments: Ancient and Modern*, 401 pp. Special Publication of the Geological Society of London **35**.

Kocurek, G.A. (1996) Desert aeolian systems. In: *Sedimentary Environments: Processes, Facies and Stratigraphy* (ed H.G. Reading), pp. 125–153. Blackwell Science, Oxford.

Pye, K. & Lancaster, N. (eds) (1993) *Aeolian Sediments Ancient and Modern*, 167 pp. International Association of Sedimentologists Special Publication **16**.

Rachocki, A.H. & Church, M. (eds) (1990) *Alluvial Fans: A Field Approach*, 391 pp. Wiley, Chichester.

9   Rivers: the fluvial environment

The world's rivers are the main arteries for the transport of clastic detritus. With the exception of material carried directly to the sea by glaciers, wind or coastal erosion, all the terrigenous clastic deposits of the oceans are supplied by rivers. In addition to acting as conduits for the transport of sediment to lakes or oceans, rivers act as depositional systems. Sediment can accumulate in the bed of the river and on the overbank areas when the river floods. The grain size and the sedimentary structures in the river channel deposits are governed by the supply of detritus, the gradient of the river, the total discharge and seasonal variations in flow. Overbank deposition consists mainly of finer-grained sediment carried by floodwaters beyond the edges of the river. Plant and animal activity on alluvial plains contributes to the formation of soils which can be recognized in the stratigraphic record as palaeosols. Rivers are possibly the best-known depositional environment because it is relatively easy to study modern river processes quantitatively and relate features seen in the stratigraphic record to them. Care must be taken in this uniformitarian approach because continental environments have changed dramatically through geological time as land plant and animal communities have evolved.

9.1 River forms and patterns

The term *fluvial* is used for anything associated with rivers. *Alluvial* is a more general term to include other features, such as a water-lain fan of detritus (an alluvial fan), which are not necessarily related to rivers. Water flow in river systems is normally confined to *channels*, depressions or scours in the land surface which contain the flow. The *overbank* area or *floodplain* is the area of land between or beyond the channels which (apart from rain) receives water only when the river is in flood. Floods occur when water is supplied into the river at a higher rate than can be carried within the channel.

9.1.1 Types of river

River channels can be categorized into four basic forms (Fig. 9.1) (Cant 1982). *Straight* channels are the simplest form; they are single, without dividing bars and have a low *sinuosity* (a value of less than 1.5). The sinuosity of

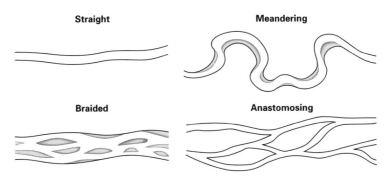

Fig. 9.1 Forms of river channel.

a channel is calculated by dividing the distance between two points measured along a channel by the straight-line distance between those two points in a downstream direction. *Meandering* rivers have a high sinuosity (greater than 1.5); they characteristically have bends in the river which change shape through time by erosion of one bank and by deposition on the other *(9.2.2)*. *Anastomosing* rivers consist of a number of separate channels which divide and join along the river. *Braided* rivers also have divisions, *bars* in the river which are only emergent when the level of water in the river is low; when the level of water in the channel is up to the tops of the banks, the bars are covered with water. Braided rivers typically have a low sinuosity. These four river types can be regarded as end members, with many intermediate forms existing (Schumm 1981).

From the point of view of sedimentary geology, meandering and braided rivers are the most important because these are the types most frequently recognized in the geological record. Straight channels are sometimes recorded and a few examples of anastomosing rivers have been reported.

9.1.2 Controls on river types

The most important controls on the river type are the gradient, local vegetation and the proportions of bedload and suspended load. On steeper slopes (greater than 0.1°) rivers tend to be braided and relatively straight, whilst at lower gradients meandering rivers predominate. Anastomosing rivers form on shallow gradients and it is likely that bank stability plays an important role in causing the channels to split *(9.2.3)*. A high bedload content leads to the development of bars in rivers (Schumm 1981). Fluvial systems can be made up of rivers of all or some of these types of channel. The upper reaches of some rivers are braided because of the steeper slopes. Larger river systems tend to have more extensive meandering tracts—for example, the Mississippi River is highly sinuous for the lower parts of its course. Where mountainous areas are close to the sea river systems are short and steep throughout and braided forms exist down to the coast.

9.1.3 Catchment and discharge

The area of ground which supplies water to a river system is the *catchment area*. Rivers and streams are mainly fed by surface run-off and groundwater from subsurface aquifers in the catchment area following periods of rain. Soils act as a sponge, soaking up moisture and gradually releasing it out into the streams. A continuous supply of water can be provided if rainfall is frequent enough to stop the soils drying out. Two factors are important in controlling the supply of water to a river system. First, the size of the catchment area: a small area has a more limited capacity for storing water in the soil and as groundwater than a large one. Second, the climate: catchment areas in temperate or tropical regions which have regular rainfall remain wet throughout the year and keep the river supplied with water.

A large river system in a tropical rainforest area is constantly supplied with water, and the *discharge* (the volume of water flowing in a river in a time period) remains constant. In contrast, rivers which have much smaller drainage areas with seasonal rainfall have highly variable discharge. *Ephemeral* rivers are dry for long periods of time and only experience flow after there has been sufficient rain in the catchment area. These rivers may show braided or meandering forms, but if the flow is of short duration the deposits may be considered to be wadi gravels *(8.3.1)*.

When the water level is well below the level of the banks it is at *low flow stage*. A river with water flowing close to or at the level of the bank is at *high flow stage* or *bank-full flow*. When the volume of water being supplied to a particular section of the river exceeds the volume which can be contained within the channel at that point, the river *floods* and *overbank flow* occurs beyond the limits of the channel.

As water flows in a channel it is slowed down by friction with the floor of the channel, the banks and the air above. These frictional effects decrease away from the edges of the flow to the deepest part of the channel where there is the highest-velocity flow. The line of the deepest part of the channel is called the *thalweg*. The existence of the thalweg and its position in a channel are important to the scouring of the banks and the sites of deposition in all channels.

9.1.4 Patterns of rivers

Most rivers exhibit a *tributary* drainage pattern in which small streams converge to form larger trunk channels. The pattern on a map is dendritic (Fig. 9.2), with a single channel reaching the sea or lake, sometimes via a delta. Tributary river systems are efficient at carrying sediment through the system, and the accumulation of great

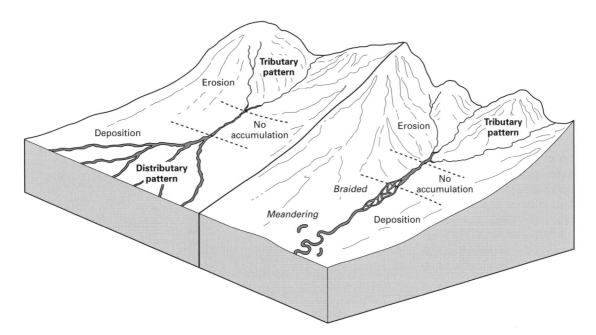

Fig. 9.2 Tributary river pattern and distributary river pattern and the changes in river form downstream in some river systems.

thicknesses of sediment does not usually result unless the lower reaches of the river are areas of continued subsidence. In basins of internal drainage (where rivers do not reach the sea) the rivers may be initially tributary in the areas of erosion, but upon reaching the area of deposition they become distributary. In distributary systems the rivers split or *bifurcate*, branching into smaller rivers and streams in a downstream direction (Fig. 9.2). The

distributary parts of a river system are less efficient at carrying sediment. With each bifurcation the channels become smaller, transmit less water and are less able to carry sediment. Most of the bedload of a river is deposited along the distributary tract of the river.

9.2 Modern rivers

The ease with which present-day river processes and deposits can be measured and observed has led to a considerable amount of work on the hydrology and geomorphology of rivers. Much of this work is relevant to

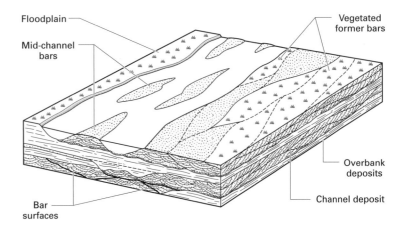

Fig. 9.3 Main morphological features of a braided river.

the recognition and interpretation of fluvial deposits in the stratigraphic record. Only three river types are considered here, and it must be emphasized that these are end members: intermediate forms exist today and doubtless many ancient deposits were the products of rivers which did not fit neatly into these categories.

9.2.1 Braided rivers

Rivers with a high proportion of sediment carried by rolling and saltation along the channel floor are referred to as *bedload rivers*. The bedload is deposited as bars of sand or gravel in the channel which are exposed at low flow stages, giving the river its braided form (Figs 9.3 & 9.4). Flow in the channel is divided around these bars except when the discharge is at bank-full level, when flow may occur over the whole width of the channel. Mid-channel bars can be exposed for long enough for vegetation to take hold (Bridge 1993).

Flow occurs between the bars and will usually be strongest along the thalweg in the deepest part of the channel. Along the thalweg the coarsest material will be transported and deposited. Most of the bedload of a braided river accumulates on bars within the channel or at the channel margins (Fig. 9.5). *Longitudinal bars* are elongate along the axis of the channel. Bars which are wider than they are long, spreading across the channel, are *transverse bars*. Crescentic bars with their apex pointing downstream are *linguoid bars*. Combinations of sandy and gravelly bar forms commonly exist (Allen 1983). This situation most frequently occurs when a river

Fig. 9.4 A sandy braided river at low flow stage with channel bars exposed, near Morondava, Madagascar.

Fig. 9.5 Gravel bars in a braided river, Alberta, Canada.

regularly experiences two strengths of flow: one of high enough velocity to transport gravel clasts, for example during monsoon or spring meltwater floods or other high flow stages; and a lower-velocity flow which can only transport sand-sized material. *Compound bars* are built up consisting of cross stratified gravels with lenses of cross bedded sands or lenses of gravel in sandy bar deposits result from this situation.

Bars in a braided river are not stationary for a prolonged period (Bristow 1987). As water flows over and around them, material is picked up from the upstream side of the bar and deposited mainly on the downstream side, but some accretion may also occur on the bar flanks. Movement of the bedload occurs mainly at high flow stages when the bars are submerged in water. In bars composed of gravelly material the clasts accumulate in parallel layers on the bar faces as cross stratification. In sandy braided rivers bars are seen to comprise a complex of subaqueous dunes over the bar surface. These subaqueous dunes migrate over the surface of the bar in the stream current to build up stacked cross bedded sands. Arcuate (linguoid) subaqueous dunes normally dominate creating trough cross bedding, but straight-crested subaqueous dunes producing planar cross stratification also occur.

The orientation of cross stratification in gravels deposited on bars in braided rivers will be related to the direction of flow in the river but is not necessarily directly downstream because accretion often occurs on the sides of bars as well as on the downstream edge. Where cross stratification is not seen palaeocurrent information can be determined from imbrication of discoid clasts *(2.3.3)* although the orientation may be oblique to the axial flow in the channel. In sandy braided rivers the direction of subaqueous dune migration is often not directly downstream but follows the orientation of local currents over and around the bar form. Palaeoflow directions determined from cross beds in braided river bar deposits can show a variance of around 60° either side of the mean channel flow.

Bars continue to migrate until the channel moves sideways, leaving the bar out of the main flow of the water. Alternatively, a bar may be overlain by other bar deposits which bury it. In most cases it is a combination of these effects: a bar is abandoned due to channel migration or avulsion, and is subsequently covered by overbank deposits or the bars of another channel. If a bar is abandoned due to lateral migration of the channel it becomes covered with finer deposits, culminating in overbank deposits which cap the bar sequence (Figs 9.3 & 9.6). In regions where braided rivers repeatedly change position on the alluvial plain, a broad, extensive region of gravelly bar deposits many times wider than the river channel may result. These *braidplains* are found in areas with very wet climates or where there is little vegetation to stabilize the river banks (e.g. glacial outwash areas: *7.3.3*).

A fining-upward succession can sometimes be recognized in braided river deposits (Fig. 9.7). The coarsest deposits at the base of the channel are those carried in the thalweg in the areas between bars. This *basal lag* on the scour surface will be overlain by bar deposits as the bars migrate in the channel. In a gravelly braided river the bar deposits will commonly consist of cross stratified

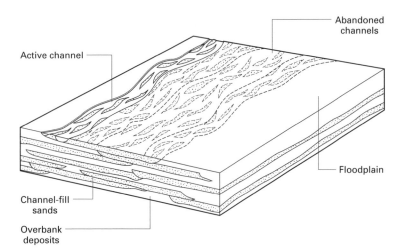

Fig. 9.6 Depositional architecture of a braided river: lateral migration of the channel and the abandonment of bars leads to the build-up of channel-fill successions.

Abandoned channels

Active channel

Floodplain

Channel-fill sands

Overbank deposits

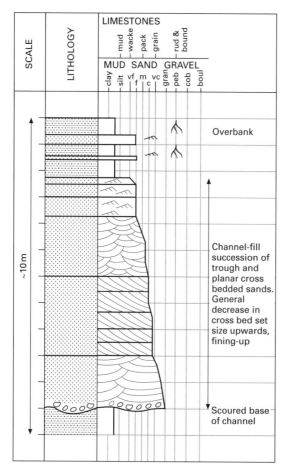

deposited on the lower parts of the bar, and at the top finer sands or silts represent the abandonment of the bar when it is no longer actively moving. However, some braid bars show little variation in grain size or cross bed set thickness in vertical section.

9.2.2 Meandering rivers

In plan view the thalweg in a river is usually not straight even if the channel banks are straight and parallel (Fig. 9.8). It may follow a sinuous path, moving from side to side along the length of the channel. When the thalweg comes close to the river bank it creates an asymmetry in the channel cross-section. The bank closest to the thalweg has relatively fast-flowing water against it whilst the opposite bank has slower-flowing water alongside. It is this asymmetry which causes river channels to meander. Meanders develop by the erosion of the bank closest to the thalweg accompanied by deposition on the opposite side of the channel. Deposition occurs where the flow is sluggish and the bedload can no longer be carried. With continued erosion of the outer bank and deposition of bedload on the inner bank the channel develops a bend and meander loops are formed (Figs 9.9 & 9.10).

Meandering rivers may dominantly carry suspended load, or there may be a mixture of suspended load and bedload (referred to as *mixed load*) (Schumm 1981). The bedload is carried by the flow in the channel, with the coarsest material carried in the central, deepest parts of the channel. Finer bedload may be saltating at higher levels in the flow and along the inner bend of a meander loop where friction reduces the flow velocity. The deposits of a meander bend develop a characteristic profile of coarse material at the base, normally sands and fine gravels, becoming progressively finer up the inner bank (Fig. 9.9). The faster flow in the more central, deeper parts of the channel forms subaqueous dunes in the sediment, which develops trough or planar cross bedding as the sand accumulates. Higher up on the inner bank where the flow is slower, ripples form in the finer sand, producing cross lamination (Allen 1970b).

Fig. 9.7 Graphic sedimentary log through a succession of sandy braided river deposits. For key to symbols see Fig. 5.7.

granules, pebbles or rarely cobbles in a single set. A sandy bar composed of stacked sets of subaqueous dune deposits will form a succession of cross bedded sands. As the flow is stronger in the lower part of the channel the subaqueous dunes, and hence the cross beds, tend to be bigger at the bottom of the bar, decreasing in set size upwards. There is a tendency for the coarser sands to be

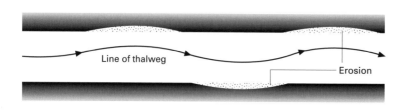

Fig. 9.8 Flow in a river follows the sinuous thalweg resulting in erosion of the bank in places.

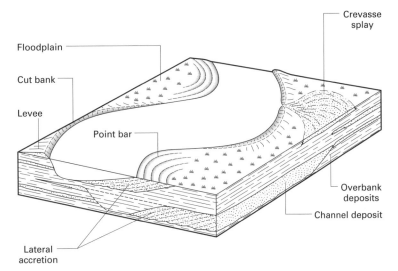

Fig. 9.9 Main morphological features of a meandering river.

Fig. 9.10 A meandering river at low flow stage exposing the point bars on the inner banks of the channel bends, near Morondava, Madagascar.

A channel moving sideways by erosion and deposition on the outer and inner banks, respectively, is undergoing *lateral migration*. The deposit on the inner bank is referred to as a *point bar* (Allen 1964b, 1965). A vertical section through a point bar deposit exhibits a gradation from coarser material at the base to finer at the top, that is, it will fine upward (Fig. 9.11). It may also show larger-scale cross bedding at the base and smaller sets of cross lamination nearer the top. As the channel migrates the top of the point bar becomes the edge of the floodplain. The fining-upwards succession in the point bar deposits will be capped by overbank deposits.

The lateral migration of the meander loop may be a steady continuous process or it may occur in a series of steps produced by changes in the flow strength in the river. If there are periods of non-deposition on the point bar a recognizable surface, sometimes draped by mud, may be preserved within the bar. These are known as *lateral accretion surfaces* because they mark stages in the lateral migration of the river by accumulation of sediment on the point bar (Fig. 9.9). These surfaces are also referred to as *epsilon cross stratification* and were recognized and interpreted as the product of a meandering river by Allen (1965). Epsilon cross stratification is the most reliable indicator of deposition by a meandering river. The cross strata characteristically dip towards the centre of the river perpendicular to the direction of flow, and are typically of low angle, up to 15° (Puigdefabregas & van Vliet 1978). The height of the set of epsilon cross strata may be up to the depth of the river channel but is usually less.

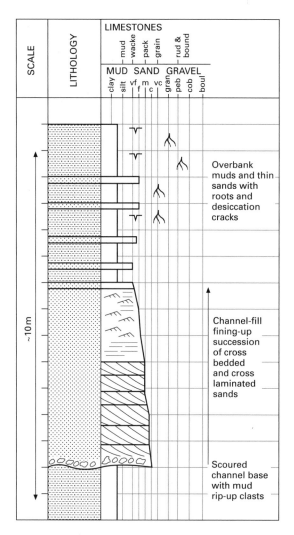

Fig. 9.11 Graphic sedimentary log through meandering river deposits.

During periods of high stage flow water may take a short-cut over the top of a point bar. This flow may become concentrated into a *chute channel* (Fig. 9.12) which cuts across the top of the inner bank of the meander. Chute channels may be semi-permanent features of a point bar but they are only active during high stage flow. They may be recognized in the deposits of a meandering river as a scour which cuts through lateral accretion surfaces.

In places where the bends in a meander belt exceed 180°, whole sections of the channel may become abandoned by the development of a new channel which takes a shorter route. This process normally occurs during flooding when water flowing over the bank streams across towards the next bend, eroding a channel. The abandoned meander loop becomes isolated as an *oxbow lake* (Fig. 9.13) and will remain as an area of standing water until it becomes silted up by floodwater deposition and/or choked by vegetation. Oxbow lake deposits are not always easy to recognize in ancient fluvial sediments; they are usually seen as channel fills of fine-grained sediment.

The scouring of the channel involves the removal of material, usually floodplain deposits. As this material is eroded it is carried downstream by the river. Sand and gravel become disseminated amongst the rest of the bedload of the river, but muddy floodplain deposits are cohesive and not broken down so easily. Consolidated mud eroded from the banks of a river frequently remains in pieces several centimetres across. These *mud clasts* become incorporated in the bedload of the river and are deposited along with the sandy material. Mud clasts remain angular if they are not moved far, but they are rapidly abraded and rounded by the interaction with other bedload material during transport.

9.2.3 Anastomosing rivers

Multiple river channels with frequent branches and interconnections are seen in places where the gradient is very low. These anastomosing (Smith & Smith 1980; Smith 1983) tracts of rivers occur most commonly where the banks are stabilized by vegetation which inhibits the lateral migration of channels but are also known from more arid regions with sparse vegetation. The positions of channels tend to remain fixed with time and most of the sedimentation occurs in the overbank regions during flooding. New channels may develop as a consequence of flooding as the water makes a new course across the floodplain, leaving an old channel abandoned. Recognition of anastomosing rivers in the stratigraphic record is problematic because the key feature is that there are several separate active channels. In ancient deposits it is not possible to demonstrate unequivocally that two or more channels were active at the same time. A similar pattern may form as a result of a single channel repeatedly changing position (avulsing: *9.4.2*).

9.2.4 Proximal to distal variations in fluvial systems

The character of a river usually undergoes marked

Fig. 9.12 A chute channel can be distinguished cutting across the point bar surface on the inner bend of this seasonally dry river course, near Morondava, Madagascar.

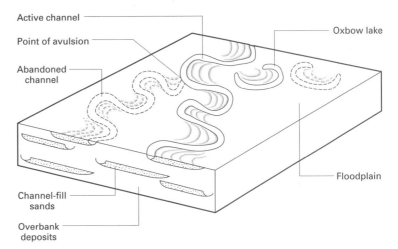

Fig. 9.13 Depositional architecture of a meandering river: sandstone bodies formed by the lateral migration of the river channel remain isolated when the channel avulses or an oxbow lake forms.

changes downstream as it passes from the upland, erosional tract to different parts of the depositional tract before reaching a lake or sea, or drying up completely (Fig. 9.2). These changes are principally due to the reduction in gradient and grain size along the river course. The slope of the river and the discharge affect the velocity of the flow, which in turn controls the ability of the river to scour and the size of the material which can be carried as bedload and suspended load.

In upland areas river courses are steep and may include rapids and waterfalls. Rivers in these areas are erosional. At lower gradients the river course may no longer be erosional but neither is it depositional, all the bedload and suspended load being carried downstream through this tract or only temporarily stored. Gravelly

braided rivers have the steepest depositional gradient (although the angle is typically less than 0.5°) and bars of pebbles, cobbles and boulders form. Finer debris is mostly carried through to the lower reaches of the river. At lower gradients the sandy bedload is deposited on bars in braided rivers, the flow having decreased sufficiently to deposit most of the gravel upstream. A meandering pattern tends to develop at very gentle gradients (*c.* 0.01°) in rivers carrying fine-grained sediment as bedload and suspended material (Collinson 1986, 1996). It is worth noting that the best-developed meander belts occur at these low gradients in homogeneous floodplains and that obstacles and irregularities in the floodplain appear to hinder meander belt formation. These tracts of different erosional and depositional characteristics are not present in all rivers. In areas of steep topography or high discharge, rivers may be erosional or non-depositional throughout or will be braided down to the mouth of the river where they will form coarse-grained deltas *(12.3)*.

9.3 Floodplains

The areas between and beyond the river channels are as important as the channels themselves from the point of view of sediment accumulation. When the discharge exceeds the capacity of the channel water flows over the banks and out on to the floodplain where *overbank* or *floodplain* deposition occurs.

9.3.1 Floodplain deposition

Most of the sediment carried out on to the floodplain is suspended load which will be mainly clay- and silt-sized debris but may include fine sand if the flow is rapid enough to carry sand in suspension. As water leaves the confines of the channel it spreads out and loses velocity very quickly. The drop in velocity prompts the deposition of all the sandy and silty suspended load, leaving only clay in suspension (Hughes & Lewin 1982). The sand and silt are deposited as a thin sheet over the floodplain. The remaining suspended load will be deposited as the floodwaters dry out and soak away after the flow has subsided.

Sandy and silty floodplain deposits are characterized by high-velocity flow as the water initially spreads out over the floodplain, followed by rapid sedimentation as the flow velocity quickly falls and most of the sediment load is deposited. Sandy and silty suspended load becomes bedload at the lower velocity. The initial deposits commonly display planar lamination produced by rapid flow. As flow velocity decreases current ripples form and migrate, and climbing ripples *(4.3.1)* will form because the rate of sedimentation will be comparable to the rate of ripple migration.

Sheets of sand and silt deposited during floods are thickest near to the channel bank because coarser suspended load is dumped quickly by the floodwaters as soon as they start flowing away from the channel. Repeated deposition of sand close to the channel edge leads to the formation of a *levée*, a bank of sediment at the channel edge which is higher than the level of the floodplain (Fig. 9.9). With time the level of the bottom of the channel can become raised by sedimentation in the channel and the level of water at bank-full flow becomes higher than the floodplain level. When the levée breaks, water laden with sediment is carried out on to the floodplain to form a *crevasse splay* (Fig. 9.9), a low cone of sediment formed by water flowing through the breach in the bank and out on to the floodplain (O'Brien & Wells 1986). The breach in the levée does not occur instantaneously but as a gradually deepening and widening conduit for water to pass out on to the floodplain. Initially only a small amount of water and sediment will pass through, and the volume of water and the grain size of the detritus carried will increase until the breach reaches full size. Crevasse splay deposits are therefore characterized by an initial upward coarsening of the sediment particle size. They are typically lenticular in three dimensions. Crevasse channels may develop into new river channels and carry progressively more water until avulsion occurs.

The primary depositional structures commonly observed in floodplain sediments are:

1 thin beds normally graded from sand to mud;
2 evidence of initial rapid flow (plane-parallel lamination) quickly waning and accompanied by rapid deposition (climbing ripple lamination);
3 thin sheets of sediment, often only a few centimetres thick but extending for hundreds of metres, even kilometres, away from the channel;
4 erosion at the base of the sheet sandstone beds, normally localized to areas near the channel where the flow is most vigorous;
5 desiccation cracks in muds at the tops of the normally graded beds;
6 evidence of soil formation *(9.7)*.

9.3.2 Floodplains and lakes

Floodplains are periodically inundated with water. Under conditions where the water does not evaporate quickly large areas of standing water develop and persist for months or years. These ephemeral ponds will have the characteristics of a lake whilst they are full of water. It is difficult to draw a sharp distinction between areas where floodwaters periodically pond and lakes which are semi-permanent on the alluvial plain. This is an instance where the 'pigeon-holing' of sedimentary environments is unsatisfactory and in any consideration of floodplain deposits the characteristics of lakes should also be considered *(10.7)*.

9.4 Ancient fluvial deposits

Fluvial deposits in the stratigraphic record can be considered at all scales, from the products of deposition on bars within a channel to the pattern of sedimentation of whole river systems tens to hundreds of kilometres across. For the most part direct comparison can be made with modern rivers but the possibility of ancient fluvial environments which have no direct modern analogues must also be considered.

9.4.1 Fluvial lithofacies

Lithofacies in fluvial deposits can be broadly divided into coarser sediments which occur in river channels and finer material deposited on the floodplain. Gravels may occur in cross bedded units representing bar deposits in gravelly braided rivers or as gravel lags, thin layers of coarse debris lying on the erosional scours at the bottoms of river channels. Sands deposited in channels are typically cross bedded, formed by subaqueous dune migration on bars in sandy braided rivers and on the lower parts of point bars in meandering rivers. Lateral accretion surfaces may be superimposed on cross bedded sands and are important because they indicate point bar formation in a meandering river. Cross laminated and parallel laminated sands may occur on the tops of braid bars or on point bars and they also occur in overbank deposits. Other overbank facies include thin-bedded sands deposited as thin sheets and mudstone which may display desiccation cracks and evidence of soil formation (*9.7.2*). Climbing ripple laminated sands are typical of levée deposits and crevasse splays.

The proportions of individual lithofacies and lithofacies associations are important in characterizing different fluvial environments. Floodplain lithofacies may be present in braided river tracts but in small proportions relative to the channel deposits (Fig. 9.6). Meandering rivers occur on the lower slopes where the alluvial valley is wider, providing a broad area for overbank facies to be deposited and preserved. The proportion of muddy lithofacies with soils and thin sheets of sand is hence higher in meandering river deposits.

9.4.2 The fill of river channels

Channels can change course in two ways. They may move sideways (*lateral migration*) by the gradual or stepwise erosion of either bank, creating a body of sand at the accretionary margin of the channel. Alternatively, rivers may *avulse* in which case the old river course is completely abandoned and a new channel is scoured into the land surface (Fig. 9.13). When this avulsion occurs the flow in the old river course reduces in volume and slows down. As the velocity falls the bedload will be deposited. A decrease in the amount of water supplied limits the capacity of the channel to carry sediment and the water gradually becomes sluggish, depositing its suspended load. Abandonment of the old river channel will leave it with sluggish water containing only suspended load as all the bedload is diverted into the new course. Oxbow lakes are an example of abandonment of a section of river channel as the change in the river course leaves the meander loop as a body of standing water. Abandoned and empty stretches of river channel are unlikely to remain empty for very long. When the river floods from its new course it will carry sediment across the floodplain to the old channel scour where sediment will gradually accumulate. The final fill of any river channel is therefore most likely to be fine-grained overbank sedimentation related to a different river course (Fig. 9.14). Channels entirely filled with mud may be very difficult to distinguish from overbank sediments in the stratigraphic record.

9.4.3 Recognition of channel margins

Evidence for flow in channels scoured into floodplain deposits is one of the features which is most distinctive in the fluvial environment. The cut bank at the edge of a channel is good evidence for the formation of channels but this is a feature which is not always recognized in ancient fluvial deposits. The absence of clear cut banks at the edges of channels in certain exposures of fluvial sediments may be explained by a combination of factors. First of all, rivers with a strong tendency to migrate laterally sweep across the floodplain, leaving channel margins only at the edges of the valley. In contrast, rivers which avulse frequently and occupy numerous separate courses for short periods of time tend to leave more cut banks in the floodplain. The second factor is the extent of the exposure. Areas where there are large expanses of complete exposure provide more opportunity for the complete nature of the fluvial succession to be observed, including channel margins.

Fig. 9.14 The outline of a channel can be identified by the truncation of the thin, horizontally bedded overbank deposits, but the channel is largely filled with mudstone. This is interpreted as a channel which was abandoned soon after formation, providing little opportunity for sand to accumulate within it. Oligocene–Miocene, near Huesca, northern Spain.

9.4.4 Scale in fluvial deposits

Rivers vary in size from small streams only metres in width and tens of centimetres deep to rivers tens of kilometres wide and tens of metres deep. This range in channel size over several orders of magnitude is also seen in channel-fill deposits in fluvial successions. A knowledge of the scale of channels within a fluvial succession is important to the interpretation of the internal and external geometry of channel deposits. The most significant erosion surface marks the banks and floor of the channel but there are also erosion surfaces between the bar deposits which make up the fill of the channels, especially in braided river deposits (Allen 1983). In small, isolated exposures distinguishing between an erosion surface enveloping a bar form (or *macroform*) and channel scours can be difficult (Fig. 9.15).

9.4.5 Architecture of fluvial deposits

The three-dimensional arrangement of channel and overbank deposits in a fluvial succession is commonly referred to as the *architecture* of the beds. The architecture is described in terms of the shape and size of the sand or gravel beds deposited in channels and the proportion of 'in-channel' deposits relative to the finer overbank facies.

The thickness of the channel-fill deposits is determined by the depth of the rivers. Their width is governed by the processes of avulsion and lateral migration of the channel. There is a tendency for nearly all rivers (meandering and braided) to shift sideways through time by erosion of one bank and deposition on the opposite side. Lateral migration continues until avulsion of the river causes the channel to be abandoned. If avulsion is

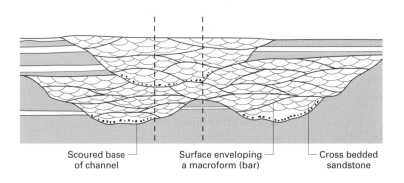

Scoured base of channel   Surface enveloping a macroform (bar)   Cross bedded sandstone

Fig. 9.15 Recognition of three separate channel-fill successions is possible in this example where there is sufficient exposure, but with limited exposure (between dashed lines) or in core, distinguishing them can be difficult.

frequent, there is less time for lateral migration to occur and the architecture will be characterized by narrow channel deposits (Fig. 9.16). Avulsion is frequent in rivers which are in regions of tectonic activity, where frequent faulting and related earthquakes affect the river course, and in settings where overbank flooding is frequent, resulting in weaker banks which make it easier for the river to change course.

Lateral migration is slowed down if the river banks are stable. Bank stability is governed by the nature of the floodplain. Muddy floodplain deposits form stable banks because clay is cohesive and is not so easily eroded as sandy beds. The type and abundance of vegetation are also important because dense vegetation, particularly grass which has fibrous roots, can very effectively bind the soils of a floodplain and stabilize the river banks. Vegetation also causes increased surface roughness which slows overland flow. In arid or cold regions where vegetation is sparse, bank stability is decreased and flows on the floodplain are faster and therefore more likely to erode.

Rates of subsidence and the quantity of sediment supplied to the floodplain also affect the architecture of fluvial deposits. With rapid subsidence and high sediment supply aggradation on the floodplain will result in a high proportion of fine deposits. In regions of slow subsidence and reduced sediment supply to the overbank

areas relatively more in-channel deposits will be preserved (Bridge & Leeder 1979).

9.4.6 Source area control on sediment type

The distribution of clast sizes deposited in different parts of a river system is determined mainly by the availability of material of different sizes. Some source areas supply only a limited range of grain sizes as regolith. For example, a drainage basin in poorly consolidated sandstone will provide dominantly sand-sized detritus to the river system and the gravelly braided tract will not develop because of the absence of clasts of an appropriate size. Similarly, a limestone source area will yield mainly gravel and mud with very little sand-sized debris. Sandy braided and meandering facies will be poorly developed under these circumstances.

9.5 Palaeocurrents in fluvial systems

Palaeocurrent data are a very valuable aid to the reconstruction of the palaeogeography of fluvial deposits. They may be used to determine the location of the source area from which the sediment was derived and it is possible to indicate the general position of the mouth of the river and hence the shoreline. Sedimentary structures which

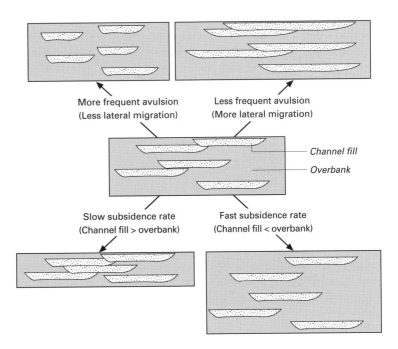

Fig. 9.16 The degree of interconnectedness between channel-fill sandstone bodies is a function of the relative frequency of avulsion and the subsidence rate.

can be used as flow indicators in fluvial deposits include the orientation of channel margins, cross bedding in sandstone and clast imbrication in conglomerate (Allen 1966; Miall 1974). An individual cross bed is formed by migration of a bar or dune bedform, but these features may be migrating obliquely to the main channel flow. Large numbers of measurements from cross bedding are therefore required to obtain a mean value which will approximate the overall flow direction in the channel. It is also important to distinguish between channel and overbank facies because flow directions in the latter will often be perpendicular to the channel.

9.6 Fossils in fluvial environments

In comparison to marine settings, the terrestrial environment has a poor potential for the preservation of fossil plants or animals. An organism which dies on the land surface is susceptible to scavenging by carrion, or its tissue will be broken down by oxidation in air. Preservation only occurs if the organism has very resilient parts (e.g. teeth and bones of vertebrates) or if the organism is covered by sediment soon after death.

The remains of animals are mostly found in floodplain sediments as sparse, scattered bones or teeth. River channels rarely contain any faunal remains. Plant fossils are more common. Broken-up leaves, branches, seeds and pollen may be locally abundant in floodplain deposits. Tree stumps may be preserved intact when a forest is inundated with sediment from floodwaters. The tree will usually be killed in the process, but the stump buried in the flood deposit may remain. Branches or trunks may be found in river channel or floodplain deposits. The more delicate parts of plants will be preserved under the more gentle depositional conditions of mud sedimentation. Pollen and some seeds are highly resistant to breakdown and can survive long periods of transport before being deposited and preserved. This makes them particularly useful for dating and correlation of terrestrial deposits *(19.3.3)*.

Trace fossils in the fluvial environment are also largely restricted to the floodplain. The footprints of animals in soft mud have a good preservation potential if the mud dries hard and is later covered with sand. The tracks of many land animals have been preserved in this way, from millipedes to dinosaurs. The presence of worms and small crustaceans is also indicated by the presence of burrows in floodplain muds; these and other organisms are also important in soil-forming processes.

9.7 Soils and palaeosols

A soil is formed by physical, chemical and biological processes which act on sediment, regolith or rock exposed at the land surface. Within a layer of sediment the principal physical processes are the movement of water down from or up to the surface and the formation of vertical cracks by the shrinkage of clays. Chemical processes are closely associated with the vertical water movement as they involve the transfer of dissolved material from one layer to another, the formation of new minerals and the breakdown of some original mineral material. The activity of plants is evident in most soils by the presence of roots and the accumulation of decaying organic matter within the soil. The activity of animals can have a considerable impact as large vertebrates, worms, insects and microscopic mites may all move through the soil, mixing the layers and aerating it.

9.7.1 Modern soils

Soils can be classified according to: (i) the degree of mineral alteration; (ii) the formation of soluble minerals; (iii) oxidizing/reducing (*redox*) conditions; (iv) the redistribution of clays, iron and organic material (*illuviation*); (v) the development of layering (horizonation); and (vi) the preservation of organic matter (Mack *et al.* 1993). Nine basic types of soil are recognized (Table 9.1). The degree of development of a soil depends upon the time available for the soil to form before the surface is inundated by another layer of sediment. Palaeosols may therefore be used as time indicators in successions of continental strata, although soil profiles can become complicated by the superimposition of a younger profile over an older one (Bown & Kraus 1987).

9.7.2 Palaeosols

A *palaeosol* is a fossil soil. Many of the characteristics of modern soils noted above can be recognized in soils which formed in the geological past. These features include the presence of fossilized roots, the burrows of soil-modifying organisms, vertical cracks in the sediment and layers enriched or depleted in certain minerals.

Seatearths are a type of palaeosol (usually a histisol, argillisol or spodosol) which is common in the Coal Measures of north-western Europe and North America (Percival 1986). They are characterized by a bed of organic matter underlain by a leached horizon of white

Table 9.1 Table of modern soil types.

Soil type	Characteristics and conditions of formation
Oxisols	Alteration of silicate minerals is extensive in wet equatorial climates, forming clays such as kaolinite and iron oxides
Calcisols	Precipitation of calcium carbonate as nodules and layers occurs in hot, dry climates
Gypsisols	Under very arid conditions calcium sulphate may form in a soil as gypsum or anhydrite
Gleysols	These are waterlogged soils with poor drainage and reducing conditions in temperate and polar regions
Argillisols	Layers and grain coatings of clay are common in these soils which form commonly in wet mid-latitude areas
Spodosols	Form under similar conditions to argillisols but the coatings are of organic material and iron oxides
Vertisols	Have poorly developed layering but strong vertical structures due to repeated desiccation of expandable clays in relatively dry climates
Histisols	High concentrations of organic material forming a peat layer; occur in moist temperate areas and occasionally in humid equatorial areas
Protosols	Poorly developed soils which can occur in any setting but are most common in polar regions

sandstone from which iron has been washed out (Fig. 9.17). Ancient vertisols are common in many floodplain deposits but are easily overlooked if they are not fully developed. Where they are more mature there is a marked horizonation and mottling of the sediment to reds, yellows and purple due to the precipitation of different iron oxides and hydroxides in gleysols and some vertisols.

A conspicuous feature of calcisols formed in climates with low rainfall is the presence of calcium carbonate precipitated within the soil profile as *calcrete* (Fig. 9.18). The evaporation of water at the surface draws

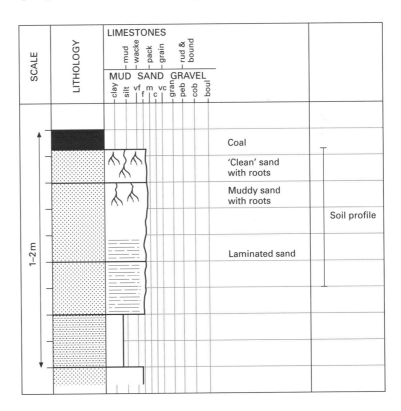

Fig. 9.17 Graphic sedimentary log through a seatearth, a palaeosol formed under relatively wet conditions.

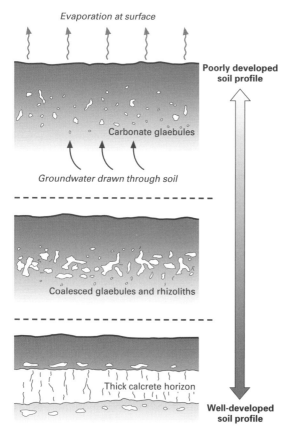

Evaporation at surface

Poorly developed soil profile

Carbonate glaebules

Groundwater drawn through soil

Coalesced glaebules and rhizoliths

Thick calcrete horizon

Well-developed soil profile

Fig. 9.18 Degrees of development of a soil profile with calcrete formation.

Fig. 9.19 Pale nodules of calcium carbonate in a mudstone deposited in an overbank environment; these soil nodules are called glaebules, and their dispersed nature indicates relatively immature calcrete development (Eocene, near Tremp, northern Spain).

groundwater up into the soil. As the water evaporates calcium carbonate dissolved in the groundwater is precipitated in the soil initially as root encrustations (*rhizocretions*) which develop into small nodules called *glaebules* (Fig. 9.19). These glaebules grow and coalesce as precipitation continues. A fully developed calcrete consists of a dense layer of calcium carbonate near the surface with *tepee structures*, domes in the layer formed by the expansion of the calcium carbonate as it is precipitated (Allen 1974). The stages in the development of a calcrete soil profile are easily recognized in palaeosols so if the rate of development of a mature profile can be measured, the time over which an ancient profile formed can be estimated (Leeder 1975). However, the time taken for a calcrete profile to develop varies considerably with temperature, rainfall and the availability of calcium carbonate, so estimates range from tens of years to tens of thousands of years.

There are a number of other soil types which consist of concentrations of a particular mineral. In arid or semi-arid regions where there is a high alkalinity to the groundwaters, silica is precipitated in the soil to form a *silcrete* crust. Coastal arid regions may have gypsisol soils with gypsum nodules developing in them which are preserved as a *gypcrete* horizon. Iron-rich soils contain displacive nodules and layers of red to black haematite forming a *ferricrete*.

Identification of a palaeosol profile is probably the most reliable indicator of a terrestrial environment. Channels are not unique to the fluvial regime because they also occur in deltas, tidal settings and deep marine environments, and thin sheets of sandstone are also common to many other depositional settings. However, recognition of a palaeosol can be made difficult by diagenetic alteration *(17.3)* which can destroy the original pedogenic structures.

9.8 Recognition of fluvial deposits: summary

Fluvial environments are characterized by flow and deposition in river channels and associated overbank sedimentation. In the stratigraphic record the channel fills are represented by lenticular to sheet-like bodies with scoured bases and channel margins, although these margins are not always seen. The deposits of gravelly braided rivers are characterized by cross bedded conglomerate representing deposition on channel bars. Both sandy braided river and meandering river deposits typically consist of fining-upward successions from a sharp scoured base through beds of trough and planar cross bedded, laminated and cross laminated sandstone. Lateral accretion surfaces characterize meandering rivers which are also often associated with a relatively high proportion of overbank facies. Floodplain deposits are mainly alternating thin sandstone sheets and mudstones with palaeosols; small lenticular bodies of sandstone may represent crevasse splay deposition. Palaeocurrent data from within channel deposits are unidirectional, with a wider spread about the mean in meandering river deposits; palaeocurrents in overbank facies are highly variable.

Other characteristics of fluvial facies include an absence of marine fauna, the presence of land plant fossils, trace fossils and palaeosol profiles in the overbank deposits. Fluvial facies are commonly associated with deposits of lacustrine, deltaic or coastal environments, and ephemeral river deposits may be found accompanying aeolian facies.

Further reading

Best, J.L. & Bristow, C.S. (eds) (1993) *Braided Rivers*, 432 pp. Geological Society of London Special Publication **75**.

Collinson, J.D. (1996) Alluvial sediments. In: *Sedimentary Environments: Processes, Facies and Stratigraphy* (ed H.G. Reading), pp. 37–82. Blackwell Science, Oxford.

Collinson, J.D. & Lewin, J. (eds) (1983) *Modern and Ancient Fluvial Systems*, 575 pp. International Association of Sedimentologists Special Publication **6**.

Ethridge, F.G., Florez, R.M. & Harvey, M.D. (eds) (1987) *Recent Developments in Fluvial Sedimentology*, 387 pp. Society of Economic Paleontologists and Mineralogists Special Publication **39**.

Fielding, C.R. (ed.) (1993) *Current Research in Fluvial Sedimentology*, 656 pp. Sedimentary Geology **85**.

Marzo, M. & Puigdefabregas, C. (eds) (1993) *Alluvial Sedimentation*, 586 pp. International Association of Sedimentologists Special Publication **17**.

Miall, A.D. (ed.) (1978) *Fluvial Sedimentology*, 859 pp. Canadian Society of Petroleum Geologists Memoir **5**.

Miall, A.D. (1992) Alluvial deposits. In: *Facies Models: Response to Sea Level Change* (eds R.G. Walker & N.P. James), pp. 119–142. Geological Association of Canada, St Johns, Newfoundland.

Retallack, G.J. (1990) *Soils of the Past: An Introduction to Paleopedology*, 520 pp. Unwin Hyman, Boston.

Wright, V.P. (ed.) (1986) *Paleosols: Their Recognition and Interpretation*, 315 pp. Blackwell Scientific Publications, Oxford.

10 Lacustrine environments: fresh and saline lakes

Lakes form where there is a supply of water to a topographic low on the land surface. They are fed mainly by rivers and lose water by flow out into a river and/or evaporation from the surface. The balance between inflow and outflow and the rate at which evaporation occurs controls the level of water in the lake and the water chemistry. Under conditions of high inflow the water level in the lake may be constant, governed by the spill point of the outflow, and remain fresh. Low input coupled with higher evaporation rates results in the concentration of dissolved ions in a saline lake. Lakes are therefore very sensitive to climate and climate change. Many of the processes which occur in seas also occur in lakes: deltas form where rivers enter the lake, beaches form along the margins, density currents flow down to the water bottom and waves act on the surface. There are, however, important differences: the fauna and flora are distinct, the chemistry of lake waters varies from lake to lake and certain physical processes of temperature and density stratification are unique to lacustrine environments. Lacustrine deposits occur throughout the stratigraphic record but tend to be less commonly represented than some other continental and marine environments of deposition.

10.1 Modern lakes

Lakes are inland bodies of water which are unconnected to the open ocean except by outflow from the lake to the sea in some instances. *Lacustrine* environments occur on about 1% of the land surface today. They are sites of deposition of clastic, carbonate and evaporite material in a wide variety of climatic and tectonic settings.

10.1.1 Types of lake

Fresh water lakes have waters with low concentrations of dissolved salts. They occur in areas of moderate to high rainfall where the input of fresh water from rivers and direct precipitation greatly exceeds the rate of evaporation from the water surface. The water in *saline lakes* has a relatively high concentration of dissolved salts, defined as greater than $5\,g\,l^{-1}$ of solutes. It may have a higher concentration of dissolved ions than sea water. Saline lakes develop in arid or semi-arid regions where the rate of evaporation matches or exceeds the input from rivers, and the salts brought into the lake become more concentrated with time (Hardie *et al.* 1978).

This chapter is concerned only with *perennial lakes*, those which are permanent bodies of water, and excludes the ephemeral or playa lakes which are temporary features of arid regions discussed in Section 8.5.

10.1.2 Dimensions and settings of modern lakes

Modern lakes are highly variable in size and shape. Lake Superior in North America is $83\,000\,km^2$ in area and has a depth of hundreds of metres. Other large lakes, such as Lake Chad in western central Africa, are much shallower, whilst some relatively small lakes may have greater depths. Lakes occur in a wide variety of tectonic and climatic settings. The East African Rift Valley lakes, such as Lake Tanzania, have formed as a result of active crustal extension along the rift valley over the last few million years. In contrast, Lake Eyre in central Australia lies in the centre of a stable cratonic area which has been undergoing very slow subsidence. Lake Eyre is also an example of a lake in an arid region where evaporation greatly exceeds inflow and the lake contains water only after periods of heavy rainfall. The Dead Sea in Jordan lies in a similarly dry area, is deep and permanent, but the high evaporation rate makes the waters permanently saline with a higher concentration of salts than in normal sea water. In higher latitudes fresh water lakes have formed in over-deepened glacial valleys such as Lake Geneva in the Swiss Alps. These lakes currently have abundant supplies of water from rivers fed by the thaw of winter ice and snow in nearby mountain regions.

10.2 Morphology and processes in lakes

There are certain differences between the physical, chemical and biological processes in lakes and those in seas. These processes control the nature and distribution of sediments and produce some characteristics which distinguish lacustrine environments. There is an absence of tidal and geostrophic currents in lakes, making wind-driven currents most important. Water in lakes develops a density stratification which is rarely seen in oceans, and this influences deposition of fine-grained sediment. Changes in water level may be dramatic in response to climate changes.

10.2.1 Physical processes in lakes: currents and waves

One of the main characteristics of the physical environment of lakes is the absence of strong permanent currents. The flow from rivers entering lakes rapidly loses energy as the current encounters the static water body and most of the coarse sediment carried by the river is deposited close to the shore. Even the largest of modern lakes is too small to experience any significant tidal effect. The main force which can move water in a lacustrine basin is the action of wind on the water surface.

Wind-driven currents affect the surface water in lakes and may generate flows with velocities up to $30 \, cm \, s^{-1}$

(Talbot & Allen 1996). Lakes in narrow valleys where the wind is funnelled by the topography are most likely to develop significant currents. However, these flows are insufficient to move anything more than silt and fine sand and will not redistribute coarser sediment. The effects of wind on the lake water will be most commonly seen as wave ripples *(4.4.1)* in the lake sediment. In the absence of other mechanisms to rework sediment, wave ripples are well preserved in sands and silts in the shallower parts of the lake. The presence of wave rippled sediment is a useful water depth indicator in lake sediments because the effects of waves decrease with depth *(4.4.1)*.

10.2.2 Density stratification in lakes

In areas where the climate is seasonal one of the consequences of the absence of strong currents in many lakes is the development of a thermal layering in the lake water. During the summer months the surface of the lake warms up by heat from the Sun, and wave action distributes this heat in the near-surface layer affected by waves (normally a few metres). This upper surface layer of water is sometimes referred to as the *epilimnion*, whilst the lower, cold waters (typically about 4°C) are called the *hypolimnion* (Talbot & Allen 1996). The boundary between these two layers is called the *thermocline* (Fig. 10.1). This layered condition is stable because the warmer surface water is less dense than the cooler water below. Influxes of water containing suspended, fine sediment may have a density lower than that of the hypolimnion and will remain above the thermocline. This enhances the distribution of suspended load in the surface waters of the lake by wave action before deposition in a wide area of the lake floor. Organisms living in the surface waters of the lake, particularly algae, will fall to the lake floor when they die. The lake bottom waters are essentially devoid of oxygen (*anoxic*) because of the lack of circulation with the oxygenated surface waters, and the organic matter remains dark and unoxidized.

Fig. 10.1 Poor circulation allows stratification to develop in lakes: the upper, warmer waters (epilimnion) are separated from the colder waters below (hypolimnion) by the thermocline and the bottom waters may become anoxic.

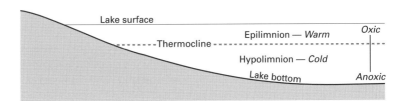

10.2.3 Variations in water depth

Wholly enclosed lake basins are extremely sensitive to climate fluctuations. A decrease in the supply of water to the basin due to lower rainfall, perhaps coupled with increased evaporation in a period of warmer weather, will result in a fall in the water level in the lake basin. The area of the lake will contract, shifting the lake shoreline towards the basin centre and leaving a peripheral area exposed to subaerial conditions (Fig. 10.2). Around the lake margin there will hence be a change from shallow subaqueous conditions with, perhaps, wave rippled sand, to a dry land surface where desiccation cracks may form in mud. Over a longer period soils will develop in the dried-out lake sediments. If wetter climatic conditions return the water level will rise once again and the lake area will expand, flooding the margins. Hence, climate changes resulting in lake level changes will be seen in a vertical section through lake margin sediments as changes from shallow subaqueous facies to subaerial facies (Fig. 10.2). Water depth changes may also be recognized in lake sediments by the presence or absence of wave ripples in sand.

Fig. 10.2 Changes in the water level in a lake can result in facies changes, which, if repeated, result in a succession of alternating beds on a graphic sedimentary log.

The effects of climate changes are not seen as lake level fluctuations in lakes which are *hydrologically open*. In these lakes the level of water in the lake is maintained by the rate of flow out of the lake into the sea. During periods of high river discharge into the lake this is matched by outflow and any reduction in input is compensated by a reduced outflow, keeping the overall lake level constant. These lakes are stable with permanent fresh water conditions.

10.3 Lacustrine sediments and facies

A wide range of depositional facies occur in lakes and they are controlled by water depth, sediment supply and water chemistry. The shallower lake margins are generally the sites of the coarsest sediment deposition, although biogenic carbonate and evaporite deposits may also form in shallow lake waters under the appropriate conditions. Deeper lake environments are sites of deposition of suspended material and turbidity currents from the lake margins.

10.3.1 Lake margin facies

Where a sediment-laden river enters a lake a delta forms as coarse material is deposited at the river mouth (Fig. 10.3). The form of and processes in a lake delta will be

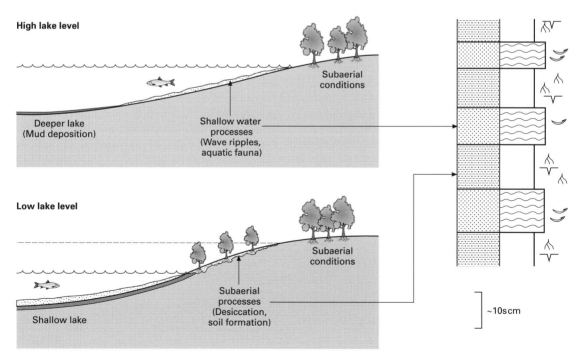

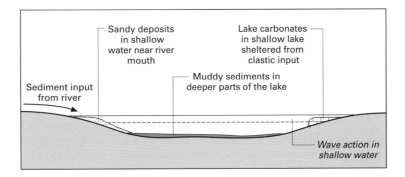

Fig. 10.3 Facies distribution in lakes.

similar to those seen in river-dominated deltas *(12.2.2)*, with some wave reworking of sediment also occurring if the lake experiences strong winds. Away from the river mouth the nature of the lake shore deposits will depend on the strength of winds generating waves and currents in the lake basin. If winds are not strong, lake shore sediments will tend to be fine-grained as there is no mechanism for the redistribution of anything much coarser than very fine sand and silt. Strong, wind-driven currents can redistribute sandy sediment around the edges of the lake where it can be reworked by waves into sandy beach deposits (Reid & Frostick 1985).

In situations where the slope into the lake is very gentle the edge of the water body is poorly defined as the environment changes from wet alluvial plain into a lake margin setting. Plants and animals living in this lake margin setting live in and on the sediment in a wet soil environment. Lake margin sediments modified by soil (pedogenic) processes *(9.7.2)* often have a nodular texture and may be calcareous. This is sometimes referred to as a *palustrine* environment.

10.3.2 Sediment dispersal in lakes

Away from the margins, clastic sedimentation occurs in the centre of a lake by two main mechanisms: dispersal as suspended sediment and transport by density currents (Fig. 10.4) (Sturm & Matter 1978). The suspended load brought by the rivers entering the lake will remain in the water column for some time before finally settling out to the bottom. Silt-sized detritus is deposited quite close to the river mouth, but clay-sized material can be redistributed by the weak lake currents before depositing as mud on the lake floor. These hemipelagic sediments (cf. *15.5*) are hence mostly very fine, clay-sized material but may also include larger particles such as organic debris which are relatively buoyant.

Coarser material may reach far out into the centre of a lake if it is carried by a density current. Dense mixtures of sediment and water can flow along the floor of the lake, depositing sediment as they travel. Density currents with low to moderate concentrations of sediment are turbidity currents *(4.6.2)* which deposit layers of sediment which grade from coarse material (deposited first) to finer sediment (which settles out last). The trigger for the redistribution of material from the river mouth by a turbidity

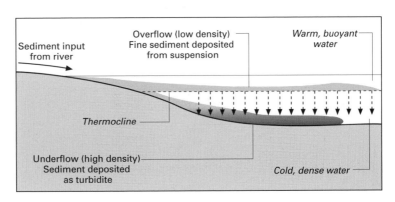

Fig. 10.4 Sediment distribution in lakes related to lake stratification.

current may be an earthquake, or a rainstorm abruptly bringing large quantities of sediment down the river to the lake. Carbonate deposits in the shallow lake margin may also be reworked and redeposited by turbidity currents in deeper parts of the lake basin.

10.3.3 Deep lake facies

This density stratification is upset by cold water entering the lake after the winter thaw. The influx of cold water will also be sediment-laden and will deposit a distinct layer of fine sediment. A colour banding between dark, organic-rich deposits formed in the summer months when organic productivity is higher and paler sediments introduced by the meltwaters is commonly seen in lakes formed in seasonal climates adjacent to mountain areas. The bands of dark and light laminae may only be a millimetre or so thick but are laterally continuous; in glacial lakes these annual layers are called *varves*, and when recognized in layers of lake sediment they can be used in chronostratigraphy *(20.4.5)*. The occurrence of very thin, fine laminae in sediments is one of the characteristics of lake sedimentation but it should be noted that varves are not so likely to form in lakes in tropical (nonseasonal) climates or in shallow lakes which have a good vertical circulation.

10.3.4 Lacustrine carbonates

In the absence of clastic input in fresh water lakes carbonate sedimentation occurs as an inorganic precipitate or due to the growth of carbonate-producing organisms. Inorganic precipitation as lime mud results from evaporation, temperature changes in the lake reducing the solubility of calcium carbonate, or the mixing of fresh and more saline water. *Tufa* is an inorganic precipitate of calcium carbonate which may form sheets or mounds in lakes. Springs along the margins or in the floor of the lake can be sites of quite spectacular build-ups of tufa (Fig. 10.5). In agitated lake waters ooids may form and build up oolite shoals in shallow water (Einsele 1992).

Biogenic sources of calcium carbonate in lakes are the skeletal remains of organisms such as bivalves, gastropods and calcareous algae. This coarse skeletal material may be deposited in shallow water or redistributed around the lake by wave-driven currents. Stromatolites formed by cyanobacteria *(3.1.2)* occur as bioherms and biostromes in lakes. The breakdown of calcareous algal filaments is an important source of lime mud. Carbonate muds from inorganic and biogenic sources is deposited in shallow parts of the lake or redeposited by density currents into deeper parts of the lake.

10.4 Saline lakes

Perennial saline lakes are most common in tropical and subtropical regions where rainfall is relatively low. There are examples of saline lakes today in east Africa, southwest North America and central Australia. They are *hydrologically closed* systems without an outflow river connecting to an ocean.

The chemistry of saline lake waters depends on the

Fig. 10.5 Pillars formed by calcium carbonate precipitating around mineral springs in a saline lake environment, Mono Lake, California, USA.

nature of the waters brought in by rivers: all rivers contain salts dissolved from the bedrock of the drainage basin of the river system. As these rivers feed into a lake they supply salts in proportions unique to that drainage system. The chemical composition of every lake is therefore unique, unlike marine waters which all have the same composition of salts, but in variable concentrations. The main ions present in modern saline lake waters are the sodium, calcium and magnesium cations and the carbonate, chloride and sulphate anions. The balance between the concentrations of different ions determines the minerals formed (Fig. 10.6) in three main saline lake types (Eugster & Hardie 1978). *Soda lakes*, such as Mono Lake, California, today, are *bitter lakes* dominated by sodium carbonate which is precipitated as the mineral trona. Searles Lake, California, is an example of a *sulphate lake* in which sodium sulphate precipitates as mirabilite. Salt lakes or *chloride lakes* such as the Dead Sea are similar in mineral composition to marine evaporites. Seasonal temperature variations in permanent saline lakes such as the Dead Sea result in a fine layering. This is due to direct precipitation of calcium carbonate as aragonite needles in the hot summer to form white laminae. These alternate with darker laminae formed by clastic input when sediment influx is greater in the winter.

Organisms in saline lakes are very restricted in variety but large quantities of blue-green algae and bacteria may bloom in the warm conditions. These form part of a food chain which includes higher plants, worms and specialized crustaceans. Organic productivity may therefore be high even in places like the Great Salt Lake of Utah (Eugster 1985) resulting in sedimentary successions which contain both evaporite minerals and black, organic-rich shales. The organic layers (sometimes oil shales: *3.7.2*) represent periods of wetter climate when the lake is relatively fresh and stratified. As the lake waters evaporate under warmer, drier conditions evaporite minerals precipitate.

10.5 Life in lakes

Palaeontological evidence is often a critical factor in the recognition of ancient lacustrine facies. Fresh water lakes may be rich in life, with a large number of organisms, but they are of a limited number of species and genera when compared to an assemblage from a shallow marine environment. Certain taxa which can only tolerate fully marine conditions (*stenohaline* fauna) are restricted to marine environments (echinoderms, corals, brachiopods, cephalopods) and their absence from a succession may be used as a pointer to a lacustrine environment of deposition. Fauna commonly found in lake deposits include gastropods, bivalves, ostracods and arthropods, sometimes occurring in monospecific assemblages. Some organisms, such as the brine shrimp arthropods, are tolerant of saline conditions and may flourish in perennial saline lake environments.

Fig. 10.6 Evaporite minerals formed in chloride lakes ('normal' waters), sulphate lakes and soda (sodium-enriched) lakes. (After Eugster & Hardie 1978.)

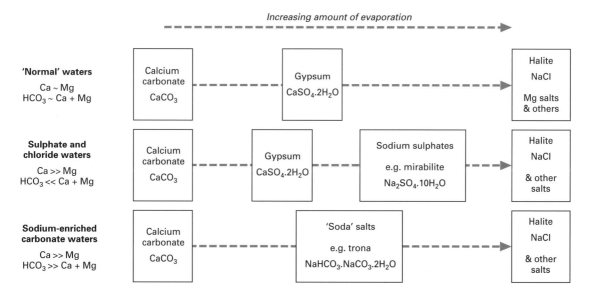

Algae are an important component of the ecology of
lakes and also have sedimentological significance. One
of the commonest types of organism found in lake deposits are *charophytes*, algae belonging to the Chlorophyta
(3.1.2) which are seen in many ancient lacustrine
sediments in the form of calcareous encrusted stems and
spherical reproductive bodies. Charophytes are generally
thought to be intolerant of high salinities and therefore a
good indicator of fresh or possibly brackish water
conditions. Cyanobacteria form stromatolite bioherms in
shallow (less than 10 m) lake waters. These carbonate
deposits in lakes form extensively where there is a low
clastic input such that the biogenically produced structures are not swamped by mud and silt. A common
feature of lakes which are areas of active carbonate
deposition is coated grains. Green algae and cyanobacteria form *oncoids*, irregularly shaped, concentrically
layered bodies of calcium carbonate several millimetres
or more across, formed around a nucleus. Oncoids form
in shallow, gently wave-agitated zones.

Cold, sediment-starved lakes in mountainous or polar
environments may be sites of deposition of siliceous
oozes. The origin of the silica is diatom phytoplankton
which can be very abundant in glacial lakes. These deposits are typically bright white cherty beds which are
called *diatomites* as they are basically made up entirely
of the silica from diatoms.

10.6 Lake environments: summary

There are a number of features which allow lacustrine
deposits to be distinguished from deposits formed in
marine environments. In lakes the fauna is characteristically limited to a few taxa, if organisms are present at
all and stenohaline marine organisms absent. Lake
deposits in temperate, seasonal climates may show the
distinctive fine laminae of lake varves. Facies variations
due to small lake level changes may occur on a fine scale
in lake margin deposits. Saline lake deposits include
evaporite minerals which are characteristic of local water
chemistry and will be different from those found in sea
water. The context of lacustrine facies is also important
as they are usually in association with other continental
facies, typically fluvial deposits.

Further reading

Anadon, P., Cabrera, L. & Kelts, K. (eds) (1991) *Lacustrine
Facies Analysis*, 318 pp. International Association of
Sedimentologists Special Publication **13**.
Kelts, K. & Hsü, K.J. (1978) Freshwater carbonate
sedimentation. In: *Lakes: Chemistry, Geology, Physics*
(ed. A. Lerman), pp. 295–323. Springer-Verlag, Berlin.
Matter, A. & Tucker, M.E. (eds) (1978*) Modern and Ancient
Lake Sediments*, 290 pp. International Association of
Sedimentologists Special Publication **2**.
Picard, M.D. & High, L.D. (1974) Criteria for recognising
lacustrine rocks. In: *Recognition of Ancient Sedimentary
Environments* (eds J.K. Rigby & W.K. Hamblin), pp.
108–145. Society of Economic Paleontologists and
Mineralogists Special Publication **16**.
Picard, M.D. & High, L.D. (1981) Physical stratigraphy of
ancient lacustrine deposits. In: *Recent and Ancient
Nonmarine Depositional Environments* (eds F.G. Ethridge
& R.M. Flores), pp. 233–259. Society of Economic
Paleontologists and Mineralogists Special Publication **31**.
Talbot, M.R. & Allen, P.A. (1996) Lakes. In: *Sedimentary
Environments: Processes, Facies and Stratigraphy* (ed. H.G.
Reading), pp. 83–124. Blackwell Science, Oxford.

11   The marine realm: morphology and processes

Nearly three-quarters of the surface of the Earth is covered by sea water. The area of the marine realm varies with time because of changes in the configuration of the continents due to plate movements. The degree to which continental margins are flooded by sea water varies but most of the surface of the Earth has been covered by sea since early in the history of the planet. Almost all of the sub-sea environments are potentially areas of sedimentation, the only exceptions being steep submarine slopes. These environments can be separated partly on the basis of the water depth. Currents within the seas are driven by winds, water density, temperature and salinity variations and tidal forces. Wave action affects water to a depth of a few tens of metres, reaching deeper levels during storms when higher-energy waves impact on parts of the sea floor. Tides profoundly affect coastal regions and tidal currents influence sedimentation on wide areas of continental shelves and epicontinental seas. The seas also teem with life: long before there was life on land, organisms evolved in the marine realm, and they continue to occupy habitats within the waters and on the sea floor. The remains of these organisms and the evidence for their existence provide important clues in the understanding of palaeoenvironments. The morphology of the marine realm, the physical and chemical processes which occur in the seas and the ecological niches which exist need to be considered before individual environments are described.

11.1 Marine environments

Marine environments include all the areas within and on the shores of the world's oceans and seas. The *bathymetry*, the shape and depth of the ocean, is determined by tectonic and thermal subsidence and the patterns of sediment accumulation. The proximity of the shore, the water depth and the gradient of the sea floor are important factors which are used to classify the environment. Coastal environments lie at the interface between land and sea. They are influenced by marine processes but are not necessarily always under water. Sediment accumulation occurs at points along the coast where rivers meet the sea at deltas and estuaries (Chapter 12) and along shorelines where deposition occurs on beaches, tidal flats and in lagoons (Chapter 13).

Away from the coastline are shallow seas (Chapter 14) which are mostly *continental shelves*, areas of continental crust covered by sea water (Fig. 11.1). Continental shelves are almost flat or very gently sloping, with gradients ranging from steep shelves of 1 in 40 to more typical gradients of 1 in 1000. They may extend for tens to hundreds of kilometres from the coastline to the shelf edge break, the line along which the slope abruptly increases at the top of the continental slope, typically at between 100 m and 200 m water depth. Where there is no distinct break of slope the 200 m bathymetric contour is taken as the edge of the shelf. Shelf seas are important areas of carbonate and/or terrigenous clastic deposition.

The *continental slope* is a relatively steep sub-sea slope, usually between about 2° and 7°, between the edge of the continental shelf and the *continental rise*, a lower-angle slope which passes down into the abyssal plain. Sediment accumulation occurs on these slopes and a lot of material by-passes the slope through canyons which channel flows down into deeper marine areas (Reineck & Singh 1973). The *deep seas* (Chapter 15) which cover most of the globe are floored by oceanic crust. On average, the water depth is around 4000 m, but is significantly shallower along mid-ocean ridges (around 2000 m) and

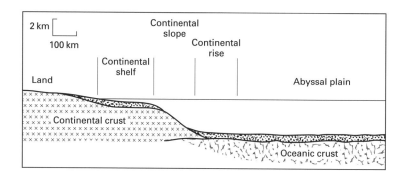

Fig. 11.1 A cross-section across the continental shelf to the continental slope, continental rise and abyssal plain.

deeper in oceanic trenches (over 8000 m). Other shallower regions include volcanic edifices such as individual volcanoes (seamounts) and volcanic plateaux.

In addition to the oceans and their margins there are large seas floored by continental crust. These *epicontinental* or *epeiric* seas are connected by straits to the oceans but have their own water circulation systems in depths of a few tens to hundreds of metres. The North Sea in north-west Europe and Hudson Bay in North America are modern examples of epicontinental seas.

The movement of water around the oceans and seas is fundamental to the physical and chemical processes of deposition and is an important factor in biogenic activity. Of the processes which move water around, the most important sedimentologically are *tidal currents* driven by the gravitational effects of the Moon and the Sun, *waves* and *storms* driven by atmospheric circulation, and *thermo-haline currents* resulting from temperature and salinity contrasts.

11.2 Tides

The surface of the Earth is affected by the gravitational pull of other bodies in the solar system, particularly the Moon, which is closest to the Earth, and the Sun, which is the largest body. Both the Sun and the Moon exert forces which move water in the world's oceans and seas in patterns known as *tides*. The significance of tides in sedimentary processes is that they involve the movement of water at velocities which are capable of moving sediment. The effects of tides can therefore be recognized in sedimentary deposits.

11.2.1 Tidal cycles

The gravitational effect of the Moon creates a bulge in the ocean waters on the side closest to the Moon but there is also a slightly smaller bulge on the opposite side of the Earth (Fig. 11.2). The second bulge on the far side is because the gravitational effect of the Moon on the water there is less than it is on the solid Earth. If the land areas are ignored the effect of these bulges is to create an asymmetric ellipsoid of water with its long axis orientated towards the position of the Moon. As the Earth rotates about its axis the bulges move around the planet. At any point on the surface the level of the water will rise and fall twice a day as the two bulges are passed in each rotation. This creates the daily or *diurnal tides*. One high

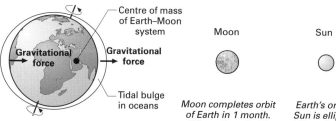

Earth rotates under each tidal bulge once each day — results in two tides each day

Moon completes orbit of Earth in 1 month. When aligned with Sun (twice a month), spring tides occur. When perpendicular, neap tides occur

Earth's orbit of Sun is elliptical. When the Earth is closest (spring and autumn equinoxes), tides are highest

Fig. 11.2 The forces on the surface of the Earth resulting from the gravitational forces acting between the Earth, the Moon and the Sun; these result in tides in the oceans and seas.

tide will always be slightly higher than the other because of the asymmetry of the ellipsoid. Each of the two tides in the diurnal tidal cycle is just over $12\frac{1}{2}$ hours because the Moon is orbiting the Earth as the planet is rotating, changing its relative position each day.

The Moon rotates around the Earth in the same plane as the Earth's orbit around the Sun. The Sun also creates a tide about half the strength of that created by the Moon, and when the Sun and Moon are in line with the Earth the gravitational effects of these two bodies are added together to increase the height of the tidal bulge. When the Moon is at 90° to the line joining the Sun and the Earth, the gravitational effects of the two on the water tend to cancel each other. During the four weeks of the Moon's orbit, it is twice in line and twice perpendicular. This creates *neap–spring tidal cycles* with the highest tides in each month, the *spring tides*, occurring when the three bodies are in line. A week either side are the *neap tides*, which happen when the Moon and Sun tend to cancel each other and the tidal effect is smallest.

Superimposed on the diurnal and neap–spring cycles is an *annual tidal cycle* caused by the elliptical nature of the Earth's orbit around the Sun. At the spring and autumn equinoxes, the Earth is closest to the Sun and the gravitational effect is strongest. The highest tides of the year occur when there are spring tides in late March and late September. In mid-summer and mid-winter the Sun is at its furthest away and the tides are smaller. This pattern of three superimposed tidal cycles (diurnal, neap–spring and annual) is complicated further by the shapes of the world's oceans and seas.

11.2.2 Tidal range and tidal currents

The tidal bulge can be considered as a wave of water which passes over the surface of the Earth. In any wave form, resonance effects are created by the shape of the boundaries of the 'vessel' through which the wave is moving. In oceans and seas the shape of the continental shelf as it shallows towards land, indentations of the coastline and narrow straits between seas can all create resonance effects in the tidal wave form. These can increase the amplitude of the tide. Locally, the mean oceanic tidal range of 50 cm is increased to several metres. The highest tidal ranges in the world today are in bays on continental shelves. The Bay of Fundy, on the Atlantic seaboard of Canada, has a tidal range of 15 m, and in Europe, the Rance Estuary, France, has a tidal range which is sufficient to drive an electricity generating

station. Within a body of water the pattern of tides can be complex. In the North Sea the tidal range varies from less than 1 m to over 6 m (Fig. 11.3).

Tidal ranges are grouped as follows (Dalrymple 1992): up to 2 m tidal range the regime is microtidal; between 2 m and 4 m mesotidal; and over 4 m macrotidal. In microtidal regimes the change in the level of water is small and hence there is little horizontal movement of water between high and low tides. This horizontal movement of water induced by tides is the *tidal current* which is more pronounced in mesotidal regimes and is capable of carrying large quantities of sediment in macrotidal regimes.

11.2.3 Characteristics of tidal currents

There are a number of features of nearshore tidal currents which produce easily recognizable characteristics in sediment deposited by them (Fig. 11.4). First, tidal currents regularly change direction from the *flood tide* current which moves water onshore between the low and high tide, and the *ebb tide* current which flows in the opposite direction as the water level returns to low tide. These currents are *bipolar*, acting in two opposite directions. Second, the tidal flow varies in velocity in a cyclical manner. At the peaks of high and low tides, the water

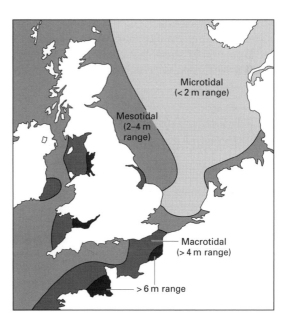

Fig. 11.3 The variation in tidal ranges around the North Sea. (Data from Komar 1976.)

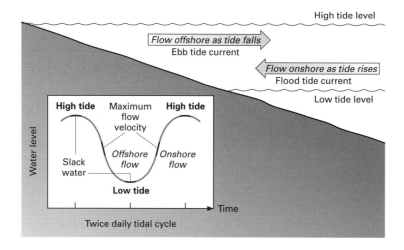

Fig. 11.4 Changes in direction and velocity of flow during a tidal cycle.

is still. As the tide turns, the water starts to move and increases in velocity up to a peak between high and low tides in each direction. Third, the strength of the flow is directly related to the difference between the levels of the high and low tides. As the tidal range varies according to a series of cycles the velocity of the current varies in the same pattern. The strongest tidal currents occur when the highest tides occur at the spring and autumn equinoxes.

Offshore tidal currents are affected by the *Coriolis force* (the rotation of the Earth causes deflection of flows of water or air over the surface) which results in an elliptical pattern to the tidal flow. The flow becomes a *rotary tide* over an area of the shelf, varying in strength during the tidal cycle, but without a period of slack water (Dalrymple 1992).

11.2.4 Sedimentary structures generated by tidal currents

BIPOLAR CROSS STRATIFICATION

An analysis of current directions recorded by cross bedding in sands deposited by tidal currents may show a *bimodal* (two main directions of flow) and bipolar (two opposite directions of flow) pattern. It should be noted, however, that this characteristic is not ubiquitous in tidal sediments because one half of the current (the ebb or the flood tide) may be much stronger than the other and because the ebb and flood currents may follow different paths. Asymmetric tidal currents result in the bulk of the sediment transport in one direction.

Under certain circumstances, bipolar cross stratifica-

tion may be seen in a single vertical section produced by alternating directions of migration of ripples or dunes. This is known as *herring-bone cross stratification*. It is characteristic of tidal sedimentation, but its formation requires ebb and flood currents to have occurred at different times in a place where the rate of sedimentation is high enough to preserve the cross stratification. These special circumstances do not occur in every tidally influenced regime (Figs 11.5 & 11.6).

MUD DRAPES ON CROSS STRATA

At the time of high or low tide when the current is changing direction there is a short period when there is no flow. Whilst the water is still some of the suspended load may be deposited. When the current becomes

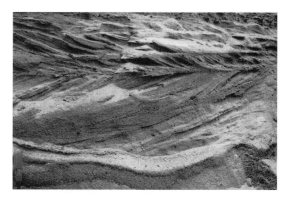

Fig. 11.5 Herring-bone cross stratification in sandstone beds (width of view 1.5 m).

Herring-bone cross stratification

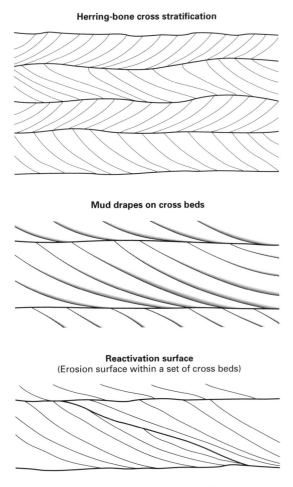

Mud drapes on cross beds

Reactivation surface
(Erosion surface within a set of cross beds)

Fig. 11.6 Features of tidal sedimentation: herring-bone cross stratification, mud drapes on cross bedding formed during slack water stages of tidal cycles, and reactivation surfaces in cross stratified sandstone.

Fig. 11.7 Cross bedded sandstone in sets 35 cm thick with the surfaces of individual cross beds picked out by thin layers of mud. Mud drapes on cross beds are interpreted as forming during slack water stages in the tidal cycle. Variations in the thickness of the sand deposited between each mud drape are interpreted as being due to changes in strength of the flow during the lunar tidal cycle (Eocene, La Puebla de Roda, northern Spain).

stronger during the next tide the suspension deposit may not be removed because of the cohesive nature of clay-rich sediment *(2.5.5)*. This effect can be seen within cross beds: a lamina of sand is deposited on the lee slope of the subaqueous dune during the tidal flow but as the tide changes direction mud falls out of suspension and drapes the subaqueous dune to form a *mud drape* or *clay drape* (Visser 1980; Dalrymple 1992). Mud drapes can form in other depositional regimes — for instance, in rivers which have only seasonal flow — but they are most common in tidal settings and abundant, regular mud drapes are a good indicator of a tidally influenced environment (Figs 11.6 & 11.7).

REACTIVATION SURFACES

In places where there is one dominant current in the tidal pair the bedforms migrate in one direction, producing unidirectional cross stratification. These bedforms can be modified by the reverse current, principally by the removal of the crest of a subaqueous dune. When the bedform recommences migration in the direction of the dominant flow the cross strata build out from the eroded surface. This leaves a minor erosion surface within the cross stratification which is termed a *reactivation surface* (Fig. 11.6) (Visser 1980; Dalrymple 1992).

TIDAL BUNDLES

The strength of the tidal current varies cyclically, changing its capacity to carry sediment. At the highest tides the current is strongest, enabling more transport and deposition of sand on the bedforms in the flow. When the difference between high and low tide is smaller the current will carry a reduced bedload or none at all. There are cases where the cyclical variation in the thickness of foreset laminae has been attributed to a control by the neap–spring cycle; these are referred to as *tidal bundles*. In an idealized case, the laminae would show thickness variations in cycle in multiples of 7 or 14, but there is often no sedimentation or bedform migration during the weaker parts of the tidal cycle, so this ideal pattern is rarely seen. A regular variation in the thickness of foreset laminae has been recognized in a number of sandstones interpreted as tidal deposits (Yang & Nio 1985).

Sediments affected by rotary tidal currents in offshore areas will show patterns of bedform migration which vary in direction across the shelf, but the deposits typically do not show herring-bone cross stratification or reactivation surfaces. Variations in the flow strength may be indicated by clay drapes or tidal bundles.

11.3 Wave and storm processes

The depth to which surface waves affect a water body is referred to as the *wave base*. On continental shelves two levels can be distinguished. The *fair weather wave base* is the depth to which there is wave-influenced motion under normal weather conditions. The depth of the fair weather wave base is normally between 5 m and 15 m. This depth varies according to the size of the waves on the shelf or in the sea and the size increases with the distance the wave forms are driven (the fetch of the waves). The *storm wave base* is the depth waves reach when the surface waves have a higher energy due to stronger winds driving them. This may typically be 20–30 m but may be 200 m or more where storms are very strong (Komar *et al.* 1972). Below the storm wave base the sea bed is not affected by surface waves.

11.3.1 Storms

Storms occur all over the world. Pressure gradients in the atmosphere due to the development of patterns of high and low pressure result in strong winds. In a severe storm the wind may reach velocities of $200\,\mathrm{km\,h^{-1}}$ or more. Storms are important in aeolian transport *(8.2)* of material over land and out to sea, and at sea they are responsible for strong, wind-driven currents. These currents may be very effective at moving sediment in suspension and as bedload in the direction of the prevailing wind, which may be onshore, offshore or along the shoreline. The effects of storms are therefore important on continental shelves, especially in places where the tidal currents are weak. Shallow marine environments are commonly divided into those which are tide-dominated and those which are storm-dominated *(14.3, 14.4)*.

The effects of storms can be dramatic in coastal areas where a *storm surge* is experienced. The force of an onshore wind piles up water along the coast, raising the elevation of the sea surface. The effect is compounded by extremely low atmospheric pressure which also raises the sea level. A severe storm surge combined with large storm waves and a high tide can result in catastrophic flooding of coastal regions.

11.3.2 Storm-related processes and products

TEMPESTITES

A *tempestite* is a storm deposit formed during a storm event (Aigner 1985). The strong currents may scour the coast and the shallow parts of the shelf, and move eroded material in the direction of the prevailing wind. As the strength of the wind subsides the storm flow decreases and sediment is deposited across the shelf area. Storm deposits (Fig. 11.8) are similar to turbidity current deposits as both are the products of single waning flow events of relatively dense mixtures of sediment and water. A tempestite often shows a scoured base overlain by a normally graded succession of material. The grain size decreases in the direction of the current as material is progressively deposited. However, there are important differences between turbidites and tempestites: the former often show current ripple cross lamination in the sandy part of the deposit whereas tempestites are more likely to display wave ripple lamination *(4.4.2)* and hummocky cross stratification (below).

HUMMOCKY AND SWALEY
CROSS STRATIFICATION

A storm creates conditions for the formation of bedforms and sedimentary structures which may be exclusive to storm-influenced environments (Hamblin & Walker 1979; Dott & Bourgeois 1982). *Hummocky cross stratification* (HCS) is distinctive in form, consisting of

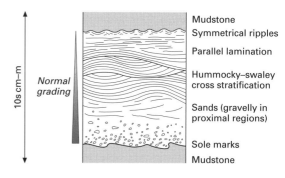

Fig. 11.8 The characteristics of storm deposits.

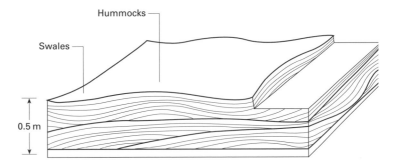

Fig. 11.9 Hummocky and swaley cross stratification.

rounded mounds of sand on the sea floor a few centimetres high and tens of centimetres across. Internal stratification of these hummocks is convex upwards and dips in all directions, a feature which is not seen in any other form of cross stratification (Figs 11.9 & 11.10). Between the hummocks lie swales and where concave layers in them are preserved this is sometimes called *swaley cross stratification* (SCS). Hummocky and swaley cross stratification are believed to form as a result of *combined flow* which only occurs when a current is generated by a storm at the same time as high-amplitude waves reach deep below the surface. The strong current takes sand out into the deeper water in temporary suspension and as it is deposited the oscillatory motion caused by the waves results in deposition in the form of hummocks and swales. One of the characteristics of HCS/SCS is that these structures are normally only seen in fine- to medium-grained sand, suggesting that there is some grain size limitation involved in this process.

Fig. 11.10 Hummocky cross stratification (HCS) in fine-grained sandstone beds, Carboniferous, Northumberland, England. The convex-up surfaces are characteristic of HCS.

11.3.3 Tsunami

Waves with periods of 10^3–10^4 s are generated by events such as earthquakes, meteorite impacts, large volcanic eruptions and submarine landslides. These events can set up a surface wave with an amplitude of a few tens of centimetres in deep ocean water and a wavelength of several hundred kilometres. As it reaches the shallower waters of the continental shelf, the amplitude is increased to 10 metres or more, producing a wave which can have a devastating effect on coastal areas (Tarbuck & Lutgens 1997). These waves are sometimes incorrectly referred to as 'tidal waves' (they have nothing to do with tidal forces) but are properly called *tsunami*.

The effects of tsunami are dramatic, with widespread destruction occurring near coasts. They also have a serious effect on shallow marine environments causing disruption and redeposition of foreshore and shoreface sediments. It has been suggested that beds of poorly sorted debris containing a mixture of deposits and fauna from different coastal and shallow marine environments may form as a consequence of tsunami (Pilkey 1988). It may be possible to distinguish them from ordinary storm deposits by their larger size, but in practice it is very difficult to show that a deposit is generated by a specific mechanism.

11.4 Thermo-haline currents

Thermo-haline currents flow in the world's oceans and are mainly driven by cold, dense water sinking at high latitudes and by relatively saline (also dense) waters spilling out of marginal seas (Stow 1985) (Fig. 11.11). Temperature and salinity contrasts create differences in density of water masses and generate flow of the denser fluid beneath the less dense water. The effects of these currents on sedimentation are most noticeable in deeper

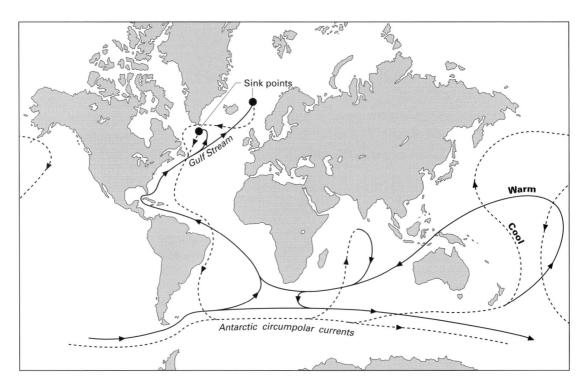

Fig. 11.11 The main geostrophic ocean currents affecting the world's oceans at the present day.

waters *(15.4)* as their effects are masked in shallower water by the influences of tides, waves and storms. Thermo-haline currents are typically weaker than storm and tidal currents but are of larger volume. They mainly move clay and silt in suspension and fine sands as bed-load. In addition to driving water bodies through the oceans, thermo-haline currents are important in the creation of bottom currents which can move nutrients large distances across oceans to areas where upwelling occurs. Organic productivity is high in regions of upwelling bottom currents *(11.6.2)*.

11.5 Divisions of the marine realm

A number of different sets of terms are used to describe and define the parts of the marine realm. A commonly used scheme is illustrated in Fig. 11.12 (McLane 1995) and two basic divisions are recognized.
1 The *neritic* environment is the shelf area down to the shelf edge break at 200 m water depth. It is subdivided into the *littoral* zone, the region between high and low tide levels, and the *sublittoral* zone which comprises the remainder of the shelf. A *supralittoral* zone above the high tide is also recognized.
2 The *oceanic* environment is divided into the region of the continental slope, the *bathyal* zone, and the *abyssal* depths of ocean floor between 4000 m and 5000 m below sea level. Deeper regions are sometimes referred to as *hadal.*

In schemes which further subdivide the shelf area (neritic environment) the main criterion used is water depth, which in turn determines the processes occurring on the sea floor (Reading & Collinson 1996). It should be noted that only ranges of depths can be given because the depths to which tidal processes, waves and storms affect the shelf vary considerably.

The *foreshore* is the littoral zone, the region between mean high water and mean low water marks of the tides, and depending on the tidal range this may be a vertical distance of anything from a few tens of centimetres (microtidal) to many metres (macrotidal). The extent of the foreshore is governed also by the slope and it may be anything from a few metres to a kilometre or more in places where there is a high tidal range and a gently sloping shelf. The foreshore is part of the beach environment *(13.2.1)*.

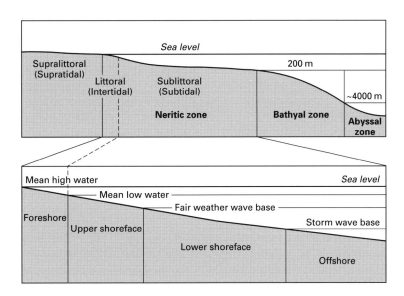

Fig. 11.12 Divisions of the marine realm.

Below the low tide level the shelf is always covered with water and a region of the shelf called the *upper shoreface* is defined between low tide mark and the depth to which waves normally affect the sea bottom (the fair weather wave base). The depth to which the shoreface reaches depends on the energy of the waves in the area but is typically somewhere between 5 m and 20 m. Some authors refer to this part of the shelf profile as the *shoreface*.

In deeper water the sea bed is only affected by waves during storm periods. The depth to which this occurs is the storm wave base and this is very variable on different shelves. In some places it may be as little as 20 m water depth but can be 50–200 m. This deeper shelf area affected by storms is called the *lower shoreface* or the *offshore transition* zone. The *offshore* zone is the area beneath the storm wave base out to the shelf edge break at around 200 m depth. It is mainly a region of pelagic sedimentation. Alternative schemes divide the shelf into *nearshore* (foreshore and upper shoreface), *inner shelf* (lower shoreface) and *outer shelf* (offshore). The characteristics of the depositional environments in these different depth zones on the continental shelf are considered in Chapters 13 and 14.

11.6 Chemical and biochemical sedimentation in oceans

The most important chemical and biochemical sediments in modern seas and ancient shelf deposits are carbonate sediments and evaporites, and in the oceans plankton provide large quantities of carbonate and siliceous sediment. In addition, there are other less abundant but significant chemical and biochemical deposits.

11.6.1 Glauconite

This green phyllosilicate mineral is moderately common in sedimentary rocks and is important to sedimentology and stratigraphy for a number of reasons. Glauconite is one of the few *authigenic* minerals (minerals formed in the sedimentary environment as opposed to detrital grains which are eroded from elsewhere) chemically precipitated in marine sedimentary environments. It is found on modern shelves and occurs associated with shallow marine sediments (Bornhold & Giresse 1985). It forms by the transformation of clay minerals and within pores in the substrate, the insides of shells and faecal pellets. Its formation appears to be sensitive to geochemical conditions (Berner 1981) and it is only found associated with shelf sea sediments. The restricted occurrence of glauconite makes it useful in the stratigraphic record as an indicator of shallow marine environments, although care must be exercised to ensure that it has not been reworked.

Glauconite forms relatively slowly and is only found concentrated in sediments where there is low clastic and carbonate sediment supply. High background sedimentation rates tend to swamp the glauconite. It can therefore be used as an indicator of low sedimentation rates, a feature which is useful when carrying out a sequence stratigraphic analysis *(21.6)*. Glauconite may be associated with ichnofacies which are characteristic of slow rates of sedimentation on firmground and hardground surfaces *(11.7.2)*.

A further use of glauconite in sedimentary rocks is that as an authigenic mineral it forms at the time of sedimentation. Dating of glauconite by potassium–argon radiometric methods *(20.1.6)* provides one of the few opportunities to determine the age of sedimentary rocks directly.

11.6.2 Phosphorites

The origins of sedimentary phosphates are disputed but are thought to be related to regions of high organic productivity where there is deposition of organic material rich in phosphates. Continental margins may be regions of high organic productivity where there are regions of upwelling of nutrient-rich waters at continental margins (Froelich *et al.* 1982; Tucker 1991). There are occurrences of phosphorites today off the coast of Chile and Peru and off west Africa where Antarctic water comes to the surface. These nutrient-rich waters coming up into warm regions promote blooms of plankton which are at the bottom of the food chain. Phosphates are more concentrated in higher organisms such as fishes which have high phosphate contents in their bones.

11.6.3 Organic-rich sediments

Organic material from dead plants, animals and microbial organisms is abundant in the oceans and becomes part of the material which falls to the sea floor. Where the sea floor is oxygenated by currents bringing water down from the surface scavengers consume much of this organic material or it is oxidized chemically, leaving little of it to become part of the sediment. Poor circulation reduces the oxygen in the waters at the sea floor and the conditions become anoxic. Breakdown of the organic matter is slower or non-existent in the absence of oxygen and the conditions are not favourable for scavenging organisms. The organic matter accumulates under these anoxic conditions and contributes to the pelagic sediment to form a *black shale*, a mudrock which typically contains 1–15% organic carbon (Stow *et al.* 1996).

Black shales are commonly deposited in deep oceans where geostrophic currents are not circulating the sea bottom waters. They also occur in shallow seas with

restricted circulation, shelves with rims or epicontinental seas with poor connection to the open ocean. The Black Sea is an example of a modern restricted epicontinental sea with anoxic bottom waters. Black shales have an economic importance in sedimentology and stratigraphy as they are hydrocarbon source rocks.

11.7 Ecology of the seas

Shelves are areas of oxygenated waters periodically swept by currents to bring in nutrients. As such they are habitable environments for many organisms which may live swimming in water (*planktonic*) or on the sea floor (*benthic*) either on the surface of or within the sediment. Plants and animals living in the marine realm contribute detritus, modify other sediments and create their own environments. Modern shelf environments teem with life and it is rare to find an ancient shelf deposit which does not contain some evidence for the organisms which lived in the seas at the time.

11.7.1 Body fossils

In shallow seas with low clastic input the calcareous hard parts of dead organisms make up the bulk of the sediment, either as the loose detritus of mobile animals or as biogenic reefs, whole sediment bodies built up as a framework by organisms such as corals and algae. Terrigenous clastic sandy and muddy shelf deposits may also contain a rich flora and fauna, the type and diversity of which depend on the energy on the sea bed (fragmentation can occur in high-energy environments) and the post-depositional history (Chapter 17) which affects preservation of material.

Many plants and animals occupy ecological niches which are defined by such factors as water depth, temperature, nutrient supply, nature of substrate and so on. If the ecological niche of a fossil organism can be determined this can provide an excellent indication of the depositional environment. In the younger Cenozoic strata the fossils may be of organisms so similar to those alive today that determining the likely environment in which they lived is quite straightforward. Farther back through geological time this task becomes more difficult. Groups of organisms such as trilobites and graptolites which were abundant in the Lower Palaeozoic seas have no modern representatives for direct comparison of lifestyle. Clues as to the ecological niche occupied by a fossil organism are provided by considering the *functional morphology* of the body fossil. All organisms are in some way adapted to their environment, so if these adaptations can be recognized the lifestyle of the organisms can be determined to some extent. In trilobites, for example, it has been recognized that some types had well-developed eyes whilst in others they were very poorly developed: one interpretation of this would be that the trilobites with eyes needed them to help move around on the sea floor but those which lived buried in the sediment had no need of sight.

Some organisms are thought to have occupied very specific niches and can provide quite precise information about the environment of deposition. Some algae and hermatypic corals require clear water and sunlight to thrive, so they are indicators of shallow, mud-free shelf environments. Other organisms (certain bivalves, for instance) are more tolerant of different environments and can live in a range of conditions and water depths provided that a supply of nutrients is available. In general, the abundance of benthic organisms decreases as the water depth increases. Shoreface environments usually have the most diverse assemblages of benthic fauna and flora due to the well-oxygenated conditions of the wave-agitated water and the availability of light (provided that it is not too muddy). The abundance of organisms living on the sea floor decreases down in the offshore transition and offshore parts of the shelf. In the deep oceans only a few very specialized organisms live on the sea floor adjacent to areas of hydrothermal activity.

The abundance of planktonic organisms is not much influenced by water depth but is controlled by the supply of nutrients and the surface temperature of the water. The hard parts of planktonic organisms may be distributed in sediments of any water depth, although dissolution of calcium carbonate occurs in very deep water *(15.5.1)*. One approach to the problem of determining the depth at which a sediment was deposited is to consider the ratio of benthic to planktonic organisms present: if the propor-

tion of benthic organisms is high the water was probably shallow whereas a high count of planktonic organisms indicates deeper water. This method normally only provides a very rough guide to relative water depth but is applied in a semi-quantitative way in Cenozoic and Mesozoic strata by considering the proportions of benthic and planktonic forms of Foraminifera.

11.7.2 Trace fossils

In addition to the body fossils, the organisms living on or in the sediment leave a record of their activities as *trace fossils*, also known as *ichnofossils*. Trace fossils are the tracks, trails or burrows and borings left by organisms as they move around, create holes for protection or sift nutrients from sediment. The study of these features is *ichnology* and the trace fossils themselves are referred to as *ichnofauna*. The general term for evidence of disturbance of the sediment by moving organisms is *bioturbation*.

Trace fossils have considerable value in the interpretation of palaeoecology and palaeoenvironments because they provide information about the behaviour of organisms. Methods of locomotion, the gait of an animal as it moves, and the methods used to search for food can be determined from trace fossils. In this way, they can tell us more about animal behaviour and the environment than assemblages of body fossils and they are not subject to removal and reworking in the way that the hard parts of fossils are. However, there is a limitation to the extent of these interpretations because there are only a few cases where the trace fossil can be confidently matched to a particular organism (an example where it is possible is *Cruziana*, the tracks of certain trilobites). Many burrows and tracks were almost certainly made by soft-bodied animals (worms, for example) which are not preserved in the fossil record.

Trace fossil assemblages in sedimentary rocks are

Table 11.1 Classification of ichnofossils according to the inferred manner in which they were formed. (After Seilacher 1953; Simpson 1975.)

Name	Description	Mode of formation
Cubichna	Resting traces	The impressions of an organism resting on soft sediment
Domichna	Dwelling traces	Burrows and borings made as a place to live
Fodinichna	Feeding traces	Marks made by an organism searching for food
Pascichna	Grazing traces	Marks made by organisms sifting nutrients from the surface
Repichna	Locomotion traces	Tracks and trails left by animals moving
Fuichna	Escape burrows	Made by animals moving to the surface after burial by sediment

normally considered in terms of intensity of bioturbation, diversity of forms and the recognition of specific types of trace fossil. This provides information about the water depth, nature of the substrate and palaeoecology of the environment of deposition.

CLASSIFICATION OF TRACE FOSSILS

Ichnofossils are classified according to the inferred manner in which they were formed—for example, by movement of an animal over a surface, feeding, creation of a shelter, and so on (Table 11.1) (Seilacher 1953; Simpson 1975; Ekdale *et al.* 1984). However, there is a lot of variation within these categories, as dinosaur footprints and trilobite tracks classify as the same type of trace fossil. There is also a lot of overlap between them: who is to say whether an animal was just walking or walking and feeding at the same time? Different types of locomotion can be distinguished in some instances—for example, the trace formed by an animal or bird swimming in shallow water is distinct from that made by a walking animal or bird (Fig. 11.13).

The most common trace fossils are burrows made for dwelling or feeding or both. Escape burrows are common in settings where there is rapid sedimentation by storms or turbidity currents. Trace fossils are usually found on or within sediment which was unconsolidated but with sufficient strength to retain the shape of the animal's trace. Contrasts in sediment type between a burrow and the host sediment are a considerable aid to recognition. A distinction is made between *burrows* formed in soft sediment and *borings* made by organisms into hard substrate. A *bioturbated* sediment is material which has been to some extent modified by organisms moving

Fig. 11.13 A trace fossil formed by a reptile or bird walking (Oligocene, northern Spain).

through it. Many shallow marine sediments exhibit bioturbation, the degree of which can be so intense as to destroy any primary sedimentary structures such as stratification.

TRACE FOSSILS AND SEDIMENTARY ENVIRONMENTS

Trace fossils are a valuable aid to the interpretation of ancient sedimentary environments because they provide information on the conditions at the time of deposition. Given an understanding of the behaviour of organisms which are sensitive to environmental conditions, it is possible to develop interpretations based on *ichnofacies*, that is, the associations of trace fossils in a sedimentary deposit. An example of this approach is the use of trace fossils as water depth indicators and to define ichnofacies for different water depth ranges (Fig. 11.14) (Pemberton *et al.* 1992; Pemberton & MacEachern 1995).

Trace fossil assemblages which occur along shorelines may be subdivided according to the degree of consolidation of the substrate. Along sandy shorelines *Skolithos* ichnofacies are characteristic. This facies is named after simple vertical tubes formed by organisms which lived in the high-energy region of the foreshore. In this ichnofacies also occur *Ophiomorpha*, a larger, mainly vertical burrow lined with faecal pellets, and *Diplocraterion*, a U-shaped burrow. The animals which formed *Skolithos*, *Ophiomorpha* and *Diplocraterion* are thought to have moved up and down in the sediment with the changing water level of the foreshore. Where the sediment is semi-consolidated the *Glossifungites* ichnofacies assemblage occurs: the burrows are similar in form to those of the *Skolithos* facies but they tend to have sharp, well-defined margins to the tubes and may extend into excavated dwelling cavities. Some organisms (such as bivalves, echinoids and some sponges) are able to bore into rock to create dwelling traces: this assemblage is called *Trypanites* (Fig. 11.15a).

In the shoreface zone of the shelf, *Cruziana* ichnofacies include *Cruziana* itself in Palaeozoic rocks; *Rhizocorallium*, an inclined U-shaped burrow; *Chondrites*, a vertically branching small burrow; *Planolites*, a horizontal branching burrow; and *Thalassanoides*, larger burrows (more than 1 cm in diameter) in a complex three-dimensional network (Fig. 11.15b).

In the deeper waters of the outer shelf/offshore zone the *Zoophycos* assemblage is the characteristic ichnofacies. *Zoophycos* has a rather variable, partly radial form which may be tens of centimetres across. Few other

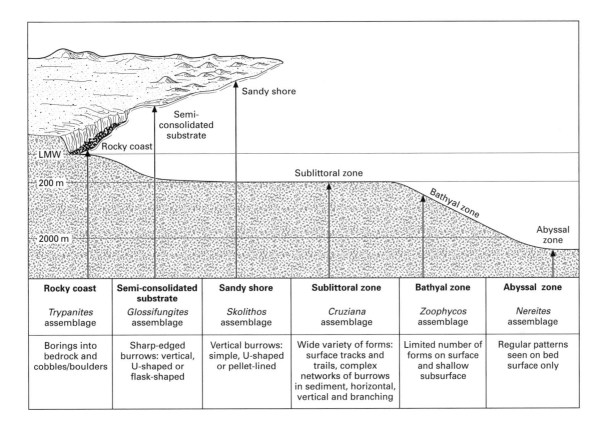

Rocky coast	Semi-consolidated substrate	Sandy shore	Sublittoral zone	Bathyal zone	Abyssal zone
Trypanites assemblage	*Glossifungites* assemblage	*Skolithos* assemblage	*Cruziana* assemblage	*Zoophycos* assemblage	*Nereites* assemblage
Borings into bedrock and cobbles/boulders	Sharp-edged burrows: vertical, U-shaped or flask-shaped	Vertical burrows: simple, U-shaped or pellet-lined	Wide variety of forms: surface tracks and trails, complex networks of burrows in sediment, horizontal, vertical and branching	Limited number of forms on surface and shallow subsurface	Regular patterns seen on bed surface only

Fig. 11.14 Trace fossil assemblages and their relationship to water depths and substrate. (After Pemberton *et al.* 1992.)

trace fossils are found in these depths. In the deeper *Nereites* ichnofacies found in bathyal to abyssal depths the traces found are characteristically feeding traces showing regular patterns. These include *Helminthoides* (Fig. 11.15c), which, like *Nereites*, is a looping surface trace, and the enigmatic *Palaeodictyon* (Fig. 11.15d) which has a regular hexagonal pattern. The regular structure of the traces of this ichnofacies is attributed to the scarcity of nutrients and the need to move efficiently; in shallower, nutrient-rich sediment more random feeding structures are the norm.

TRACE FOSSILS AND RATES OF SEDIMENTATION

Ichnofacies can be used as indicators of the degree of consolidation of the substrate, and this can be a useful tool in the analysis of a stratigraphic succession. Where rates of sedimentation are high, the sea floor is covered by loose sandy or muddy material and a variety of ichnofacies occur according to the water depth. Sediment

exposed on the sea floor starts to consolidate if there rate of sedimentation is relatively slow and a firmground forms. The characteristic ichnofacies of firmgrounds is *Glossifungites* (Ekdale *et al.* 1984). At even slower rates of sedimentation complete lithification (*17.1*) of the sea floor occurs with the formation of a hardground typified by the ichnofacies *Trypanites* (Ekdale *et al.* 1984). Recognition of hardgrounds and firmgrounds is particularly important in the sequence stratigraphic analysis of sedimentary successions (*21.2.4*).

11.8 Marine environments: summary

The physical processes of tides, waves and storms in the marine realm define regions bounded by water depth changes. The beach foreshore is the highest-energy depositional environment where waves break and tides regularly expose and cover the sea bed. At this interface between the land and sea storms can periodically inundate low-lying coastal plains with sea water. Across the submerged shelf, waves, storms and tidal currents affect the sea bed to different depths, varying according to the range of the tides, the fetch of the waves and the

(a)

(b)

(c)

(d)

Fig. 11.15 (a) A boring into the edge of a limestone boulder, *Trypanites* ichnofacies, (Miocene, Yemen); (b) *Thalassanoides* (Eocene, northern Spain); (c) *Helminthoides*, a deep water trace fossil of the *Nereites* ichnofacies (Eocene, northern Spain); (d) *Palaeodictyon*, a distinctive hexagonal trace fossil of the *Nereites* ichnofacies and characteristic of abyssal water depths (Jurassic, Oman).

intensity of the storms. Sedimentary structures can be used as indicators of the effects of tidal currents, waves in shallow water and storms in the offshore transition zone. Further clues about the environment of deposition are available from body fossils and trace fossils found in shelf sediments. More details of the coastal, shelf and deep water environments are presented in the following chapters.

Further reading

Bromley, R.G. (1990) *Trace Fossils, Biology and Taphonomy*, 280 pp. Special Topics in Palaeontology **3**. Unwin Hyman, London.

Dalrymple, R.W. (1992) Tidal depositional systems. In: *Facies Models: Response to Sea Level Change* (eds R.G. Walker & N.P. James), pp. 195–218. Geological Association of Canada, St Johns, Newfoundland.

Johnson, H.D. & Baldwin, C.T. (1996) Shallow clastic seas. In: *Sedimentary Environments: Processes, Facies and Stratigraphy* (ed. H.G. Reading), pp. 232–280. Blackwell Science, Oxford.

Pemberton, S.G. & MacEachern, J.A. (1995) The sequence stratigraphic significance of trace fossils: examples from the Cretaceous foreland basin of Alberta, Canada. In: *Sequence Stratigraphy of Foreland Basin Deposits* (eds J.C. Van Wagoner & G.T. Betram), pp. 429–476. American Association of Petroleum Geologists Memoir **64**.

Pemberton, S.G., MacEachern, J.A. & Frey, R.W. (1992) Trace fossil facies models: environmental and allostratigraphic significance. In: *Facies Models: Response to Sea Level Change* (eds R.G. Walker & N.P. James), pp. 47–72. Geological Association of Canada, St Johns, Newfoundland.

Stride, A.H. (ed.) (1982) *Offshore Tidal Sands: Processes and Deposits*, 222 pp. Chapman & Hall, London.

Tillman, R.W., Swift, D.J.P. & Walker, R.G. (eds) (1985) *Shelf Sands and Sandstone Reservoirs*, 708 pp. Society of Economic Paleontologists and Mineralogists Short Course Notes **13**.

Walker, R.G. & Plint, A.G. (1992) Wave- and storm-dominated shallow marine systems. In: *Facies Models: Response to Sea Level Change* (eds R.G. Walker & N.P. James), pp. 219–238. Geological Association of Canada, St Johns, Newfoundland.

12   Deltas and estuaries

Where the land meets the sea, the surface processes which act on the continents interact with marine processes, resulting in the most complex depositional environments. Waves, tides and ocean currents act on sediment brought down by rivers, redistributing material in estuaries and deltas at the river mouth. These environments are influenced by the size and discharge of the rivers, the climate, and the energy associated with waves, tidal currents and longshore drift, all of which control the type and distribution of sediments in deltas and estuaries. A delta is a sedimentary body formed at the mouth of a river which is building outwards through time, usually into a marine or lacustrine environment. Estuaries are sites of deposition at the mouths of rivers where there is mixing with sea water; they are commonly influenced by the tide and do not build outwards. Both are almost exclusively sites of clastic deposition. Deltas and estuaries are very sensitive to changes in sea level, caused by uplift and subsidence of the land or changes in the absolute level of the sea. Deposits formed in deltaic environments are important in the stratigraphic record as sites for the formation and accumulation of fossil fuels.

12.1 Deltas

The problem of defining a delta has vexed sedimentologists and geomorphologists for decades. The origin of the term was simple enough, coming from the plan shape of the Nile delta (Fig. 12.1), which resembled the fourth letter of the Greek alphabet, Δ. A quick glance at the shape of some of the other major deltas of the world such as the Mississippi, Rhône or Ganges (Fig. 12.1) reveals that shape is clearly not the common factor. Undoubtedly, location at the mouth of a river must form part of the definition, but is not enough in itself as not all rivers form deltas; some have an estuary at the mouth, others have no distinctive sediment body where the river meets the sea.

A definition commonly used is 'a discrete shoreline protuberance formed at a point where a river enters an ocean or other body of water' (Elliott 1986a; Bhattacharya & Walker 1992). Deltas are sites of net accumulation, with sediment being supplied more rapidly than it can be redistributed in the basin. They have a high preservation potential if the area is a region of overall subsidence.

A number of depositional sub-environments (Fig. 12.2) are common to all but the coarsest-grained deltas (12.3). Many of these may occur in other depositional settings, such as channels in river systems, sandy bars in coastal environments, and so on, but it is the association of these sub-environments which characterizes a delta. Similarly, the recognition of the association of the different depositional facies allows the recognition of deltaic deposits in the stratigraphic record.

12.1.1 Delta top environments

All deltas are fed by a river and there is inevitably a transition from the channel which is considered part of the fluvial environment to the channel which occurs on the *delta top* or *delta plain* (Fig. 12.2). Delta channels may be in the form of a single course to the delta front or may be distributary in form. The coarsest delta top facies are found in the channels, where the flow is strong enough to transport and deposit bedload material (sand in all but the finest-grained delta systems). Adjacent to the channels are subaerial overbank areas which are sites of sedimentation of suspended load when the river

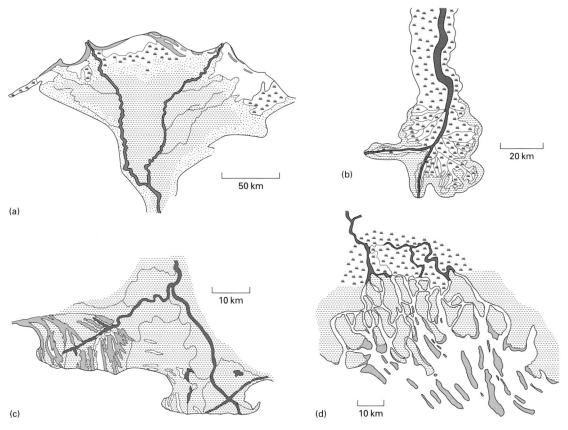

(a)

(b)

(c)

(d)

Fig. 12.1 The forms of modern deltas: (a) the Nile delta, the 'original' delta; (b) the Mississippi delta, a river-dominated delta; (c) the Rhône delta, a wave-dominated delta; (d) the Ganges delta, a tide-dominated delta.

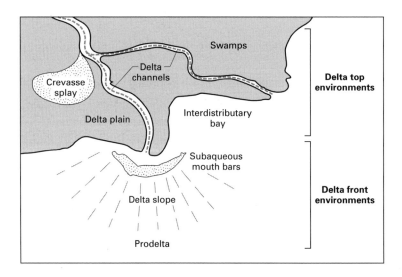

Fig. 12.2 Delta sub-environments.

floods. These may be vegetated under appropriate climatic conditions and in hot, wet tropical regions large, vegetated swamps may form on the delta top. These may be sites for the accumulation of peat, although if there is frequent overbank flow from the channel the deposit may contain enough clastic material to form a carbonaceous mud. Crevasse splays *(9.3.1)* may result in lens-shaped sandy deposits on the delta top. Sheltered areas along the edge of the delta top between the main channels are called *interdistributary bays*. These are subaqueous regions of low-energy sedimentation between the lobes built up by the channel–overbank complexes. The water may be brackish if there is sufficient influx of fresh water from the channel and overbank areas and the boundary between the floodplain and the interdistributary bays may be indistinct, especially if the delta top is swampy.

12.1.2 Delta front environments

At the mouth of the channels the flow velocity is abruptly reduced as the water enters the standing water of the lake or sea. The delta front immediately forward of the channel mouth is the site of deposition of bedload material as a *subaqueous mouth bar* (Fig. 12.2). Mouth bar deposits typically coarsen upwards as coarser

sediment can be transported and deposited in higher-energy, shallower water; they may be extensively reworked by wave and tide action. Conditions are often brackish as the fresh river water mixes with sea water at the river mouth. On cross-sections across deltas the *delta slope* is always shown as a steep incline away from the delta top. In fact, the delta slope is usually at an angle of only 1–2°, except in the case of some coarse-grained deltas. The current from the river is dissipated away from the channel mouth and wave energy decreases with depth, leading to a pattern of progressively finer material being deposited further away from the river mouth. The finest-grained deposits of the delta front occur in the deeper water (*prodelta*) area where deposition is mainly from suspension, but gravity currents may also bring coarser sediment from the delta front and deposit material as turbidites.

River-borne suspended load enters the relatively still water of the lake or sea to form a plume in front of the delta. Fresh river water with a suspended load may have a lower density than saline sea water and the plume of suspended fine particles will be buoyant, spreading out away from the river mouth. As mixing occurs deposition out of suspension occurs, with the finest, more buoyant particles travelling further away from the delta front before being deposited in the prodelta region.

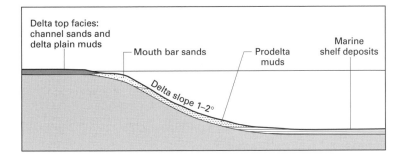

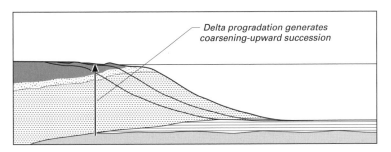

Fig. 12.3 A cross-section across a delta lobe.

12.1.3 Progradation of deltas

The characteristics in cross-section (Fig. 12.3) are formed as clastic material is deposited in all the delta sub-environments and the sediment body builds out into the lake or sea. As the delta progrades, the channel is extended out over the mouth bar, which in turn builds out over the delta slope. The most proximal sub-environments (channel and mouth bar) are areas of coarser bedload deposition and the delta slope and prodelta deposits are progressively finer distally. Progradation therefore results in a coarsening-upward succession of the finest prodelta deposit at the base and the coarsest mouth bar and channel sediments at the top (Fig. 12.3).

12.2 Controls on deltas

The factors which influence the form and character of a delta can be divided into hinterland controls (discharge of the river, nature of sediment carried, local climate) and basinal controls (strength and orientation of waves and tides, slope and bathymetry within the water body). These are summarized on Fig. 12.4. A popular way of classifying deltas in terms of processes was proposed by Galloway (1975): a triangular plot (Fig. 12.5) on which the relative importance of waves, tides and river pro-

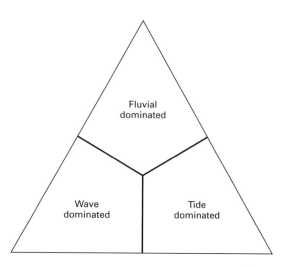

Fig. 12.5 Galloway's (1975) delta classification which only considers wave, tide and river influences and does not take into account water depth or grain size. (After Elliot 1986a.)

cesses is considered allows any modern delta to be classified in terms of these three factors. The Mississippi is the classic 'river-dominated' delta, plotting at the apex of the triangle, the Rhône lies close to the wave-dominated corner and the Ganges has strong tidal influences. The drawback of this classification is that some modern deltas plot in the same part of the triangle but have very

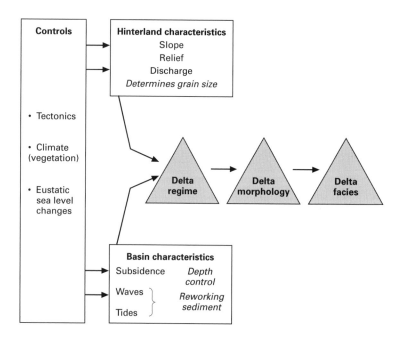

Fig. 12.4 Controls on delta environments and facies. (After Elliot 1986a.)

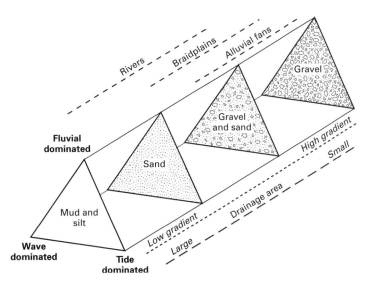

Fig. 12.6 Classification of deltas taking grain size, and hence sediment supply mechanisms, into account. (After Orton & Reading 1993.)

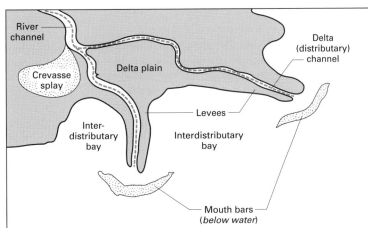

Fig. 12.7 River-dominated delta environments.

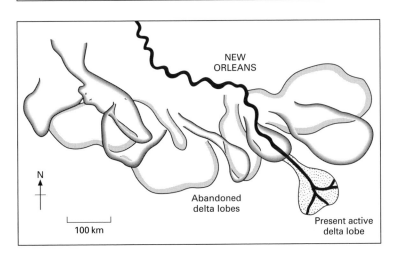

Fig. 12.8 Lobe switching on the Mississippi delta resulting in a pattern of overlapping lobes of deltaic sediments.

different morphologies and characteristics. The 'Tobler-one' plot shown in Fig. 12.6 is a modification of the Galloway triangle, adding another axis to show the grain size. However, even this does not provide a comprehensive classification scheme as factors such as the slope and depth of the 'receiving basin' (the sea or lake into which the delta is building) are not taken into account (Postma 1990). It may not be possible to represent all deltas on a single classification diagram, but by considering the effects of the different hinterland and basinal controls there is a means to describe adequately both modern and ancient deltas.

12.2.1 Grain size and hinterland slope controls

Rivers which have traversed a long, low-gradient tract of land before reaching the coast are likely to have deposited most of their bedload before the water arrives at the delta. Deltas adjacent to broad coastal plains fed by meandering rivers are hence predominantly made up of fine-grained sediment. This fine material is relatively easily reworked and dispersed by waves, currents and tides. With increasing river gradient, the flow velocity is greater and the proportion and size of the bedload increases. Deltas at the mouths of braided rivers may be sandy or pebbly, requiring more energetic basinal processes to remobilize them.

If the gradient of the hinterland adjacent to a lake or sea is very steep, an alluvial fan may form and build out as a partially subaqueous feature. In these cases the sediment body may be referred to as a *fan delta*. The term 'fan delta' is often loosely applied to any coarse-grained delta, although the usual definition is 'an alluvial fan which progrades into a standing body of water' (Wescott & Ethridge 1990). Along steep coasts a scree cone may build out into the water; such features would not be regarded as a true delta, but as very coarse, subaqueous gravity deposits which may form a talus cone deposit.

12.2.2 River-dominated deltas

A delta is regarded as river-dominated where the effects of tides and waves are minor. This requires a microtidal regime *(11.2.2)* and a very gently sloping shelf where wave energy is effectively dissipated before the waves

Fig. 12.9 Schematic graphic sedimentary log through a river-dominated delta. For key to symbols see Fig. 5.7.

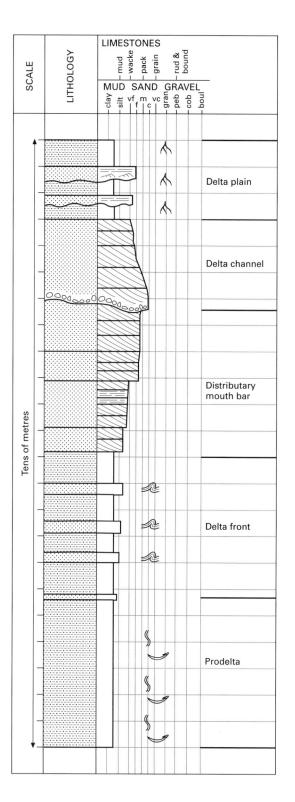

reach the coastline. Under these conditions, the form of the delta is largely controlled by fluvial processes of transport and sedimentation. The unidirectional fluvial current at the mouth of the river continues into the sea or lake as a subaqueous flow. The channel form is maintained, with well-defined subaqueous levees and overbank areas (Fig. 12.7). Bedload and suspended load carried by the river are deposited on the subaqueous levees, building up to sea level and extending the front of the delta basinwards as thin strips of land either side of the main channel to form the characteristic 'bird's foot' pattern of a river-dominated delta (Bhattacharya & Walker 1992). A characteristic of large, fluvially dominated

deltas such as the Mississippi (Fig. 12.1) is channel instability due to the very low gradient on the delta plain causing frequent avulsion of the major and minor channels. The course of the river changes as one route to the sea becomes abandoned and a new channel is formed, leaving the former channel, its levees and overbank deposits abandoned. Repeated switching of the channels on the delta top builds up a pattern of overlapping abandoned lobes (Fig. 12.8).

The deposits of river-dominated deltas have well-developed delta top facies, consisting of channel and overbank sediments. The characteristics of these facies will be essentially the same as those of a similar fluvial

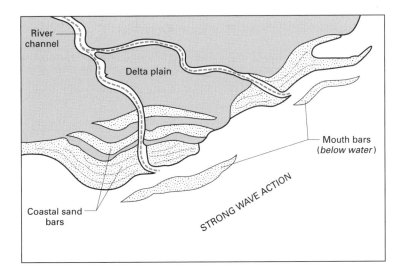

Fig. 12.10 Wave-influenced delta environments.

Fig. 12.11 Sand bars at the mouth of a wave-influenced delta (western Madagascar).

system. The overbank areas of a delta top may be sites of prolific growth of vegetation, leading to the formation of peat and eventually coal. The channels build out to form the 'toes' of the 'bird's foot', between which there are large interdistributary bays. These bays are relatively sheltered and are sites of fine-grained, subaqueous sedimentation. Crevasse splays from the distributary channels supply sediment into these bays and they gradually fill to sea level to become the vegetated part of the delta plain. The filling of interdistributary bays results in small-scale (a few metres thick) coarsening-upward successions (Fig. 12.9).

In front of the channels, mouth bars form and are localized to areas in front of the individual delta lobes. Little redistribution of mouth bar sediments by wave or tidal processes occurs, so individual mouth bar bodies are relatively small.

12.2.3 Wave-influenced deltas

Waves driven by strong winds agitate the top few metres of the water column and have the capacity to rework and redistribute any sediment deposited in shallow water, especially under storm conditions. The river mouth and mouth bar areas of a delta are susceptible to the action of waves, resulting in a modification of the patterns seen in river-dominated deltas (Fig. 12.10). Progradation of the channel outwards is limited because the subaqueous levees do not form and bedload is acted upon by waves as quickly as it is deposited. Any obliquity between the wind direction and the delta front causes a lateral migration of sediment as the waves wash material along the coast to form beach spits. Beaches and mouth bars build up as elongate bodies parallel to the coastline, as seen on the Rhône delta (Oomkens 1970). Wave action is effective at sorting the bedload into different grain sizes and the mouth bar deposits of a wave-influenced delta may be expected to be better sorted than those of a river-dominated delta.

Progradation of a wave-dominated delta occurs because the wave action does not transport all the material away from the region of the river mouth. A net supply of bedload by the river results in a series of shore-parallel sand ridges forming as mouth bars build up and out to form a new beach (Figs 12.10 & 12.11). In the stratigraphic record wave-dominated delta deposits display well-developed mouth bar and beach sediments,

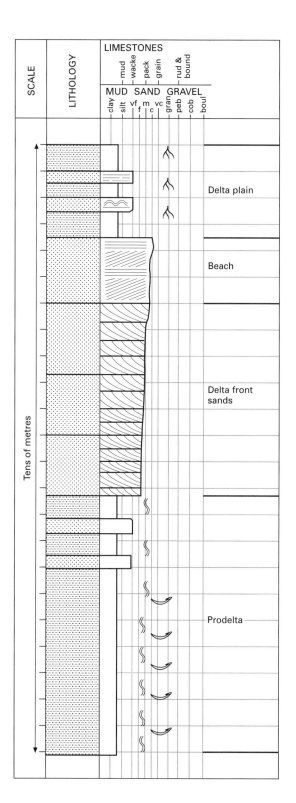

Fig. 12.12 Schematic log through a wave-influenced delta.

occurring as elongate coarse sediment bodies approximately perpendicular to the orientation of the delta river channel. This is in contrast to the river-dominated delta deposits, which would be expected to show less continuous mouth bars, and a higher proportion of channel and overbank deposits forming the delta lobes. Delta slope and prodelta deposits may not significantly differ between these two delta types (Fig. 12.12).

12.2.4 Tide-influenced deltas

Coastlines with high tidal ranges experience onshore and offshore tidal currents which move both bedload and suspended load. A delta building out into a region with strong tides will be modified into a pattern distinct from both river- and wave-dominated deltas (Fig. 12.13). First, the delta top channels are subject to tidal influence and the channel is subject to either reverses of flow or periods of stagnation as a flood tide balances the fluvial discharge. This may be seen in strata as reversals of palaeocurrent indicated by cross stratification, and layers of mud deposited in-channel due to flow stagnation. Overbank areas on the delta top may be partially tidal flats, and all of the delta top will be more susceptible to flooding during periods of high fluvial discharge coupled with high tides. A large part of Bangladesh experiences frequent floods because the land forms the top of the tidally influenced Ganges–Brahmaputra delta (Coleman 1969).

The tidal currents rework sediments at the river mouth into elongate bars which are perpendicular to the shoreline. These are modified mouth bars, which coarsen upwards, but may show bidirectional cross stratification and mud drapes on the cross bed foresets due to the reversing nature of the ebb and flood tidal currents (Fig. 12.14).

The deposits of tide-influenced deltas can be distinguished from other deltas by the presence of sedimentary structures and facies associations which indicate that tidal processes were active (reversals of palaeoflow, mud drapes, and so on: *11.2.4*), and subaqueous bars will be elongate parallel to the river channels. The overall succession of strata will display the characteristic coarsening upward of deltas, a feature which allows them to be distinguished from other tidally influenced environments such as estuaries, which have much in common in terms of depositional processes. The main distinguishing feature is that a delta is always a progradational feature, whereas an estuary commonly forms as part of a retrogradational, or transgressive, succession *(21.1.2)*.

12.2.5 Effects of climate

In a climate which favours extensive development of plant cover on the delta top, channel behaviour will be influenced by the stabilizing effect of vegetation on the channel levees and the overbank areas. This will encourage the progradation of the individual lobes in a fluvial-dominated delta system.

More indirect effects of climate on a delta include the controls which rainfall and evaporation have on the efficiency with which the river can supply detritus to the

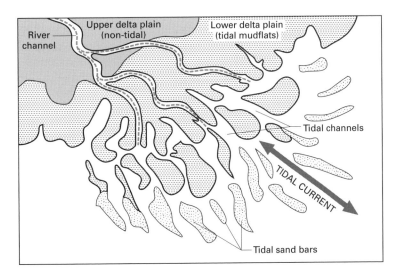

Fig. 12.13 Tide-influenced delta environments.

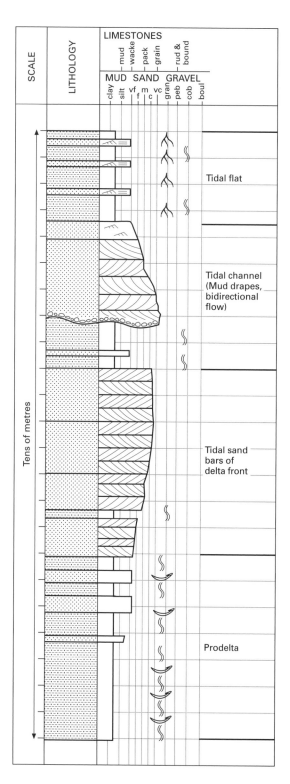

delta. These are important factors in the hinterland controls on the delta morphology and facies (Fig. 12.4). Low rates of sediment supply to the delta favour reworking of the delta front by basinal processes.

12.2.6 Basin bathymetry

A delta progrades by sediment accumulating on the sea floor at the delta front and building up to sea level to increase the area of the delta top. For a given supply of sediment, the rate at which the delta progrades will depend on the thickness of the sediment pile which must be created to reach sea level. Delta progradation will hence be faster if the sea or lake into which it is building is shallow (Fig. 12.15). In deeper water, a greater volume of sediment will be required to achieve the same amount of progradation. The effects of basin bathymetry are seen in the stratigraphic record as a contrast between thin, widespread coarsening-upward successions, formed by delta progradation across a shallow shelf, and thick, laterally restricted successions, formed where the adjacent basin was deeper (Elliott 1989).

12.3 Coarse-grained deltas

Any delta of gravelly material may be referred to as a *coarse-grained delta*. Coarse-grained deltas are bodies of gravelly detritus which form on the margins of lakes and seas where a braided river or alluvial fan supplies predominantly coarse material. Deltas of this type supplied by a braided river are sometimes called *braid deltas* although it is tempting to use the term 'fan delta' because of the distinct fan-shaped geometry that they may possess. *Gilbert-type deltas* are a class of coarse-grained delta with steep foreset deposition.

The morphology of a coarse-grained delta is particularly sensitive to the offshore slope and depth of water, and three distinct forms of coarse-grained delta can be recognized in this respect (Fig. 12.16). In relatively shallow water with a gentle slope, waves rework the sediment brought to the delta front, and as there is a decrease in wave energy with depth, there is a decrease in clast size offshore into deeper water. Sub-environments recognized within these *shelf-type coarse-grained deltas* comprise (Fig. 12.16):

1 a delta top setting of pebbly braided river or alluvial fan deposits;

Fig. 12.14 Schematic log through a tide-influenced delta.

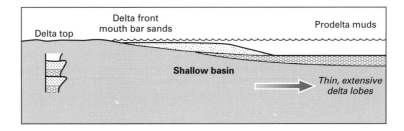

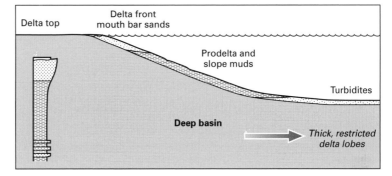

Fig. 12.15 The depth of water into which a delta progrades determines the extent of the lobes and the facies on the delta slope.

2 a delta front and mouth bar area of pebbly and sandy deposits;

3 progressively finer-grained deposition offshore to sandy and muddy material below wave base in the prodelta environment.

Progradation of this delta type generates a coarsening-upward succession. At the top of the succession there may be a sharp contrast between the relatively poorly sorted deposits of the delta top and better-sorted delta front material which has been reworked by wave action.

The second class is the *slope-type coarse-grained delta* (Fig. 12.16) which has similar delta top and nearshore facies to the shelf type, but a distinct increase in slope offshore gives rise to a different style of deposition. With a steeper gradient, sediment transport is by debris flows and turbidity currents which carry detritus into the deeper parts of the basin.

In places where there is a sharp bathymetric break at the coast (created by a steep fault, for example) a Gilbert-type delta forms. Named after G.K. Gilbert, who recognized and described them from Lake Bonneville (Gilbert 1885), these coarse-grained deltas display a distinct tripartite division (Fig. 12.16). A Gilbert-type delta consists of a *topset* deposit of coarse fluvial or alluvial fan sediments which may be partly modified by wave activity along the fringes of the topset. Coarse sediment accumulating at the delta front is in an unstable position at the top of a steep slope and periodically this sediment

moves down the delta front as a mass flow. If the mixture of material involved is relatively muddy, the mass flow will be a debris flow: well-sorted clasts move in an avalanche, or grain flow, down the slope. In either case, the sediment will be deposited at, or close to, the angle of rest for coarse aggregates, up to 30° from the horizontal, to form a *foreset*. Gilbert-type delta foresets can be tens to hundreds of metres high, and are unique as an example of decimetre to metre beds of coarse sediment deposited at a steep angle: there is no need for vertical exaggeration in the cross-section through a Gilbert-type delta (Fig. 12.16). At the toe of the foreset, deposition occurs on horizontal *bottomsets*, which are often made up of finer material, deposited by turbidity currents. Slumping is often seen on Gilbert-type deltas because the steep slopes of deposition can become unstable.

Coarse-grained or fan deltas are found in settings where there is a very steep topography adjacent to a body of water. The relief in the hinterland is normally the consequence of active tectonics causing uplift, as is the case in the Gulf of Corinth (Fig. 12.17) (Ori *et al.* 1991). Fan deltas are documented from both marine (e.g. Postma 1984) and lacustrine (e.g. Laird 1995) settings.

12.4 Delta 'cycles'

When the channel on the delta top changes course, the

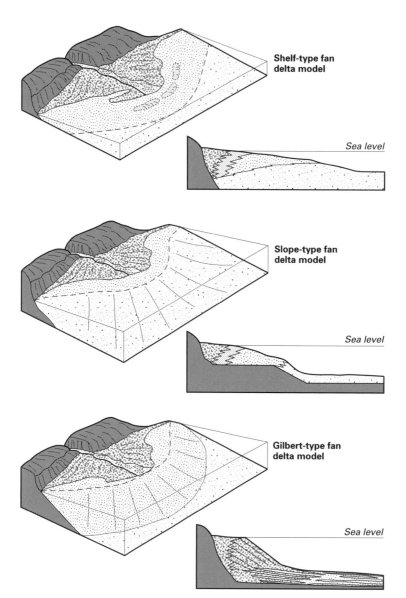

Shelf-type fan
delta model

Sea level

Slope-type fan
delta model

Sea level

Gilbert-type fan
delta model

Sea level

Fig. 12.16 Types of coarse-grained delta or fan delta: shelf-type fan delta, slope-type fan delta and Gilbert-type fan delta. (After Wescott & Ethridge 1990. Copyright John Wiley & Sons Ltd. Reproduced with permission.)

former lobe is abandoned as a new site of deposition is occupied. River-dominated deltas, such as the Mississippi, tend to have the most frequent changes in position of the active lobe, but avulsion of channel course also occurs in other delta types, for example the wave-influenced Rhône delta. An abandoned lobe will subside as the sediments lose water and compact under their own weight. The fall below sea level of the abandoned lobe will be accelerated if the delta is located in a region of overall subsidence or if there is a general rise in sea level.

The beds which mark the end of sedimentation on a delta lobe are known as the *abandonment facies*. In the upper part of the delta plain these will be peats or palaeosols which represent a low clastic supply to this part of the plain now that active lobe progradation has moved elsewhere on the delta. The fringes of the delta lobe will be areas of slow, fine-grained deposition in shallow

Fig. 12.17 A Gilbert-type coarse-grained delta in Pleistocene deposits in the southern side of the Gulf of Corinth, Greece. The cliff is over 500 m high and is made up mostly of foreset deposits dipping at around 30°. Horizontal topset strata form the top of the cliff and the toes of the foreset beds pass into gently dipping bottomset facies.

water, whilst further offshore, carbonate facies may form over the toe of the delta. Abandonment facies may show intense bioturbation because of the slow sedimentation rate.

After a number of changes in channel position the active delta lobe may reoccupy an earlier position and prograde over an older, submerged delta lobe succession. In cross-section the result is one coarsening-upward delta lobe succession built up on top of another. Repetitions of this pattern have been recognized in the stratigraphic record and are referred to as *delta cycles*, each 'cycle' representing the progradation of an individual delta lobe. (To be semantically correct, they should be referred to as 'rhythms', because the pattern of repetition is deep–intermediate–shallow–deep–intermediate–shallow, etc., whereas a true cycle would show a pattern of deep–intermediate–shallow–intermediate–deep–intermediate–shallow, etc.). The thickness of a delta 'cycle' will be controlled by the depth of water in the receiving basin (see above): cycles in Upper Carboniferous deltaic strata in northern England range from a few metres to tens or hundred of metres in thickness (Elliott 1986a).

Variants on the idealized delta cycle are frequently encountered (Fig. 12.18). A complete succession from offshore fine-grained deposits up to the delta channel fill will only be seen at the point where the axis of the lobe has built out basinward. In other positions, the top of the cycle will vary from delta plain carbonaceous mudstones, to interdistributary bay deposits or mouth bar

sands. In a hinterland direction, subsidence will not be great enough for fully marine conditions to develop at the base of each delta cycle, and only the upper parts of the typical succession may be seen (Fig. 12.18) (Elliott 1976, 1986a).

In addition to the trends which represent the progradation of delta lobes, smaller-scale grain size patterns are also present. The filling of an interdistributary bay results in a coarsening-upward succession, but this will normally be on a scale which is an order of magnitude smaller than the main delta cycle. Small-scale fining-upward trends are formed by the filling of distributary channels when they are abandoned.

12.5 Post- and syndepositional effects on deltas

Progradation of a delta lobe involves relatively rapid sedimentation over a small area. A delta lobe succession consists of poorly consolidated material with undercompacted mud at the base overlain by denser sands. This unstable sediment body is susceptible to diapiric movements of the low-density mud up through the overlying sands *(17.1.2)* and failure of the delta slope, producing collapse scars and slumps at the surface. These deformation features disrupt the stratal pattern of a deltaic succession (Coleman *et al.* 1983; Elliott 1989). The instability of the pile of sediment in a delta body results in internal deformation as it develops. Faults occur, offsetting and rotating the bedding. These can be

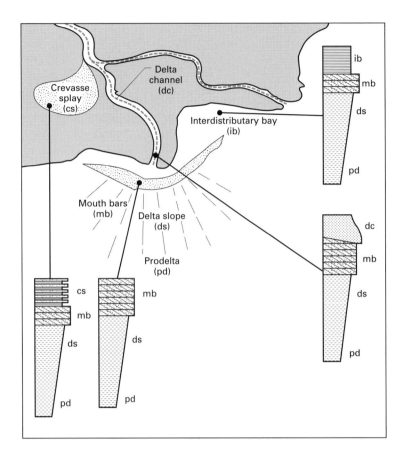

Fig. 12.18 Delta cycles: the facies succession preserved depends on the location of the vertical profile relative to the depositional lobe of a delta.

recognized with deltaic successions as *syndepositional faults* because they affect only parts of the sediment pile, with younger beds unaffected.

12.6 Recognition of deltaic deposits: summary

A key feature of many deltas is the close association of marine and continental depositional environments. In delta deposits this association is seen in the vertical arrangement of facies. A single delta cycle may show a continuous vertical transition from fully marine conditions at the base to a subaerial setting at the top. This transition is typically within a coarsening-upward succession from lower-energy, finer-grained deposits of the prodelta to the higher-energy conditions of the delta mouth bar where coarser sediment accumulates. The shallower water deposits may be extensively reworked by wave and/or tidal action. Under these condi-

tions the mouth bar may show cross bedding and cross lamination.

The delta top contains both relatively coarse sediment of the distributary channel as well as finer-grained material in overbank areas and interdistributary bays. The channel may be recognized by its scoured base, a fining-upward pattern and evidence of flow which will be unidirectional unless there is a strong tidal influence resulting in bidirectional currents. The delta top will show signs of subaerial conditions, including the development of a soil. Deposits in the sheltered interdistributary bays may show thin bedding resulting from influxes of sediment from the delta top and symmetrical ripples due to wave action. In a stratigraphic succession delta deposits show a variety of sediment body geometries in channels and as bar deposits with forms depending on the relative importance of river, tidal and wave processes.

Palaeontological evidence from fauna and flora can be

important in the recognition of the marine and continental sub-environments of a delta. A distinct fauna tolerant of brackish water may be found near the mouths of channels and in the interdistributary bays where fresh and marine water mix. The mixture of shallow marine, brackish and fresh water fauna plus coastal vegetation is also characteristic of deltaic environments. The contrast between fresh and saline water is not present in deltas formed at the margins of fresh water lakes. In these settings, recognition of the delta setting will be reliant on the evidence of the physical processes influencing sedimentation.

Of particular importance in the recognition of delta deposits is the relationship to other facies in the stratigraphic section. In a succession which is grossly progradational, delta facies will be underlain by fully marine deposits of the shelf and overlain by fluvial and coastal plain sediments. This sequence of environments will be reversed in a setting where there is an increase in water depth with time.

12.7 Estuaries

An *estuary* is a semi-enclosed coastal water body where there is a mixture of river and sea water and where there is a mixture of fluvial and marine processes (Boyd *et al.* 1992; Dalrymple *et al.* 1992). The mouths of many rivers are estuaries at present as a result of the relative sea level rise which has occurred since the last ice age (Fig. 12.19). The river mouths have been inundated by sea water, exposing them to tidal processes. In areas of low

tidal range, a stable shoreline is established and there is a clear boundary between areas of fluvial processes within the river channel and wave activity in the marine environment. Under macrotidal regimes, the rise and fall of the tide and the ebb and flood tidal currents establish two sub-environments within the estuarine setting: tidal channels and tidal mudflats (Fig. 12.20).

12.7.1 Tidal channels

In an estuary, the flow of water in the ebb and flood tides is funnelled into single or multiple channels. Currents may be powerful enough to cause scouring at the base, and gravelly debris, including bioclastic debris brought in by onshore currents, may be left as a lag in the channel floor (Elliott 1986b; Reinson 1992). The flow within a tidal channel will move sand as a bedload, with the formation of subaqueous dunes which can build up into bars. These bar deposits will show the characteristics of tidal sedimentation—for example, mud drapes on the cross beds. However, evidence of a reversing flow is not as common as might be expected. This is because one of the characteristics of estuaries with multiple channels is that the ebb and flood tidal currents commonly follow different pathways. A particular section of an estuary may therefore experience only one dominant flow, which may be ebb or flood. Herring-bone cross stratification only develops in regions where the ebb and flood flow pathways overlap (Fig. 12.20).

Like river channels, tidal channels do not remain fixed: they may avulse, or migrate laterally in a gradual

Fig. 12.19 A modern estuarine environment: Barmouth estuary, Wales.

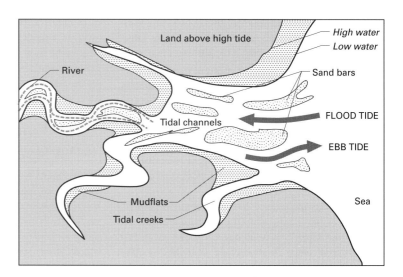

Fig. 12.20 Environments in a tide-dominated estuary.

manner. In either case, sandy bar deposits remain as a channel-fill succession, but where lateral migration has been gradual, a point bar on the inner bend of a channel forms. Tidal point bar deposits are similar to those found in rivers, but with one important difference: the episodic nature of tidal flows allows fine material to come out of suspension on to the gently sloping bar surface. Alternations of sand and mud as bar deposits produce a pattern of bedding referred to as *inclined heterolithic stratification* (Thomas *et al.* 1987). Inclined heterolithic stratification is seen in tidal channel deposits as alternating layers of sand and mud which dip into the axis of the channel, perpendicular to flow (Fig. 12.21).

12.7.2 Tidal mudflats

Away from the areas of strong tidal currents lie tidal mudflats (Elliott 1986b; Reinson 1992). These regions are flooded at high tide, and exposed when the tide falls, with the extent of flooding and exposure depending on the height of the tide. Fine sediment is carried in suspension across the mudflats on the rising tide, depositing as the tide turns and falls. A network of creeks across the mudflats partially channelizes the water flow, leaving areas in between where vegetation is established. Species tolerant to inundation by saline water at every tide develop in the lower part of the mudflats, whilst

Fig. 12.21 Large-scale cross stratification in Carboniferous estuarine deposits, north Mayo, Ireland: the cross bedded units are alternations of sandstone and mudstone as inclined heterolithic cross stratification.

upper parts are colonized by plants which can survive sea water flooding during the highest spring tides. Vegetation on the tidal flats traps sediment, and mudflats are commonly sites of net accumulation. The succession built up is predominantly mud with thin sand sheets present if very high tides or storms wash coarser material across the flats. Evidence of vegetation is normally abundant, as is bioturbation by the fauna which live on the nutrient-rich mudflats.

12.7.3 Estuarine successions

Where subsidence allows the accumulation of a succession of estuarine deposits, the vertical profile may be expected to be similar to that shown in Fig. 12.22. The base of a tidal channel is marked by a scour and lag, followed by a fining-upward succession of cross bedded sands, which may show mud drapes, inclined heterolithic stratification and bidirectional palaeocurrent indicators. Abandonment of the channel allows muddy tidal flat deposits to accumulate, often rich in organic material. When a new channel is established in the same place, the succession is repeated. Tidal channels are typically a few metres deep, occasionally tens of metres, and the thickness of tidal flat deposits will depend on the subsidence rates. Areas around the edges of an estuary will develop thick successions of muddy tidal flat deposits, with little or no coarser sediment if the channels do not reach that area. It should be noted that estuarine deposits typically show marked lateral variability (Dalrymple *et al.* 1992) and many do not conform to the model presented in Fig. 12.22. The recognition of estuarine sedimentation is important to assessing relative sea level rises in the stratigraphic record *(21.3.1)*.

12.7.4 Wave-dominated estuaries

Along coasts with smaller tidal ranges wave action is important in estuarine settings (Fig. 12.23) (Reinson 1992; Roy 1994). The sand barrier built up at the mouth of the estuary will have the characteristics of a strand plain or barrier island complex *(13.3)*. An inlet in the barrier allows water to pass through into the lagoon, a region of shallow, low-energy sedimentation *(13.3.5)*. At the mouth of the river a *bay head delta* forms. This consists of delta top facies (channel and delta plain) which may be similar to those found in other deltaic environments, but are confined by the incised valley of the estuary.

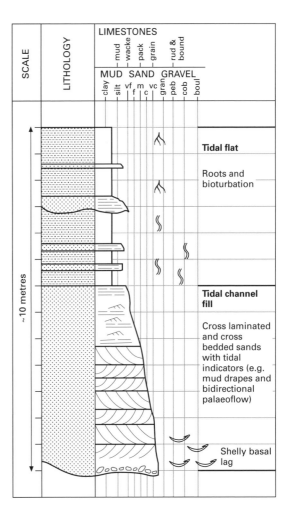

Fig. 12.22 Schematic graphic sedimentary log through tidal estuary deposits.

12.8 Recognition of estuarine deposits: summary

The deposits of deltas and estuaries in the stratigraphic record have many features in common. Both are sedimentary bodies formed at the interface between marine and continental environments and consequently display evidence of physical, chemical and biological processes which are active in both settings (for example, an association of beds containing a marine shelly fauna with other units containing rootlets). The key difference is that a delta is a progradational sediment body, that is, it builds out into the sea and will show a coarsening-upward succession produced by this progradation. In

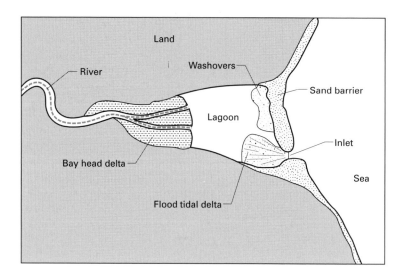

Fig. 12.23 Wave-dominated estuary. (After Dalrymple *et al.* 1992; courtesy of the Society for Sedimentary Geology.)

contrast, estuaries are mainly aggradational, building up within a drowned river channel. The base of an estuarine succession is therefore commonly an erosion surface scoured at the mouth of the river, for example, in response to sea level fall *(21.3.1)*. Estuaries are commonly, but not exclusively, tide-dominated environments and it may be difficult to distinguish between the deposits of a tidal estuary and a tide-dominated delta if there is limited information and it is difficult to establish whether the succession is aggradational and valley-filling or progradational.

Further reading

Bhattacharya, J.P. & Walker, R.G. (1992) Deltas. In: *Facies Models: Response to Sea Level Change* (eds R.G. Walker & N.P. James), pp. 157–178. Geological Association of Canada, St Johns, Newfoundland.

Colella, A. & Prior, D.B. (eds) (1990) *Coarse-grained Deltas*, 357 pp. International Association of Sedimentologists Special Publication **10**.

Dalrymple, R.W. (1992) Tidal depositional systems. In: *Facies Models: Response to Sea Level Change* (eds R.G. Walker & N.P. James), pp. 195–218. Geological Association of Canada, St Johns, Newfoundland.

Dalrymple, R.W., Zaitlin, B.A. & Boyd, R. (1992) Estuarine facies models: conceptual basis and stratigraphic implications. *Journal of Sedimentary Petrology* **62**, 1130–1146.

Davis, R.A. Jr (ed.) (1985) *Coastal Sedimentary Environments*, 716 pp. Springer-Verlag, Berlin.

Morgan, J.P. (ed.) (1970) *Deltaic Sedimentation, Modern and Ancient*. Society of Economic Paleontologists and Mineralogists Special Publication **13**.

Oti, M. & Postma, G. (eds) (1995) *Geology of Deltas*, 315 pp. A.A. Balkema, Rotterdam.

Reading, H.G. & Collinson, J.D. (1996) Clastic coasts. In: *Sedimentary Environments: Processes, Facies and Stratigraphy* (ed. H.G. Reading), pp. 154–231. Blackwell Science, Oxford.

Smith, D.G., Reinson, G.E., Zaitlin, B.A. & Rahmani, R.A. (eds) (1991) *Clastic Tidal Sedimentology*, 307 pp. Canadian Society of Petroleum Geologists Memoir **16**.

Whateley, M.K.G. & Pickering, K.T. (eds) (1989) *Deltas: Sites and Traps for Fossil Fuels*, 360 pp. Geological Society Special Publication **41**.

13 Coastlines: beaches, barriers and lagoons

Away from deltas and estuaries there are stretches of coastline which may be sites of either erosion or sediment accumulation. Erosional coastlines typically have cliff lines which may occasionally be preserved in the geological record as unconformity surfaces. Constructional coastlines have a higher preservation potential if the area is a region of subsidence. They may be sites of accumulation of clastic sediments, carbonate deposits or evaporites. Terrigenous clastic sediment is supplied to the coast by wave, storm and tide activity transporting detritus along the coast and then reworking it on beaches and associated environments. Carbonate-dominated coasts occur where clastic supply is low: the calcareous sediment is either biogenic or precipitated chemically in the shallow marine waters. Barrier islands form where sand or carbonate debris is built up by wave action as a shore-parallel bar a little way offshore on a gently sloping shelf. Behind this barrier a lagoon forms as a sheltered environment where fine clastic sediment, carbonate muds or evaporite deposits form. In arid climates evaporite minerals precipitate within coastal environments as sabkha deposits.

13.1 Coastal environments

Deltas and estuaries occupy only a small proportion of coastlines. Between them lie long stretches which may be erosional or constructional coastlines. The former act as sources of detritus whilst the latter are a range of depositional environments whose character is determined by the nature of the sediment available and the processes which sculpt the depositional bodies (Boyd *et al.* 1992).

13.1.1 Erosional coastlines

The energy of waves, tides and storms is a powerful erosive mechanism which can remove material from coastlines at rates which are significant on a human time-scale. In places attempts are made to halt the processes of erosion, but sea defences can only provide a temporary respite and often simply move the problem further along the coast. Physical weathering of cliffs includes the effects of the impact of water and loose debris on the cliff face and the growth of salt crystals within cracks where sea water has penetrated *(6.5.1)*. Material is removed in solution, as fine particles, and occasionally as large blocks as whole sections of the cliff face are removed. *Wave cut platforms* of bedrock eroded subhorizontally at beach level occur along many erosional coastlines. Eroded debris temporarily accumulates at the foot of the cliff but is ultimately removed by marine currents to contribute to the sediment deposited elsewhere in the marine realm (Fig. 13.1).

13.1.2 Constructional coastlines

A coastline which is a site of accumulation of sediment must have an adequate supply of material to counter erosive forces along the coastline and build up a deposit. Terrigenous clastic sediment accumulating on a beach may come from rivers and from the erosion of other coastal regions. Tidal, wind-driven and geostrophic currents are the main mechanisms for moving the material along the coastline, across continental shelves and throughout epicontinental seas *(11.1)*. Material formed within the sea is also an important source of sediment in the form of bioclastic debris and other carbonate material. Direct precipitation out of sea water also occurs along low-energy coasts in arid environments. There is

Fig. 13.1 An erosional coastline in Pembrokeshire, South Wales: wave action has eroded the cliff and left a wave cut platform of eroded rock on the beach.

often a mixture of sources for the material deposited along a coastline as clastic coasts often include carbonate debris and vice versa; both terrigenous mud and carbonate deposition occur along evaporitic (arid) coasts.

13.2 Morphological features of coastlines

The form of a constructional coastline is determined by the supply of sediment, the wave energy, the tidal range and the climate (Heward 1981; Boyd *et al.* 1992). Where there is an abundant supply of clastic sediment and high wave energy a well-developed beach forms. Sediment accumulation occurs along the shore where the waves break, on a beach dune ridge and in backshore areas of the coastal plain. Lagoons and tidal flats may develop behind a beach barrier. In contrast, low-energy coastlines in an arid environment may have a very poorly developed beach as the coastal plain merges gradually with the foreshore.

13.2.1 Beaches

The beach is the area washed by waves breaking on the coast (Figs 13.2 & 13.3). Sediment accumulating on a beach is either clastic or carbonate material or very commonly a mixture of the two. Where wave energy is

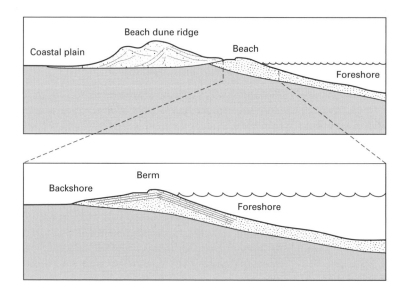

Fig. 13.2 Morphological features of a beach.

Fig. 13.3 A mixed sandy and pebbly beach near Mukalla, Yemen.

sufficiently strong, sand- and gravel-grade material may be continuously reworked on the foreshore, abrading clasts of all sizes to a high degree of roundness, and effectively sorting sediment into different sizes. Sandy sediment is deposited in laminae parallel to the slope of the shoreface, dipping offshore at only a few degrees to the horizontal (much less than the angle of repose). This low-angle stratification of well-sorted, well-rounded sediment is especially characteristic of wave-dominated beach environments, and is seen in both sandy and gravelly settings.

Similar low-angle stratification may also occur in beaches with a stronger tidal influence, although it will be associated with features formed in the intertidal zone. Beaches with very low gradients are covered with very shallow water as the tide rises and falls, shallow enough for waves generated by wind blowing across the water to form well-developed wave ripples in the sediment. Mud deposited out of suspension at the turn of the tide may result in the formation of flaser bedding *(4.5)*. Wave-formed sedimentary structures on the beach may be obliterated by organisms living in the intertidal environment and burrowing into the sediment. This bioturbation may obscure any other sedimentary structures.

At the top of the beach, a ridge, known as a *berm*, marks the division between the foreshore and *backshore*, or back-berm, area (Fig. 13.2). Water only washes over the top of the berm under storm surge conditions. Sediment carried by the waves over the berm crest is deposited on the landward side, forming layers which dip gently landward. These low-angle strata are typically truncated by the foreshore stratification, to form a pat-

tern of sedimentary structures and textures which may be considered to be typical of the beach environment (Figs 13.4 & 13.5).

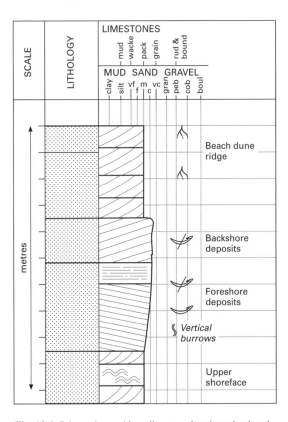

Fig. 13.4 Schematic graphic sedimentary log through a beach deposit.

Fig. 13.5 Sandstone and conglomerate made up of well-sorted, well-rounded material deposited on a beach, Pleistocene, Gulf of Corinth, Greece.

Behind some beaches a series of ridges parallel to the coastline are present (Fig. 13.6). These *chenier ridges* are the relicts of former beaches which have been left inland as the shoreline prograded (that is, the sea retreated) (Augustinus 1989). Chenier ridges are composed of sandy or shelly debris and are separated by areas of finer-grained sediment which forms a marshy area. To leave a series of distinct ridges the progradation of the shoreline must have been episodic, with the formation of a beach ridge in front of an existing beach. This occurs as a result of pulses of sediment supplied by longshore drift or the formation of new barrier islands *(13.3.2)*.

13.2.2 Beach dune ridges

Sand washed up the beach into the back-berm area is subject to reworking by wind action (Fig. 13.7). Onshore winds may pick up sand from the top of the beach and redeposit it as aeolian dunes *(8.2.3)* forming a band of aeolian sand deposits which lies parallel to the coast and extends hundreds of metres to kilometres inland. The processes of transport and deposition are the same as for desert dunes, but vegetation is likely to play a more important role in stabilizing the dunes and disrupting well-developed dune cross bedding. In a sedimentary succession these aeolian deposits may be seen as well-sorted sand at the top of the beach succession, with possible preservation of roots of plants which colonized the dune field. Aeolian deposits associated with a beach succession also occur in carbonate-dominated coastlines consisting of dunes composed of bioclastic and other carbonate detritus.

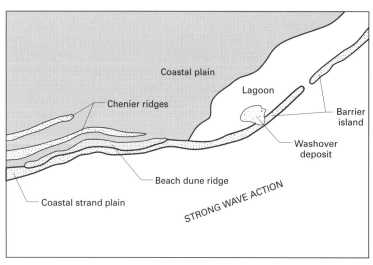

Fig. 13.6 Morphological features of a wave-dominated coastline.

Fig. 13.7 A beach dune ridge formed by sand blown by the wind from the shoreline on to the coast to form aeolian dunes, here stabilized by grass, Pembrokeshire, Wales.

13.2.3 Lagoons

Sheltered from the power of the waves by a barrier island, a reef or a carbonate shoal, *lagoons* are areas of quiet conditions and low-energy sedimentation. They are permanent bodies of water with limited connection to the open ocean. Waves will form on the surface, but will be relatively small because of the short fetch across the body of water *(4.4)*. Along coasts with abundant clastic supply, terrigenous clastic muds dominate lagoons developed behind barrier islands. Lagoons behind reefs and carbonate sand shoals on carbonate platforms are areas of lime mud deposition. Arid coastline lagoons are sites of evaporite precipitation.

13.2.4 Coastal plains

Coastal plains are low-lying areas adjacent to seas (Fig. 13.6). They are part of the continental environment where fluvial, alluvial or aeolian processes of sedimentation and pedogenic modification may be evident *(9.7.2)*. Coastal plains are influenced by the adjacent marine environment when storm surges result in extensive flooding by sea water. A deposit related to the storm flooding can be recognized by features such as the presence of bioclastic debris of a marine fauna amongst deposits which are otherwise wholly continental in character.

13.3 Clastic coastlines

The morphology of a clastic coastline is primarily controlled by the relative importance of tides and waves (Figs 13.6 & 13.8). The form of the beach, and the development of offshore constructional features, barrier islands, is dependent on whether the coastline is in a micro-, meso- or macrotidal regime *(11.2.2)*, and the amplitude of the waves which impact on the coast. Infrequent but high-energy storms also play an important role in distributing detritus and building up deposits in coastal settings.

13.3.1 Strand plains

A beach ridge formed along the coastline without a lagoon behind it is a *strand plain* (Figs 13.6 & 13.8) (Reading & Collinson 1996). Along coasts supplied with sediment, strand plains form sediment bodies tens to hundreds of metres across and tens to hundreds of kilometres long. Progradation of strand plains can produce extensive sandstone bodies covering hundreds of square kilometres. The strand plain is composed of the sands and gravels deposited on the foreshore, a berm and a backshore region, the extent of which depends on how often storm surges wash sediment over the berm crest. The backshore area merges into the coastal plain and may show evidence of subaerial conditions such as the formation of aeolian dunes and plant colonization.

13.3.2 Barrier islands

A *barrier island* is a beach detached from the main coast to form a ridge of sediment parallel to the coast (Oertel 1985). Long, linear barrier islands form most readily

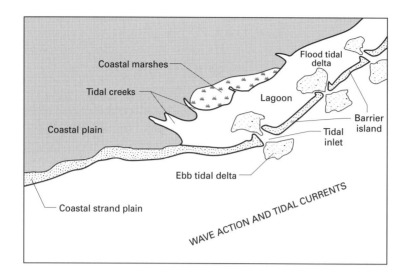

Fig. 13.8 Morphological features of a wave- and tide-influenced coastline.

along coasts with a low tidal range and high wave energy (Fig. 13.6). These conditions are achieved opposite areas of open ocean where the continental shelf is relatively narrow: waves driven across the ocean by winds lose little of their energy as they cross the narrow shelf, and there is little amplification of the low oceanic tidal range across the shelf.

The seaward margin of a barrier island has a beach at the back of which lies a ridge which marks the edge of the barrier platform. This platform is a few metres above sea level, with aeolian processes reworking sand into dunes. Vegetation helps to stabilize the dunes. Aggradation occurs by the addition of sand and bioclastic debris washed on to the top of the barrier island during storms. On the landward side of the island the layers of sand deposited during storms pinch out into the muddy marshes of the edge of the lagoon.

Coasts with higher tidal ranges and lower-amplitude waves may also be the sites for the formation of barrier islands, but without an associated lagoon. In their place, a swathe of tidal flats (salt marshes) cut by creeks lies between the sand ridge and the mainland. The depositional facies are very similar to those seen in estuarine settings *(12.7)* but in the stratigraphic record the association with the barrier island sands may allow this environment to be distinguished.

13.3.3 Flood-tidal deltas, ebb-tidal deltas and washovers

Barrier islands may be tens to hundreds of kilometres in length but they are normally dissected by gaps in the ridge which allow for the passage of water between the open sea and the lagoon behind the barrier. In microtidal regimes the small tidal flow is concentrated into these channels and the currents may be strong enough to move sediment to and fro between the open sea and the lagoon. In mesotidal coastlines with barrier islands the currents in these tidal channels are strong enough to redistribute large quantities of material. On the lagoon side of the barrier, sediment washed through the channel is deposited in a *flood-tidal delta* (Fig. 13.8). The water in the lagoon is shallow, so the sediment spreads out into a thin, low-angle cone of detritus dipping very gently landwards. Bedforms on the flood-tidal delta are typically subaqueous dunes migrating landwards which result in cross bedding with onshore palaeocurrent directions (Boothroyd 1985).

Ebb-tidal deltas form on the seaward margin of the tidal channel as water flows out of the lagoon when the tide recedes. Building out into deeper water, they are thicker bodies of sediment than flood-tidal deltas and the direction of bedform migration is seawards. The size and extent of an ebb-tidal delta is limited by the effects of reworking of the sediment by wave, storm and tidal current processes in the sea.

Between the channels along the length of the barrier island *washovers* of sediment occur in the lagoon due to water spilling over the top of the barrier during storm surges (Fig. 13.6). Washover deposits are low-angle cones of stratified sands dipping landwards from the barrier into the lagoon (Schwartz 1982).

13.3.4 Barrier island and strand plain successions

The pattern of strata built up in a barrier island setting will depend on whether the shoreline is progradational or transgressive (Reinson 1992). A seaward migration of facies in a progradational setting results in a succession of offshore and shoreface deposits *(11.5)* capped by sandy beach deposits of the barrier island. Fine-grained

Fig. 13.9 Schematic graphic sedimentary logs of the successions formed along: (a) a regressive shoreline; and (b) a transgressive shoreline.

lagoonal or tidal mudflat strata will overlie the sands and these muds will in turn be covered by continental deposits of the coastal plain (Fig. 13.9a). In a transgressive regime the pattern will be in part reversed, with the sands of the barrier island migrating over the lagoon or mudflat deposits (Fig. 13.9b).

Successions built up along coasts with strand plain beaches lack lagoonal facies. In a progradational succession the shoreface deposits are overlain by the well-sorted sediments of the beach foreshore and the succession is capped by evidence of subaerial conditions of the backshore and coastal plain environments. A

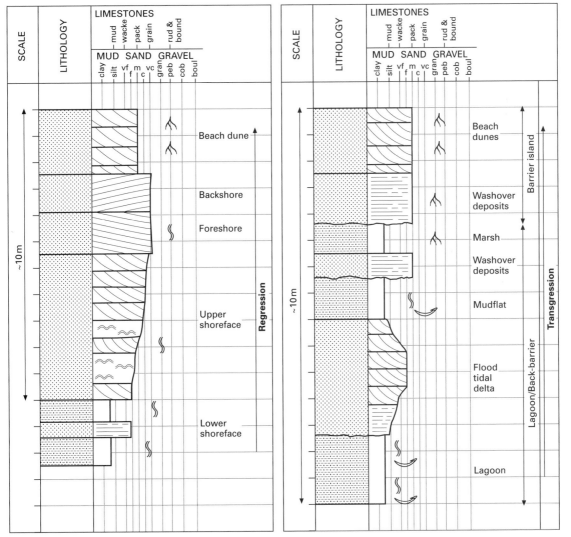

(a) (b)

transgressive succession may display a reversal of the order of facies.

13.3.5 Clastic lagoons

In the absence of any significant currents, only suspended load will be carried by the water in the lagoon. The deposits are hence almost entirely fine-grained. Coarser sediment may enter the lagoon when storms wash sediment over the barrier island as washover deposits *(13.3.3)*. A lagoonal succession is typically mudstone with thin, wave rippled sand beds. Organic material may be abundant from vegetation which grows on the shores of the lagoon (Reading & Collinson 1996). In tropical climates, trees with aerial root systems (*mangroves*) colonize the shallow fringes of the lagoon. Mangroves cause the shoreline to prograde into the lagoon as they act as sites for accumulation of sediment and organic matter along the water's edge. In more temperate climates, saline-tolerant grasses, shrubs and trees may play a similar role in trapping sediment.

Water circulation within a lagoon is restricted by the limited connection to open marine conditions, and this makes lagoons sensitive to the local climate. If rainfall is high or if there is local river water supply, the marine water of the lagoon becomes diluted and turns brackish: fauna will be restricted to organisms which are tolerant of low salinities. Similarly, fauna will be restricted in an arid climate where the evaporation of the lagoon waters increases the salinity. Although the diversity of fauna may be severely limited by brackish or hypersaline waters in the lagoon, those species which are tolerant flourish in the absence of competition in waters rich in nutrients from the surrounding vegetation. Oysters are common in many modern lagoonal environments.

In isolation, it may be difficult to distinguish the deposits of a clastic lagoon from a fresh water lake. Processes are almost identical because they are both standing bodies of water. Fauna may provide the best indicator of conditions, suggesting a marine influence, brackish or hypersaline water in a lagoon. Association with other facies is also important: lagoonal deposits occur above or below beach/barrier island sediments and fully marine shoreface deposits.

13.4 Carbonate coastlines

Carbonate-rich coastlines are most commonly found adjacent to tropical seas where organisms flourish and carbonate productivity is high. Reefs of coral and other reef-building organisms occur along some coasts *(14.7.2)*. Beach deposits of biogenic carbonate material are also found along the shore in more temperate climates away from supplies of terrigenous clastic sediment. A division into wave- and tide-dominated types of coast is again made, with a recognition that there are many cases where both processes are active in shaping the sediment bodies along the coast.

13.4.1 Wave/storm-influenced carbonate coastlines

The patterns of sedimentation along microtidal coastlines with high carbonate productivity are very similar to those of clastic, wave-dominated coastlines. Carbonate material in the form of bioclastic debris and ooids is reworked by wave action into ridges which are strand plains along the coast or barrier islands separated from the shore by a lagoon. The texture of carbonate sediments deposited on barrier island and strand plain beaches is typically well sorted and with a low mud matrix content (grainstone and packstone). Sedimentary structures are low-angle cross stratification dipping seawards on the foreshore and landwards in the backshore area.

At the top of the beach sands of carbonate debris and terrigenous debris may be reworked by wind to form aeolian dunes. When these dune sands become wet calcium carbonate is locally dissolved and reprecipitated to cement the material at the surface into a rock which is often referred to as an *aeolianite* (note that this term is sometimes used for cemented dune sand of any composition and formed in any environment) (Tucker & Wright 1990). Carbonate also precipitates around the roots of vegetation growing in the dune sands and may be preserved as nodular rhizocretions *(9.7.2)* (McKee & Ward 1983).

A feature of carbonate-rich beaches is the formation of *beachrock*. Carbonate ions in solution precipitate between sand and gravel material deposited on the beach and cement the beach sediments into fully lithified rock. Beachrock along the foreshore may act as a host for organisms which bore into the hard substrate, a feature which makes it possible to recognize early cementation of a beachrock in the stratigraphic record.

Where a barrier island ridge is cut by tidal channels, the tidal currents passing through form flood- and ebb-tidal deltas in much the same way as in clastic barrier island systems. The shape and internal sedimentary structures of these deposits are also similar on both clastic and carbonate coastlines. The nature of the carbonate material deposited on ebb- and flood-tidal deltas depends on the type of material being generated in the shallow marine waters: it may be bioclastic debris or, as is the case along the parts of the Trucial Coast of Arabia, oolitic sediment forming beds of grainstone and packstone.

13.4.2 Carbonate lagoons

Carbonate lagoons are sites of fine-grained sedimentation forming layers of carbonate mudstone and wackestone with some grainstone and packstone beds deposited as washovers near the beach barrier. In mesotidal regimes tidal channels cut the barrier and distribute relatively coarse sediment as a flood-tidal delta in the same way as sands are distributed in a similar clastic setting. Lenses of cross bedded oolitic and bioclastic packstone and grainstone are formed by subaqueous dunes on flood-tidal deltas.

The source of the fine-grained carbonate sediment in lagoons is largely algae living in the lagoon, with coarser bioclastic detritus from molluscs (Purser 1973). Pellets formed by molluscs and crustaceans are abundant in lagoon sediments. The nature and diversity of the plant and animal communities in a carbonate lagoon are determined by the salinity. Lagoons in mesotidal coastlines tend to have better exchange of sea water through tidal channels than more isolated lagoons in microtidal regimes. The local climate also has an effect. In arid regions the lagoon may become hypersaline and there will be a restricted fauna. Where the climate is more humid evaporation is lower, and as the lagoon has near-normal salinities a diverse marine fauna is present.

13.4.3 Tide-influenced carbonate coastlines

Tidal currents along carbonate-dominated coastlines transport and deposit coarse sediment in tidal channels and finer carbonate mud on tidal flats. The tidal channel sediments are similar in character to those found in tidal channels in clastic estuarine deposits *(12.7)*. The base of the channel succession is marked by an erosive base overlain by a lag of coarse debris: this may consist of broken shells and intraclasts of lithified carbonate sediment. Carbonate sands deposited on migrating bars in the tidal channels form cross bedded grainstone and packstone beds (Pratt *et al.* 1992). As a channel migrates or is abandoned the sands are overlain by finer sediment

forming beds of carbonate mudstone and wackestone. Bioturbation is normally common throughout.

In the intertidal zones deposits of lime mud and shelly mud are subject to subaerial desiccation at low tide (Fig. 13.10). Unlike terrigenous clastic mud which remains relatively wet when exposed between tidal cycles, lime mud tends to dry out. Most carbonate coastlines are in regions of warm climate where water is lost from intertidal areas by evaporation. As the water dries out carbonate cements are precipitated in the deposits of lime mud forming a crust on the surface. Repeated precipitation of cements in this crust causes the surface layer to expand and form a polygonal pattern of ridges, called *tepee structures* or *pseudoanticlines*, a few tens of centimetres across. As the pseudoanticlines grow they leave cavities beneath them which are sites for the growth of sparry calcite cements. Smaller isolated cavities and vertically elongate hollow tubes also form in lime muds in intertidal areas due to air and water being trapped in the sediment during the wetting and drying process. Patches of calcite cement which grow in the cavities in the host of lime mud give rise to a fabric generally called *fenestrae* or *fenestral cavities*. Vertical fenestrae may also result from roots and burrows. Lime mudstones with small cavities filled with calcite are sometimes referred to as *birds-eye limestones*.

A common feature of carbonate tidal flats is the formation of algal and bacterial mats which trap fine-grained sediment in thin layers to form the well-developed, fine lamination of a stromatolite *(3.1.2)*. Stromatolites may form horizontal layers or irregular mounds on the tidal flats. Their distribution is partly controlled by the activity of organisms which either feed on the microbial

Fig. 13.10 Polygonal cracks on the top surface of a fine-grained limestone bed, interpreted as the deposits of algal mudflats (Carboniferous, north Mayo, Ireland).

mats or disrupt them by bioturbation. Stromatolites tend to be better developed in the higher parts of the intertidal area which are less favourable for other organisms, particularly gastropods which graze on the mats.

13.5 Arid coastlines

Along the coasts of seas in hot dry regions chemical processes are important in determining the sediments deposited. Arid shorelines are found today in places like the Arabian Gulf where they are sites of evaporite formation within the coastal sediments. These arid coasts are called *sabkhas* (Handford 1981; Kendall 1992). They typically have a very low relief and there is not always a

well-defined beach. Along coastlines where there is a barrier formed by a beach of carbonate or clastic sediment the lagoon behind the barrier may be a region of evaporite deposition in an arid climate (Fig. 13.11).

13.5.1 Sabkhas

The coastal plain of a sabkha is occasionally wetted by sea water during very high tides or during onshore storm winds. More importantly, there is a supply of water through groundwater seepage from the sea. The hot surface of the coastal plain is an area of evaporation and water is drawn up through the sediment to the surface (Fig. 13.12). As the water rises it becomes more

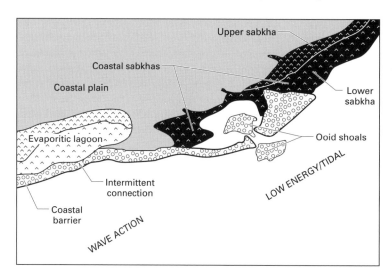

Fig. 13.11 Morphological features of arid coastal environments.

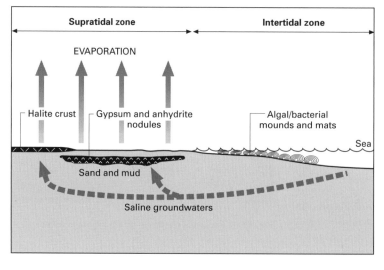

Fig. 13.12 A cross-section across a coastal sabkha: evaporation in the supratidal zone results in saline water being drawn up through the coastal sediments and the precipitation of evaporite minerals within and on the sediment.

concentrated in salts which precipitate within the coastal plain sediments. Gypsum and anhydrite grow within the sediment whilst a crust of halite forms at the surface.

In general, anhydrite forms in the hotter, drier sabkhas and gypsum where the temperatures are lower or where there is a supply of fresh, continental water to the sabkha. In some sabkhas both gypsum and anhydrite are formed. Close to the shore in the intertidal and near-supratidal zone gypsum crystals grow in the relatively high flux of sea water through the sediment. Further up in the supra-tidal area conditions are drier and nodules of anhydrite form in the sediment. The gypsum and anhydrite grow displacively within the sediment with the gypsum in clusters and the anhydrite forming amorphous coalesced nodules with little original sediment in between (Fig. 13.13). These layers of anhydrite with remnants of other sediment have a characteristic *chicken-wire* structure. Halite crusts are rarely preserved because they are removed by any surface water flows. The terrigenous

Fig. 13.13 Nodular gypsum formed in a coastal sabkha setting (Triassic, England).

sediment of the sabkha is often strongly reddened by the oxidizing conditions.

With the accumulation of sediment along an arid coast a sabkha cycle is developed (Fig. 13.14). At the base of

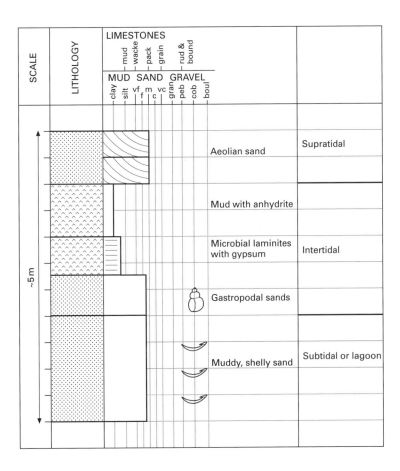

Fig. 13.14 Schematic graphic sedimentary log through a sabkha succession.

the cycle there are oolitic and bioclastic limestone beds deposited in a wave-reworked shallow subtidal setting and overlain by intertidal stromatolitic limestones. The high salinity of arid coasts permits the preservation of microbial mats as stromatolites because the organisms which normally graze on the organic matter are unable to tolerate the hypersaline conditions. Gypsum crystals formed in the upper intertidal and lower supratidal area occur next in the succession overlain by anhydrite with a chicken-wire structure. Coalesced beds of anhydrite formed in the uppermost part of the sabkha form layers contorted as the minerals have grown into an *enterolithic* bedding. The sabkha cycle may be repeated many times if there is continued subsidence along the coastal plain (Kendall & Harwood 1996). Associated beds include aeolian sands formed on the coastal plain.

Evaporites deposited in a sabkha setting can be distinguished from those formed in continental environments (8.5, 10.4) by the association with marine sediments and from lagoonal (13.5.2) or barred basin (14.9) evaporites by the displacive growth of anhydrite and gypsum within sediments. An association of sabkha sediments with marginal lagoonal or barred basin successions may occur.

13.5.2 Evaporite lagoons

In hot, dry climates the loss of water by evaporation from the surface of a lagoon is high. If it is not balanced by influx of fresh water from the land or exchange of water with the ocean the salinity of the lagoon will rise and it will become *hypersaline* — more concentrated in salts than normal sea water (Fig. 13.11). The barrier between the evaporitic lagoon or *salina* and the ocean may be terrigenous clastic sand, a carbonate shoal or reef. Connection with the ocean may be via gaps in the barrier or by seepage through it. Variations in the salinity within the lagoon may be because of climatically related changes in the fresh water influx from the land or increased exchange with open sea water during periods of higher sea level. The extent of the lagoon and the minerals precipitated in it are therefore likely to be variable, resulting in cycles of sedimentation. An alternation between laminated gypsum deposited subaqueously in a lagoon and nodular gypsum formed in a supratidal sabkha around the edges of the water body may represent fluctuations in the area of the water body. In less arid settings the supratidal facies will be carbonates which will be found in cycles with beds of gypsum. In the restricted circulation of a lagoon conditions are right for

large crystals of selenitic gypsum to form by growing upwards from the lagoon bed (Warren & Kendall 1985). Semi-enclosed bodies of water adjacent to oceans which are sites of evaporite deposition are barred basins (14.9) which have deeper water.

13.6 Recognition of coastal deposits: summary

Coastal environments of deposition are very variable. Their characteristics depend on the relative importance of wave and tide activity, the supply of detritus and the climate.

Beaches formed along strand plains or as barrier islands are characterized by deposits of well-sorted and well-rounded material in both clastic and carbonate coastlines. Stratification is typically low-angle, dipping either seawards or landwards depending on whether deposition was on the foreshore or backshore of the beach. An association with subaerial conditions may be recognized by the presence of aeolian dune bedding in sands deposited on the beach ridge and evidence of roots in soils formed on the exposed beach ridge or backshore area.

Lagoons formed behind beach barriers are sites of low-energy deposition of clastic or carbonate muds. The fauna may be restricted if the lagoon has poor connection to the sea in microtidal regimes, because either hypersaline conditions develop in arid regions or brackish waters form in humid climates. Connection to the sea is through tidal channels which are well developed in mesotidal areas. These allow coarser material to be brought into the lagoon as flood-tidal deltas preserved as thin wedges of sandstone or packstone/grainstone units in the lagoonal mudstone. Adjacent to the barrier island, coarse washover deposits are also interbedded with the lagoon sediments. Extensive beds of evaporite minerals form in lagoons on arid coastlines which are continuously recharged by influxes of sea water.

In mesotidal to macrotidal regimes extensive tidal flats are mainly areas of fine-grained sedimentation. Colonization by plants is intensive in humid regions, leaving evidence as rootlets, organic material and widespread bioturbation by animals feeding in the mudflats. Microbial mats form layered stromatolites in carbonate-rich intertidal areas. In drier regions there is evidence of desiccation, particularly in carbonate mudflats, and the formation of evaporite deposits in arid environments. Arid coastal regions are characterized by the growth of nodules of evaporite minerals within a host of carbonate or terrigenous muds in a sabkha environment.

Further reading

Davis, R.A. Jr (ed.) (1985) *Coastal Sedimentary Environments*, 716 pp. Springer-Verlag, Berlin.

Kendall, A.C. & Harwood, G.M. (1996) Marine evaporites: arid shorelines and basins. In: *Sedimentary Environments: Processes, Facies and Stratigraphy* (ed. H.G. Reading), pp. 281–324. Blackwell Science, Oxford.

Oertel, G.F. & Leatherman, S.P. (eds) (1985) Barrier islands. *Marine Geology* **63**, 1–396.

Pratt, B.R., James, N.P. & Cowan, C.A. (1992) Peritidal carbonates. In: *Facies Models: Response to Sea Level Change* (eds R.G. Walker & N.P. James), pp. 303–322.

Geological Association of Canada, St Johns, Newfoundland.

Reading, H.G. & Collinson, J.D. (1996) Clastic coasts. In: *Sedimentary Environments: Processes, Facies and Stratigraphy* (ed. H.G. Reading), pp. 154–231. Blackwell Science, Oxford.

Reinson, G.E. (1992) Transgressive barrier island and estuarine systems. In: *Facies Models: Response to Sea Level Change* (eds R.G. Walker & N.P. James), pp. 179–194. Geological Association of Canada, St Johns, Newfoundland.

Scholle, P.A., Bebout, D.G. & Moore, C.H. (eds) (1983) *Carbonate Depositional Environments*, 708 pp. American Association of Petroleum Geologists Memoir **33**.

Tucker, M.E. & Wright, V.P. (1990) *Carbonate Sedimentology*, 482 pp. Blackwell Scientific Publications, Oxford.

Wilson, J.L. (1975) *Carbonate Facies in Geological History*, 471 pp. Springer-Verlag, Heidelberg.

14   Shallow seas

The shallow marine environment is a setting for the accumulation of terrigenous clastic material washed down from the continents and of carbonate sedimentation from the remains of the organisms which live in the waters. There are extensive areas of deposition today, and probably more stratigraphic units are interpreted as shallow marine deposits than any other depositional environment. The sediments formed in shallow seas are extremely diverse. Terrigenous clastic material is distributed and separated into different depositional facies by tides, waves, storms and ocean currents. These same processes also influence the distribution of carbonate sediments, which are more abundant in this setting than any other environment. Carbonate-producing organisms are sensitive to water depth, climate and terrigenous input, resulting in very variable shallow carbonate deposits. Changes in biota through geological time have also played an important role in determining the characteristics of shallow marine sediments throughout the stratigraphic record. In some circumstances shallow marine environments can become sites for the formation of thick evaporite successions. Deposition in all shallow marine environments is sensitive to changes in sea level and the stratigraphic record of sea level changes is recorded within sediments formed in shallow seas.

14.1 Sediment supply to shallow seas

The supply of sediment to shelves is a fundamental control on shallow marine environments and depositional facies of shelves and epicontinental seas. If the area lies adjacent to an uplifted continental region and there is a drainage pattern of rivers delivering detritus to the coast the shallow marine sedimentation will be dominated by terrigenous clastic deposits. The highest concentrations of clastic sediment will be at the mouths of major rivers and the adjacent coastal regions will also be supplied by longshore movement of material. The coastlines prograde and the shelf area is supplied with sediment redistributed by tidal, geostrophic and storm currents.

Biogenic carbonate production is inhibited by the presence of clastic material so in areas of low input of detritus from rivers or by ocean currents there is the potential for carbonate deposition. Biogenic productivity is also determined by the temperature and salinity of the seas and is favoured by warm waters and normal salinity.

In shallow seas in low latitudes carbonate-producing organisms are abundant and in the absence of clastic sediment the shelves are dominated by carbonate deposition. In cooler climates where carbonate production in the seas is slower and along coasts distant from any terrigenous sediment supply the shelf area or sea is *starved*. The rate of sediment accumulation is slow and may be exceeded by the rate of subsidence of the sea floor such that the environment becomes gradually deeper with time.

Evaporite sedimentation only occurs in situations where a body of water becomes partly isolated from the ocean realm and salinity increases to the point where there is chemical precipitation of minerals. This normally only occurs in epeiric seas which are connected to the open ocean by a strait which may become blocked by changes in sea level or local tectonics.

In many cases there is a mixture of terrigenous clastic and biogenic carbonate sediment supply which is variable across the shelf or epeiric sea and changes through time. Deposition in these settings may be very long-lived and can continue uninterrupted for tens of millions of years (Tucker & Wright 1990). During these long periods the supply of terrigenous material may be subject to tectonic controls in the adjacent continents, the climate may go through short- and long-term variations and both global and local changes in sea level will have occurred. With so many interrelated controls on the supply and distribution of sediment in shallow seas conceptual models for environments and facies can only be end members in a broad, complex spectrum of possibilities.

14.2 Shallow marine clastic environments

The patterns and characteristics of deposition on shelves and epicontinental seas with abundant terrigenous clastic supply are controlled by the relative importance of wave, storm and tidal processes. The largest tidal ranges tend to be in epicontinental seas and restricted parts of shelves although in some situations the tidal ranges in narrow or restricted seaways can be very small *(11.2.2)*. Open shelf areas facing oceans are typically regions with a microtidal to mesotidal regime and are affected by ocean storms. A general classification into *storm-dominated shelves* and *tide-dominated shelves* with clastic sedimentation can be made for both modern environments and ancient facies. It must be emphasized that these are end members of a continuum, and many modern shelves and epicontinental seas show influence of both major processes. Shallow marine sediments in the stratigraphic record similarly show features which indicate the influence of both tidal and storm processes.

Fig. 14.1 Facies distribution on a storm-dominated shelf.

14.3 Storm-dominated shallow clastic seas

In the shallower part of the shelf any sediment will be extensively reworked by wave processes. Sands deposited in these settings are texturally and (usually) compositionally mature with wave ripple cross lamination and horizontal stratification. Streaks of mud in flaser beds *(4.5)* deposited during intervals of lower wave energy become more common in the deposits of slightly deeper water further offshore (Fig. 14.1). Wave ripples are less common as fair weather wave base *(11.3)* is approached in the lower part of the upper shoreface.

In the lower shoreface sands are deposited and reworked by storms. Beds of tempestites *(11.3.2)* deposited by individual storms typically taper in thickness from a few decimetres to millimetres in the outer parts of this zone several tens of kilometres offshore (Aigner 1985). Proximal tempestites have erosive bases and are composed of coarse detritus whilst the distal parts of the bed are finer-grained laminated sands. Hummocky and swaley cross stratification *(11.3.2)* occurs in the sandy parts of the normally graded storm deposits (Walker & Plint 1992). Storm conditions affect the water to depths of 20–50 m or more so hummocky and swaley cross stratification may be expected in any sandy sediments on the shelf to depths of several tens of metres. In fact, these structures are not seen in upper shoreface deposits above fair weather wave base due to reworking of the sediment by ordinary wave processes, so this characteristic form of cross stratification is only found in sands deposited between fair weather and storm wave bases on storm-dominated shelves.

In the periods between storm events the shelf is an

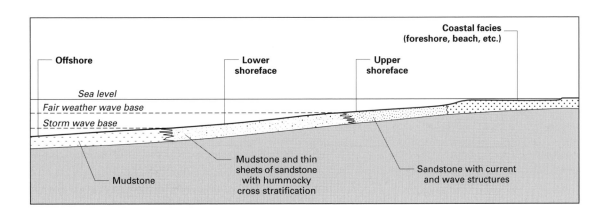

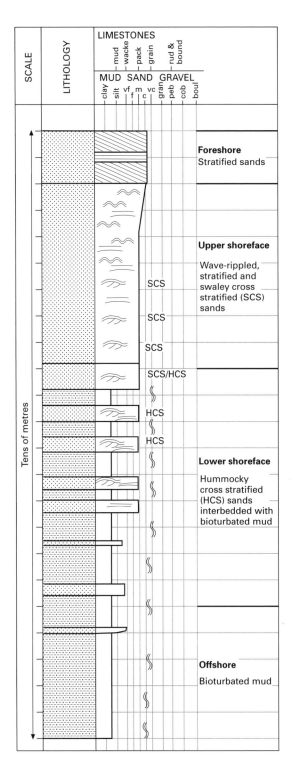

area of deposition of mud from suspension. Fine-grained clastic material is carried away from river mouths in suspension by geostrophic and wind-driven currents *(11.3, 11.4)*. Storms also rework a lot of fine sediment from the sea floor and carry it in suspension across the shelf. Storm deposits are therefore separated by layers of mud except in cases where the mud is eroded away by the subsequent storm. The proportion of mud in the sediments increases offshore as the amount of sand deposited by storms decreases. The outer shelf area below storm wave base is a region of mud deposition.

The primary sedimentary structures (wave ripples, hummocky cross stratification, and so on) are not always preserved in the shelf sediments because of the effects of bioturbation. The period between storm events provides benthic organisms with plenty of opportunity to rework the sediment, sometimes even in the deeper areas of the shelf where life is usually sparse. Small amounts of bioturbation are recognizable by the presence of individual burrows cutting across bed boundaries and sedimentary structures. If bioturbation is intense the sediment becomes homogenized, the internal structure of beds is obscured and there can be a mixing of finer and coarser material into homogeneous, structureless units.

14.3.1 Characteristics of a storm-dominated shallow marine succession

If there is a constant sediment supply to the shelf, continued deposition builds up the layers on the sea bed and the water becomes shallower. Shelf areas which were formerly below storm wave base experience the effects of storms and become part of the offshore transition zone. Similarly addition of sediment to the sea floor in the offshore transition zone brings the sea bed up into the shoreface zone above fair weather wave base and a vertical succession of facies which progressively shallow upwards is constructed (Fig. 14.2) (Walker & Plint 1992). A relative sea level fall may result in a shift of facies which superimposes upper shoreface deposits directly on offshore facies with a sharp, erosive contact (Walker & Plint 1992). In a storm-dominated shallow marine succession the upper shoreface is characterized by symmetrical (wave) ripple lamination and horizontal stratification which may be obscured by intense bioturbation. This overlies the lower shoreface facies,

Fig. 14.2 Schematic graphic sedimentary log through a storm-dominated clastic shelf succession.

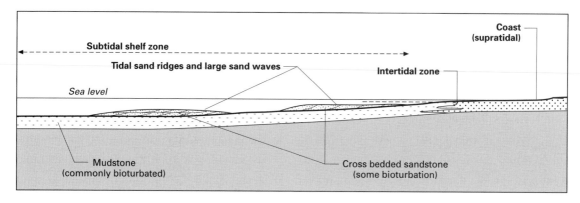

Fig. 14.3 Facies distribution on a tide-dominated shelf.

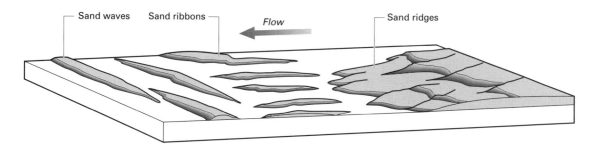

Fig. 14.4 Sand waves, sand ridges and sand ribbons in a shallow sea.

which is typically made up of sandstone beds with erosional bases with some hummocky cross stratification, interbedded with bioturbated mudstone. The thickness of the sandstone beds generally decreases down the succession. The offshore facies mainly consists of mudstone beds with some bioturbation.

14.4 Tide-dominated shallow clastic seas

The effects of tides are most obvious on coastlines but the currents generated by tides influence large areas of shelves and epicontinental seas. These tidal currents affect the sea bed up to 100 m or more below sea level and are strong enough to move large quantities of sand in shallow marine environments. The effects of waves and storms are largely removed by tidal currents reworking the material in macrotidal regimes and only the tidal signature is left in the stratigraphic record (Fig. 14.3). In seas with moderate tidal effects the influence of tides is

seen in shallower water but storm deposits are preserved in the offshore transition zone in these mixed storm/tidal shelf settings.

The form of tidal deposits in shallow marine environments depends on the velocity of the tidal current. At low to moderate near-surface tidal current velocities (less than 100 cm s^{-1}) sand waves *(4.3.2)* are typical. These bedforms are larger than subaqueous dunes having heights of between 1.5 m and 15 m and wavelengths ranging from a few tens of metres to 500 m (Fig. 14.4). The crests are straight to moderately sinuous and the lee slope is a lower angle than most subaqueous bedforms, at around 15° (Johnson & Baldwin 1996). Migration of sand waves in the direction of the predominant tidal current generates cross stratification with sets which may be 10 m or more thick (Fig. 14.5). Cross stratification on this scale is not seen in other marine environments and is only matched in size by aeolian dunes and some large bar forms in rivers. Smaller subaqueous dune bedforms are normally associated with other shallow marine tidal deposits and are preserved as decimetre-scale trough cross bedding. Mud drapes *(11.2.4)* occur on the cross strata of these bedforms.

Fig. 14.5 Large-scale cross stratification formed by the migration of subaqueous sand waves (Lower Cretaceous, Leighton Buzzard, England).

In shallow seas with higher velocity tidal currents (over $100\,cm\,s^{-1}$) sediment on the sea floor forms *sand ribbons* aligned parallel to the flow direction (Fig. 14.4). These ribbons are only a metre or so thick but are up to 200 m wide and stretch for over 10 km in the flow direction (Kenyon 1970). They are separated by a substrate of gravel. With very high-velocity currents only gravel is resident on the sea floor and extensive scouring of furrows into the sea bed occurs. These erosion surfaces are distinctive in the deposits of shallow seas with strong tidal currents; they are typically overlain by a gravel or shelly lag and thin sandstone formed as a ribbon body.

A sparse supply of sand results in isolated ribbon features and sand wave bedforms scattered over the substrate which is often mud. In areas of abundant sediment supply large banks called *tidal sand ridges* form on shelves and in epicontinental seas. These ridges of sediment may be tens of metres high and kilometres across, stretching for tens of kilometres. These may be recognized in the stratigraphic record as elongate lenses of sandstone with low-angle cross stratification. Areas of the shallow seas (lower shoreface and offshore) which are too deep for the effects of the surface tidal currents to be felt and regions where the tidal currents are weaker are sites for mud deposition and sands deposited by storm currents. Bioturbation is common in these muds and may also be identified in the coarser deposits of sand waves, ribbons and ridges.

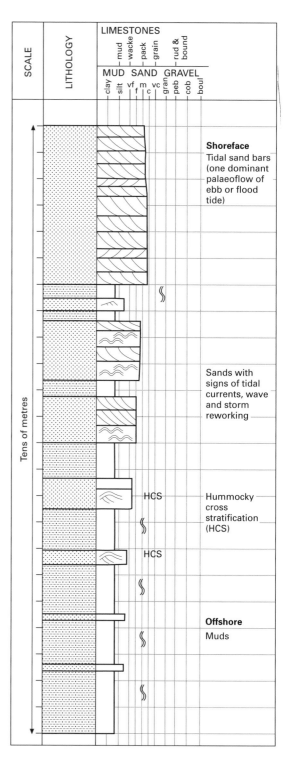

Fig. 14.6 Schematic graphic sedimentary log through a tide-dominated clastic shelf succession.

14.4.1 Characteristics of a tide-dominated shallow marine succession

Packages of cross stratified sandstone which contain a fully marine fauna and lack evidence for any subaerial exposure are normally interpreted as the deposits of tide-dominated shallow seas. In water depths of tens of metres tides are the only currents which can generate and maintain the subaqueous dune and sand wave bedforms: geostrophic currents are generally too weak and storm-driven currents are too short-lived and infrequent to create these bedforms. Features of tidal sedimentation *(11.2.4)* which may be present in these offshore tidal facies include mud drapes on some of the smaller scale cross bedding and reactivation surfaces within the sand

wave cross stratification (Allen 1982). There may be evidence of different directions of tidal currents from within a unit of tide-deposited sandstones but herringbone cross stratification is uncommon. Tidal currents on a shelf tend to follow regular patterns (rotary tides: *11.2.3*) which do not show the direct reversals seen in estuarine and coastal tidal settings. Erosion surfaces overlain by gravel or shelly lags are found representing higher-energy parts of the shelf or sea but the distinct channels found in estuarine deposits are not seen. The packages of cross bedded sandstone are typically tens of metres thick, sometimes amalgamated into even larger units, and are lens-shaped on a scale of kilometres (Fig. 14.6).

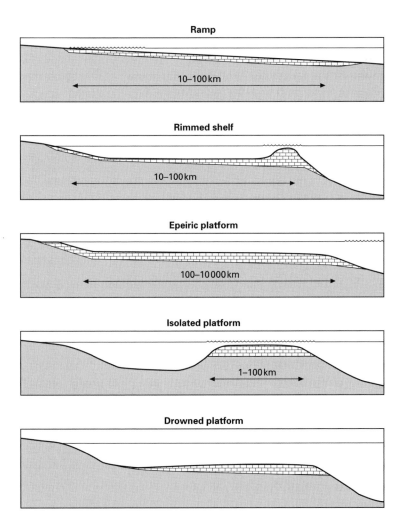

Fig. 14.7 Types of carbonate platform in shallow marine environments.

14.5 Shallow marine carbonate environments

Areas of shallow marine carbonate sedimentation are known as *carbonate platforms*. Platforms occur on areas of the continental shelf which may stretch from the shoreline out to the edge of the shelf or cover large parts of epicontinental (epeiric) seas. A number of different types of carbonate platform are recognized on the basis of their morphology (Fig. 14.7) (Tucker & Wright 1990; Wright & Burchette 1996). A gently sloping shelf analogous to a clastic shelf with water depth steadily increasing basinwards is known as a *carbonate ramp*. Also similar to a clastic depositional environment is an *epeiric platform*, a shallow epicontinental sea covering a large area. Shallow carbonate settings which do not have clastic counterparts are *rimmed shelves*, which have a reef or carbonate bank on their outer margin; *isolated platforms*, regions of shallow water sedimentation completely surrounded by deeper water; and *drowned platforms*, where relative sea level has risen too rapidly for shallow marine deposition to continue and deeper marine sedimentation takes over.

The skeletal grain associations which occur on carbonate platforms are dependent on temperature and salinity (Lees & Butler 1972). In low latitudes where the shallow sea is always over 15°C and the salinity is normal, corals and calcareous green algae are common, along with numerous other organisms, to form a *chlorozoan* assemblage. In restricted seas where the salinities

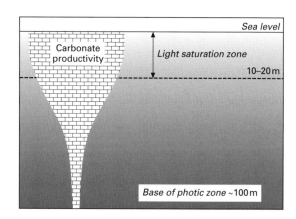

Fig. 14.8 Water depth and biogenic carbonate productivity.

are higher only green algae flourish, a *chloralgal* association. Temperate carbonates formed in cooler waters are dominated by the remains of benthic foraminifera and molluscs, a *foramol* assemblage. Ooids are most commonly associated with chlorozoan and chloralgal assemblages.

The amounts of biogenic carbonate produced in shallow seas are determined by the productivity within the food chain. Photosynthetic plants and algae at the bottom of the food chain are dependent on the availability of light. Penetration by sunlight is controlled by the water depth and the amount of suspended material in the sea. In bright tropical regions with clear waters the *photic zone* may extend up to 100 m water depth (Fig. 14.8) (Bosscher & Schlager 1992). Photosynthetic organisms typically flourish in the upper 10–15 m of the sea and it is in this zone that the greatest abundance of calcareous

Fig. 14.9 Cross-section and facies distributions across a carbonate ramp.

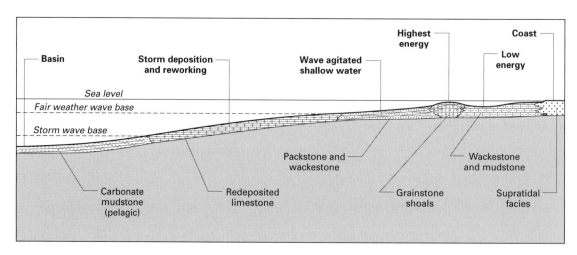

organisms is found. This shallow region of high biogenic productivity is referred to as the *carbonate factory* (Tucker & Wright 1990). Hermatypic corals dependent on symbiotic algae are most productive in shallow clear water with strong currents whilst most other benthic marine organisms prefer quieter waters.

14.6 Carbonate ramps

The bathymetric profile of a carbonate ramp (Fig. 14.9) and the physical processes within the sea and on the sea floor are very similar to an open shelf with clastic deposition. The term 'ramp' may give the impression of a significant slope but in fact the slope is a gentle one of less than 1° in most instances (Wright & Burchette 1996), in contrast to carbonate slope environments *(15.1.2)* which are much steeper. In macro- to mesotidal regimes tidal currents distribute carbonate sediment and strongly influence the coastal facies. Wave and storm processes are dominant in microtidal shelves and seas. The effects of tides, waves and storms are all depth-dependent and consequently the processes and facies on a carbonate ramp display a relatively simple pattern from nearshore, shallow water conditions to deeper water sedimentation in the more distal parts of the ramp.

14.6.1 Distribution of facies on a carbonate ramp

Coastal facies along tide-influenced shorelines will be characterized by deposition of coarser material in channels and carbonate muds on tidal flats; a beach barrier may be developed to form a lagoon (Tucker & Wright 1990; Jones & Desrochers 1992). Wave-dominated shorelines may have a beach ridge which confines a lagoon or a linear strandplain attached to the coastal plain. Ramps with mesotidal regimes will show a mixture of beach barrier, tidal inlet, lagoon and tidal flat deposition. Agitation of carbonate sediment in shallow nearshore water results in a shoreface facies of carbonate sand bodies. Skeletal debris and ooids formed in the shallow water form bioclastic and oolitic carbonate sand shoals. Benthic foraminifera are the principal components of some Tertiary carbonate ramp successions. Shoals of carbonate material may be formed by the migration of subaqueous dune bedforms which develop cross bedded grainstone and packstone facies. Wave action in shallow water results in well-sorted, stratified carbonate sands similar in character to terrigenous clastic sands deposited in shallow shelf environments.

Below fair weather wave base the extent of reworking by shallow marine processes is reduced. Storm processes transport bioclastic debris out on to the shelf to form deposits of wackestone, packstone and grainstone which may even include hummocky cross stratification. In deeper water below storm wave base in the distal parts of a ramp deposits are principally redeposited carbonate

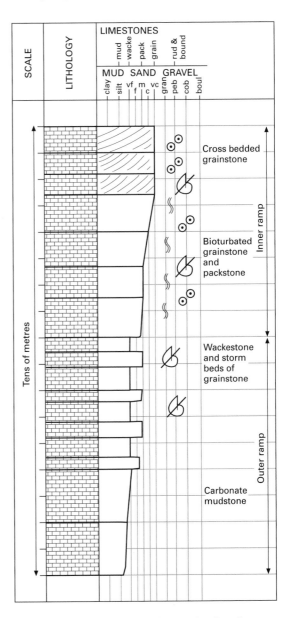

Fig. 14.10 Schematic graphic sedimentary log through a carbonate ramp succession.

mudstone and wackestone, often with the characteristics of turbidites. Redeposition of carbonate sediments is common in situations where the outer edge of the ramp merges into a steeper slope at a continental margin as a *distally steepened ramp. Homoclinal ramps* have a consistent gentle slope on which little reworking of material by mass flow processes occurs (Read 1985).

In contrast to rimmed shelves *(14.7)* reefal build-ups are relatively rare in ramp settings. Isolated *patch reefs* may occur in the more proximal parts of a ramp and mud mounds are known from Palaeozoic ramp environments. *Mud mounds* are build-ups of fine-grained biogenic carbonate material without a framework formed by reef-forming organisms.

14.6.2 Carbonate ramp succession

A succession built up by the progradation of a carbonate ramp will be characterized by an overall coarsening upward from carbonate mudstone and wackestone deposited in the outer ramp environment to wackestone and grainstone beds of the inner ramp (Figs 14.9 & 14.10) (Wright 1986). The degree of sorting typically increases upwards, reflecting the higher-energy conditions in shallow water. Outer ramp deposits include sediment reworked by storm processes and as turbidites. Inner ramp carbonate sand deposits are typically oolitic and bioclastic grainstone beds which exhibit decimetre to metre scale cross bedding and horizontal stratification.

Fig. 14.11 Cross-section and facies distribution across a carbonate rimmed shelf.

The top of the succession may include fine-grained tidal flat and lagoonal sediments. Ooids, broken shelly debris, algal material and benthic foraminifera may all be components of ramp carbonates. Locally mud mounds and patch reefs may occur within carbonate ramp successions.

On shelves and epicontinental seas where there are fluctuations in relative sea level, cycles of carbonate deposits are formed on carbonate ramps. A sea level rise results in a shallowing-upward cycle a few metres to tens of metres thick which coarsens up from beds of mudstone and wackestone to grainstone and packstone. A fall in sea level may expose the inner ramp deposits to dissolution in karstic subaerial weathering (Emery & Myers 1996).

14.7 Rimmed carbonate shelves

A bathymetric feature in many modern carbonate platforms is a distinct break in slope between a shallow inner shelf region and an outer zone of deeper water (Fig. 14.11) (Wilson 1975; Read 1982). This break in slope forms a distinct zone of high energy where wave action is important. Hermatypic corals thrive in shallow, agitated seas remote from muddy clastic input and form *barrier reefs* (e.g. the Great Barrier Reef offshore of Queensland, Australia) (Fig. 14.12). In the absence of coral reefs, skeletal or oolitic material forms carbonate sand shoals which also build up to form a distinct barrier. Landward of the barrier lies a low-energy *shelf lagoon* which is sheltered from the open ocean and may be from a few hundred metres to tens of kilometres wide.

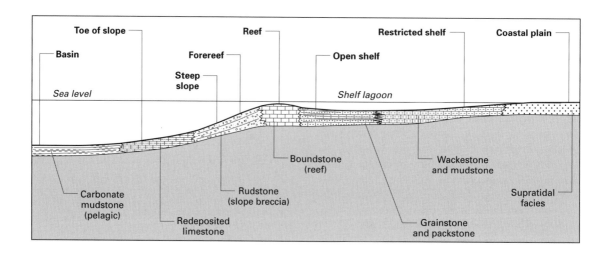

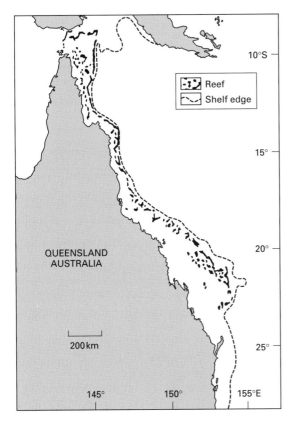

Fig. 14.12 The Great Barrier Reef, a carbonate rimmed shelf off the north-east coast of Australia. (Reproduced with the permission of Stanley Thornes Publishers Ltd from Duff (1993) *Holmes' Principles of Physical Geology*, 4th edn.)

Fig. 14.13 Facies distribution in a reef complex.

14.7.1 Shelf lagoon facies

The character of the sediments deposited in the sheltered waters of the shelf lagoon is largely determined by the extent to which the area is connected to the open sea through the reef barrier. Closed lagoons with very limited circulation are typically areas of lime mud deposition. They may be permanently or intermittently hypersaline and under conditions of raised salinities fauna tends to be very restricted: the coastline may be evaporitic, forming a sabkha *(13.5.1)*. In situations where the reef barrier is discontinuous and/or there is a moderate tidal influence the shelf lagoon will be subject to ocean waves and tidal currents. Open lagoons are therefore characterized by higher-energy facies including shoals which form lenses of grainstone and packstone. Coarse facies are commonest on the side of the lagoon nearest to the reef as bioclastic material is reworked by waves from the reef into the lagoon. Coastal facies are typically low-energy tidal flat deposits but a beach barrier may develop if the wave energy is high enough.

14.7.2 Reef facies

Reefs formed at shelf margins are areas of high biogenic productivity as a consequence of the upwelling of nutrient-rich waters from the deeper water of the continental slope. They are also high-energy environments because they are exposed to ocean waves and storms. Modern corals thrive under these conditions and present-day reef environments are used as a model for facies found in limestones in the Phanerozoic stratigraphic record. Differences between this model and ancient

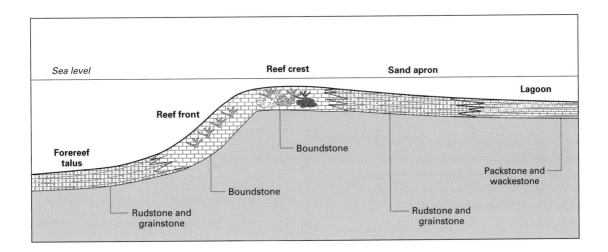

examples are to be expected as corals evolved through time from the Palaeozoic to the present (Tucker 1992: *24.4.1*) and other organisms such as bryozoa may have played a role in reef building in the past.

The *reef core* is made up of a framework of coral which forms the crest of the reef structure (Fig. 14.13) (James & Bourque 1992). The depositional facies formed by a reef core is a bioherm of boundstone which is principally bindstone with some regions of framestone *(3.1.4)*. The voids between the coral structures are filled with bioclastic debris and lime mud in sheltered pockets. Breakup of the reef core material by wave and storm action leads to the formation of a talus slope of reefal debris. This *forereef* setting is a region of accumulation of carbonate breccia to form bioclastic rudstone and grainstone facies. As these are gravity deposits formed by material falling down from the reef crest they build out as steeply sloping depositional units inclined at 10–30° to the horizontal. Behind the reef crest, the *backreef* is sheltered from the highest-energy conditions and is the site of deposition of debris removed from the reef core and washed towards the lagoon. A gradation from rudstone to grainstone deposits of broken reef material, shells and occasionally ooids forms a fringe along the margin of the lagoon.

14.7.3 Rimmed carbonate shelf successions

As deposition occurs on the rimmed shelf under conditions of static or slowly rising sea level the whole complex progrades. The reef core builds out over the forereef with backreef to lagoon facies associated with the reef bioherm (Fig. 14.13). Distally the slope deposits of the forereef prograde over deeper water facies comprising pelagic carbonate mud and calcareous turbidite deposits (Fig. 14.14). The steep depositional slope of the forereef creates a clinoform bedding geometry which may be seen in exposures of rimmed shelf carbonates large enough to show several hundred metres of vertical section and even greater lateral extent. This distinctive geometry can also be recognized in seismic reflection profiles of the subsurface *(22.2.2)* (Emery & Myers 1996). Under conditions of sea level fall the reef core may be subaerially exposed and develop karstic weathering. A distinctive surface showing evidence of erosion and solution may be preserved in the stratigraphic succession if subsequent sea level rise results in further carbonate deposition on top.

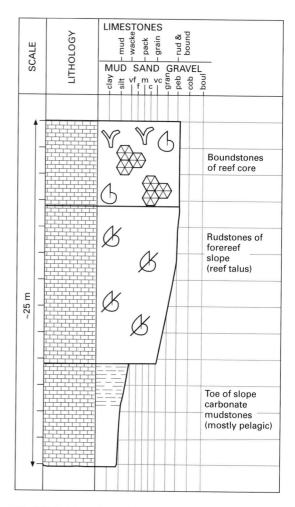

Fig. 14.14 Schematic graphic sedimentary log through a carbonate rimmed shelf succession.

14.8 Epeiric, drowned and isolated carbonate platforms

These three classes of carbonate platform are less common than ramps and rimmed shelves, both in modern seas and in the stratigraphic record. Epeiric platforms may be associated with shelves or ramps but, as their name suggests, isolated platforms are surrounded by deeper water. Drowned platforms form as a result of a rise in sea level (Fig. 14.7).

14.8.1 Epeiric platforms

There are no modern examples of large epicontinental

seas dominated by carbonate sedimentation but facies distributions in limestones in the stratigraphic record indicate that such conditions have existed in the past, particularly during the Jurassic and Cretaceous when large parts of the continents were covered by shallow seas (e.g. much of southern and western Europe) (Tucker & Wright 1990). The water depth across an epeiric platform would be expected to be variable up to a few tens to hundreds of metres. Both tidal and storm processes may be expected, with the latter more significant on platforms with small tidal ranges. Currents in broad shallow seas would build shoals of oolitic and bioclastic debris which could become stabilized into low relief islands. Deposition in intertidal zones around these islands and the margins of the sea would result in the progradation of tidal flats. The facies successions developed in these settings would therefore be cycles displaying a shallowing-upward trend which may be traceable over large areas of the platform.

14.8.2 Isolated platforms

Areas of shallow sea surrounded on all sides by deeper water are almost inevitably sites of carbonate sedimentation because there is no source of terrigenous detritus. They are found in a number of different settings, ranging from small *atolls* above extinct volcanoes in the west Pacific to horst blocks in extensional basins such as the Red Sea and larger areas of shallow seas such as the area around the Bahamas (Wright & Burchette 1996). All sides are exposed to open seas and the distribution of facies on an isolated platform is controlled by the direction of the prevailing wind. The best-developed marginal reef facies occurs on the windward side of the platform which experiences the highest-energy waves. Carbonate sand bodies may also form part of the rim of the platform. The platform interior is a region of low energy where islands of carbonate sand may develop and deposition occurs on tidal flats.

14.8.3 Drowned platforms

It was noted in Section 14.5 that carbonate productivity is related to water depth. If there is an abrupt relative sea level rise due to tectonic subsidence or eustatic sea level changes *(21.1.1)* an area which had formerly been a site of shallow marine carbonate deposition may become too deep for the production of much sediment. This is called the *drowning* of a platform. When this occurs the domin-

ant facies formed is fine-grained pelagic material which is similar in character to deep sea pelagic carbonates *(15.5.1)*. Pelagic carbonate sedimentation is considerably slower than shallow marine accumulation rates, resulting in much thinner layers in a given period of time. Successions deposited under these conditions are known as *condensed sections* and they may have as many millions of years of accumulation in them as a shallow water deposit two or three orders of magnitude thicker (Bernoulli & Jenkyns 1974).

14.9 Barred basins and saline giants

Areas of epicontinental seas with a barrier which restricts connection with the rest of the marine realm are called *barred basins*. They are distinguished from lagoons *(13.5.2)* in that they are relatively deep basins, with water depths of many hundreds of metres. The barrier may be depositional, a reef or bar as in a lagoon, or a structural feature such as an uplifted fault block. If there is a high rate of evaporation from the surface, precipitation of evaporite minerals occurs in the bottom of the basin as a *deep water evaporite* succession. A typical succession would start with normal marine deposits overlain by an organic-rich layer representing the stagnation of the basin; laminated calcium sulphates (often anhydrite) precipitate as the salinity increases, followed by halite as the salinity increases further. Evidence for the deep water origin in ancient examples comes from grading within the beds resulting from the redeposition of evaporite minerals by turbidity currents: gravity-driven flows would not be widespread in shallow basins without a gradient to drive the mass flow.

Two different patterns of deposition can be recognized (Einsele 1992). If the barred basin is completely enclosed the water body will gradually shrink in volume and area. The deposits which result will show a *bull's-eye* pattern with the most soluble salts in the basin centre. In circumstances where there is a more permanent connection a gradient of increasing salinity from the connection with the ocean to the furthest point into the basin will exist (Fig. 14.15). The minerals precipitated at any point across the basin will depend on the salinity at the point and may range from highly soluble sylvite (potassium chloride) at one extreme to carbonates deposited in normal salinities at the other. If an equilibrium is reached between the inflow and the evaporative loss then stable conditions will exist across the basin and tens to hundreds of metres thickness of a single mineral can be

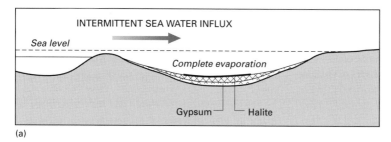

(a)

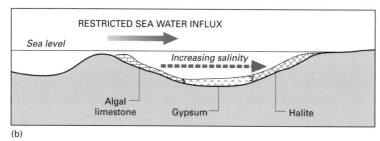

(b)

Fig. 14.15 (a) A barred basin, 'bull's-eye' pattern model of evaporite deposition; (b) a barred basin 'teardrop' pattern model of evaporite deposition.

deposited in one place. This produces a *teardrop* pattern of evaporite basin facies (Fig. 14.15). Disturbances to that equilibrium due to an increase or decrease in the interchange with the open ocean will result in changes in the concentrations of ions across the basin and the minerals deposited. In particular, a global sea level rise will reduce the salinity in the basin and may lead to widespread carbonate deposition. Cycles in the deposits of barred basins may be related to global sea level fluctuations or possibly due to local tectonics affecting the width and depth of the seaway connection to the open ocean.

Around the world today there are no examples of very large, barred evaporitic basins but evidence for seas precipitating evaporite minerals over hundreds of thousands of square kilometres exists in the geological record. These *saline giants* have evaporite sediments over 1000 m thick in them and represent the products of the evaporation of vast quantities of sea water. To produce just a metre bed of halite a column of sea water over 75 m thick must be evaporated. Examples of saline giants are the Permian Zechstein of north-west Europe (Taylor 1990) and the Salina Basin in Michigan which is Silurian–Devonian in age (Nurmi & Friedman 1977).

The mechanism for depositing such huge thicknesses of evaporite minerals relies on a continual replenishment of sea water in an epicontinental basin. Without replenishment the evaporation of the body of water in the basin would only produce a few metres of evaporites. The

volume of water in the basin is maintained by an influx of sea water from an ocean balancing the evaporation from the surface of the sea. The sea water in the basin becomes hypersaline if the connection with the ocean is limited and does not allow the salinity of the basin to balance with that of the oceans by mixing. Epicontinental seas which become evaporitic basins would be expected to have narrow, shallow seaways connecting them to the ocean to restrict circulation and mixing of waters. Flow of water effectively becomes one-way from the ocean into the basin.

Perhaps the most spectacular example of a large evaporitic basin in the geological record is known from the Tertiary history of the Mediterranean (which is not an epicontinental sea, having a basaltic crust floor over much of its area). Deep sea drilling in the western Mediterranean revealed thick evaporite deposits of latest Miocene (Messinian) age in many places (Hsü 1972). These indicate that there was a period when the Mediterranean Sea had very high concentrations of salts and was sufficiently saline for gypsum and halite to form on the sea floor. This is sometimes referred to as the *Messinian salinity crisis*. This raised salinity is assumed to be a consequence of closure of the connection to the Atlantic Ocean. In the Mediterranean Sea today the loss of water from the surface due to evaporation is greater than the supply from rivers and the level of water is maintained by influxes of water from the Atlantic Ocean. If that connection were closed today the water level would gradually fall until it was virtually zero in a matter of thousands of years. It is assumed that this happened in the Messinian, the closure of the connection being due to the northward movement of Africa towards Europe. The situation returned to normal when the barrier was breached to create the present-day Straits of Gibraltar.

14.10 Criteria for the recognition of shelf sediments

The environments of deposition on continental shelves vary according to water depth, sediment supply, climate and the relative importance of wave, tide and storm processes. The products of these interacting processes are extremely variable in terms of facies character, sediment body geometry and stratigraphic succession. There are, however, certain features which can be considered to be reliable indicators of shallow marine environments. First, extensive and diverse carbonate facies only occur in shallow marine settings. Carbonate platforms may be the sites of accumulation of thousands of metres of limestone deposited in a range of depth-sensitive environments over broad shelf areas. Other carbonate environments, such as lakes, have a restricted range of facies, and in deep marine settings deposits are all pelagic carbonate or redeposited limestone. Second, the organisms which make up most of the shallow marine carbonate facies are themselves very diverse and often distinctively shallow water forms. Benthic organisms are only abundant in shelf environments and reef-builders such as corals flourish only in a limited range of conditions. Much of the benthic fauna may also be found in clastic shelf deposits, either as body fossils, or as trace fossils which are distinctive of shallow marine conditions. Third, the physical processes of the shelf leave their mark in both carbonate and clastic shallow marine deposits. For example, extensive sheets and ridges of cross bedded sand deposited by strong currents are easily recognized and cannot be the deposits of any other environment, especially if there is evidence that the currents were tidal. Many other carbonate and clastic shelf deposits have a sheet geometry, including storm deposits which may additionally show hummocky cross stratifi-

cation as an indicator of the process of deposition. None of the above features is common in evaporitic shelf facies, but the salts in these deposits are distinctive for their chemistry, reflecting deposition from sea water, and the huge thicknesses of evaporite lithologies are characteristic of barred basins.

Further reading

Dalrymple, R.W. (1992) Tidal depositional systems. In: *Facies Models: Response to Sea Level Change* (eds R.G. Walker & N.P. James), pp. 195–218. Geological Association of Canada, St Johns, Newfoundland.
James, N.P. & Bourque, P.-A. (1992) Reefs and mounds. In: *Facies Models: Response to Sea Level Change* (eds R.G. Walker & N.P. James), pp. 323–348. Geological Association of Canada, St Johns, Newfoundland.
Johnson, H.D. & Baldwin, C.T. (1996) Shallow clastic seas. In: *Sedimentary Environments: Processes, Facies and Stratigraphy* (ed. H.G. Reading), pp. 232–280. Blackwell Science, Oxford.
Jones, B. & Desrochers, A. (1992) Shallow platform carbonates. In: *Facies Models: Response to Sea Level Change* (eds R.G. Walker & N.P. James), pp. 277–302. Geological Association of Canada, St Johns, Newfoundland.
Scholle, P.A., Bebout, D.G. & Moore, C.H. (eds) (1983) *Carbonate Depositional Environments*, 708 pp. American Association of Petroleum Geologists Memoir **33**.
Stride, A.H. (ed.) (1982) *Offshore Tidal Sands: Processes and Deposits*, 222 pp. Chapman & Hall, London.
Tillman, R.W., Swift, D.J.P. & Walker, R.G. (eds) (1985) *Shelf Sands and Sandstone Reservoirs*, 708 pp. Society of Economic Paleontologists and Mineralogists Short Course Notes **13**.
Tucker, M.E. & Wright, V.P. (1990) *Carbonate Sedimentology*, 482 pp. Blackwell Scientific Publications, Oxford.
Walker, R.G. & Plint, A.G. (1992) Wave- and storm-dominated shallow marine systems. In: *Facies Models: Response to Sea Level Change* (eds R.G. Walker & N.P. James), pp. 219–238. Geological Association of Canada, St Johns, Newfoundland.
Wilson, J.L. (1975) *Carbonate Facies in Geological History*, 471 pp. Springer-Verlag, Heidelberg.
Wright, V.P. & Burchette, T.P. (1996) Shallow-water carbonate environments. In: *Sedimentary Environments: Processes, Facies and Stratigraphy* (ed. H.G. Reading), pp. 325–394. Blackwell Science, Oxford.

15   Deep marine environments

The deep oceans are the least understood parts of the Earth. The surface waters are rich in life, but below the photic zone organisms are rarer and on the deep sea floor life is relatively sparse, apart from strange creatures around volcanic hot springs. Organisms which live floating or swimming in the oceans provide a source of sediment in the form of their shells and skeletons when they die. These sources of pelagic detritus are present throughout the oceans, varying in quantity according to the surface climate and related biogenic productivity. Around the edges of ocean basins sediment shed from the continental shelves or generated from volcanically active margins may reach hundreds of kilometres out into the basin. This material is carried by turbidity currents and debris flows down the continental slope and out on to the ocean floor to form aprons and fans of deposits. Towards the basin centre terrigenous clastic detritus is limited to wind-blown dust, including volcanic ash and fine particulate matter held in temporary suspension in ocean currents. Despite their huge capacity for the accumulation of sediment, ocean basins only develop thick successions at margins adjacent to terrigenous sources. The centres of ocean basins are sites of very slow deposition of biogenic and fine-grained material falling from the surface of the water body. These deposits are preserved in the stratigraphic record where an ocean basin has closed and the floor has been incorporated into an orogenic belt.

15.1 Modern and ancient oceans

Around 70% of the area on the globe is occupied by ocean basins floored by basaltic oceanic crust. As a result of the creation of oceanic crust at spreading ridges and destruction of it at subduction zones, oceanic crust is mostly much younger than continental crust: the oldest sea floor is Jurassic in age in the north-west Pacific. Because of these plate tectonic processes, ocean basins are continually changing in size and shape. As continental plates change position new oceans form and others are destroyed. Remnants of ancient oceans are found in orogenic belts where two continental plates have collided, closing the ocean in between. For example, along the line of the Alps there is evidence for an ocean which existed between the southern plates of Africa and India and the northern plates of Europe and Asia. This ocean existed in the Mesozoic before the Alpine Orogeny and is known as the *Tethys Ocean*.

15.1.1 Bathymetry and structure

The edge of the continental shelf makes a natural break between 'shallow' and 'deep' marine environments (see Fig. 11.1). Bathymetrically it marks a change in slope from the very gently sloping sea floor of the shelf to the steeper continental slope which leads down to the continental rise and the ocean floor. The structure and composition of the crust at this point is responsible for this profile. In *passive margins*, where there is no plate boundary between the continent and the ocean, the shelf is underlain by continental crust, which may be of normal thickness or somewhat thinned by the extension which has occurred in the development of an ocean basin. At the edge of the shelf, stretching of the continental crust has been more extensive. The crust is broken up by faults and is compositionally transitional between a continental and oceanic character. This zone of transition underlies the continental slope.

The ocean floor is underlain by oceanic crust, formed by sea floor spreading at mid-ocean ridges. Spreading centres are typically at 2000–2500 m depth in the oceans. Along them the crust is actively forming by the injection of basic magmas from below to form dykes as the molten rock solidifies, and the extrusion of basaltic lava at the surface in the form of pillows (*pillow lavas*: Fig. 15.1). This igneous activity within the crust makes it relatively hot. As further injection occurs and new crust is formed, previously formed material gradually moves away from the spreading centre. As it does so it cools and contracts and consequently the density increases. Older, denser oceanic crust is less buoyant on the mantle below, and it sinks relative to the younger, hotter crust at the spreading centre. A profile of increasing water depth away from the mid-ocean ridge results, down to a typical bathymetry of 4000–5000 m where the crust is more than a few tens of millions of years old.

Not all continental margins are passive margins. Convergent plate boundaries between an over-riding continental plate and an oceanic plate are known as *active margins*. Sites where a slab of oceanic crust is being subducted beneath another piece of oceanic or continental crust are *subduction zones* and their surface expressions form the deepest parts of oceans. The downgoing slab flexes downwards as it enters the subduction zone, creating a trough which may be over 8000 m deep (the Marianas Trench in the west Pacific is 10 000 m deep). Ocean trenches are important sites of sediment accumulation as they act as traps for sediment shed from an adjacent continental margin.

With mid-ocean ridges 2000 m above the mean bathymetry of the ocean floor and trenches 4000 m or more deeper, the relief on the sea floors is greater than that on the land surface. Additional features on the sea floor include *seamounts*, underwater volcanoes located over isolated hot-spots. Seamounts may be wholly submarine or may build up above water as volcanic islands, such as the Hawaiian island chain in the central Pacific.

15.1.2 Continental slopes and continental rises

The vertical exaggeration on diagrams which show profiles of the crust from the continent to the ocean often gives the impression of a precipitous slope down from the edge of the continental shelf. In fact, continental slopes typically only have slope angles of between 2° and 10° and the continental rise is even less. Nevertheless, they are physiographically significant, as they contrast with the very low gradients (*c.* 0.05°) of continental shelves and the flat ocean floor. Slopes of more than a couple of degrees are steep in terms of sediment stability, and unconsolidated deposits will easily move as mass flows or slumps in this setting. Continental slopes extend from the shelf edge, about 200 m below sea level, to the basin floor at 4000–5000 m depth and may be up to 100 km across in a downslope direction.

The nature and volume of sediment reaching the continental slope and rise depends on the width of the shelf and availability of material on it. In narrow shelf settings currents may transport sand or other bedload material to the edge where it may be deposited down the slope as *sand spill-over deposits* (Fig. 15.2) (Stow 1986). Sands deposited in this setting will be sheet-like bodies with little internal organization, interbedded with other slope deposits. Slopes at the edges of wide shelves will receive little, if any, coarse detritus. Pelagic and hemipelagic sediments, consisting of fine-grained terrigenous clays and biogenic material from planktonic organisms, will be the principal deposits.

Instability on the continental slope leads to the formation of slump scars (*17.1.1*), surface faults on the slope where masses of semi-consolidated sediment have moved by gravity downslope. These may remain as coherent displaced masses, *submarine slides*, or may show various degrees of breakup. A completely disaggregated mass of material will form a dense mixture of sediment and water which will move as a debris flow (*4.6.1*) down the slope. The composition of these debris flows will be determined by the nature of the material deposited on the upper parts of the slope, typically reworked fine-grained deposits in varying states of lithification with or without sand from spill-over deposits.

Fig. 15.1 Basaltic pillow lavas formed on the floor of an ocean, Upper Cretaceous, Oman.

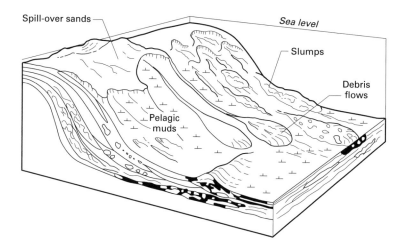

Fig. 15.2 Slope apron deposits. (After Stow 1986.)

The association of sedimentary units on a continental slope would therefore be predicted to be a mixture of fine-grained pelagic and hemipelagic deposits *(15.5)*, structureless sands, as well as debris flows and slumped strata. Such associations have been observed in the present day and are called *slope aprons* (Stow 1985), but there are few examples from the stratigraphic record. At the edges of carbonate platforms *carbonate slopes* may develop at angles ranging from a few degrees to slopes in excess of 30° (Coniglio & Dix 1992; Wright & Burchette 1996). The steeper slopes are sites of slumping and redeposition of material by debris flows; at the base of the slope an apron of carbonate turbidites forms.

An important morphological feature of continental slopes and rises is that at intervals they are dissected by steep-sided canyons which cut back into the shelf. These canyons are important conduits for the flow of material from the shelf to the ocean basin. Much of the material carried out beyond the shelf through the canyons by-passes the slope, leaving it relatively starved of sediment. These by-pass canyons act as the feeders for systems which deposit coarse sediment in the deep ocean basins.

15.1.3 Ocean floors

The only places in ocean basins where thick accumulations of sediment occur are around the margins. In the centres of ocean basins the main sources of sediment are from pelagic material *(15.5)*. Sediment bodies made up of mass flow deposits (turbidity currents and debris flows) which have travelled from the continental shelf do

not reach the more distal parts of large ocean basins and are concentrated within a few hundred kilometres of the margin. Volumetrically the most important coarse-grained sediments in ocean basins are deposited by turbidity currents *(4.6.2)* derived from the basin margins.

15.2 Deep marine mass flows

As noted in Section 4.6.2, turbidity currents can occur in a variety of environments (e.g. lakes) but they have been most frequently documented in relatively deep basins, including deeper parts of the world's oceans. Turbidites are abundant in sedimentary successions interpreted as deep water facies (Mutti 1992). Vast thicknesses of turbidite facies are recorded in the fills of sedimentary basins in almost all plate tectonic settings, most notably in peripheral foreland basins *(23.4.1)* and accretionary complexes formed of ocean trench deposits *(23.3.1)*.

In addition to turbidity currents, denser flows occur transporting material from the edge of the shelf and the continental slope as debris flows. The deposits of these flows may occur associated with turbidite deposits or localized at the base of the continental slope.

15.2.1 Composition of marine turbidites

The process of transport and deposition by a turbidity current is a rapid one and often there is little sorting of material except into the overall normal grading of the Bouma sequence. Sands deposited by turbidity currents are commonly rather muddy because the fine-grained material is partly entrained with the coarser sediment during deposition. They are therefore often classified as wackes in the Pettijohn classification *(2.4.7)*.

The composition of the sand- and mud-grade material will depend upon the source area. If the sediment is 'first cycle', derived directly from metamorphic and igneous bedrock *(2.8.2)*, there will be little separation of different grain types by the processes of transport and deposition by a turbidity current. In these cases the sandstone would be classified as a lithic wacke although it is also often called a greywacke *(2.4.7)*. Although many greywackes may be deposited by turbidity currents and many turbidites are compositionally greywackes the two terms are not interchangeable as 'greywacke' is a compositional term and 'turbidites' are a type of deposit.

In basins adjacent to volcanic provinces such as arc-related basins *(23.3)* reworked volcaniclastic material may make up all or part of the deposits. Offshore from continental margins which are supplied by the reworking of older sedimentary rocks the coarser parts of turbidites may be quartz arenites or quartz wackes, reflecting the mature composition of the margin sediments. Continental shelves dominated by carbonate deposition can also be the source areas for turbidites: *calcareous turbidites* may show the same textural characteristics as those of terrigenous clastic material with all the features of a Bouma sequence present. The grains are whole or fragmentary biogenic debris and the fine material will be a carbonate mud. Shallow marine fauna can be redeposited in deep marine environments by turbidites so recognition of the process of deposition is important in making interpretations of the depositional environment from fossil evidence in these cases.

15.2.2 Turbidity currents in modern oceans

Our understanding of the deposits on modern ocean floors and the processes which form them has improved as techniques for investigating the sea floor have become more refined. Deposits representing single turbidity current events have been mapped out—for example, in the east Atlantic off the Canary Islands a single turbidite deposit has been shown to have a volume of $125\,km^3$ (Masson 1994). Turbidity currents carrying these volumes of sediment are thought to have a flow velocity of $10–15\,m\,s^{-1}$ and occur at intervals of 25 000 years.

Direct observation of turbidity currents in natural settings is very difficult given where and how infrequently they occur. The effects of turbidity currents have, however, been monitored, the best-known instance being off the coast of Newfoundland in 1929. On 18 November 1929 at 20.32 an earthquake was recorded in the Grand Banks area. In the following hours, telegraph cables on the sea floor were cut in a sequence starting with those higher up the slope and ending with those further down. Little was known about deep ocean processes at this time, but when these data were considered decades later (Heezen & Drake 1964) the interpretation favoured was that the earthquake had triggered a flow of water and sediment down the continental slope. As the flow advanced it cut the telegraph cables, providing a unique opportunity to plot the course and velocity of a turbidity current. Interpretation of the data indicates that in the upper part of its path, the turbidity current travelled at $100\,km\,h^{-1}$ ($27.7\,m\,s^{-1}$), slowing to around $20\,km\,h^{-1}$ (c $5.5\,m\,s^{-1}$) 500 km away from the source (Fig. 15.3).

15.2.3 Debris flows

The collapse of a pile of sediment on the continental slope can result in remobilization of the material as a debris flow. The thickness and extent of a debris flow deposit in an ocean basin will be governed by the volume of material involved and the velocity the flow gains as it moves downslope. Unlike a debris flow on land, an underwater flow has the opportunity to mix with water and in so doing it becomes more dilute. This can lead to a change in the flow mechanism *(4.6)*. The top surface of a submarine debris flow deposit will typically grade up into finer deposits due to dilution of the upper part of the flow.

There are both modern and ancient examples of an association between debris flow deposits and turbidites. Images of the floor of the eastern Atlantic off north-west Africa show the patterns of debris flows which have closely associated turbidites (Masson *et al.* 1992). Thick debris flow units immediately overlain by turbidites are also known from the stratigraphic record. Many of these are considered to be genetically related, with the two flows having been triggered by the same event. In the Eocene deposits of the southern Pyrenees of Spain there are over 20 separate units consisting of debris flow deposits tens of metres thick each with a turbidite deposit on top (Johns *et al.* 1981). These are called *megabeds* and there are similar examples of very thick redeposited beds in the Tertiary of the North Sea (Pauley 1995). The volume of material involved in the megabeds indicates that the events were very large and catastrophic.

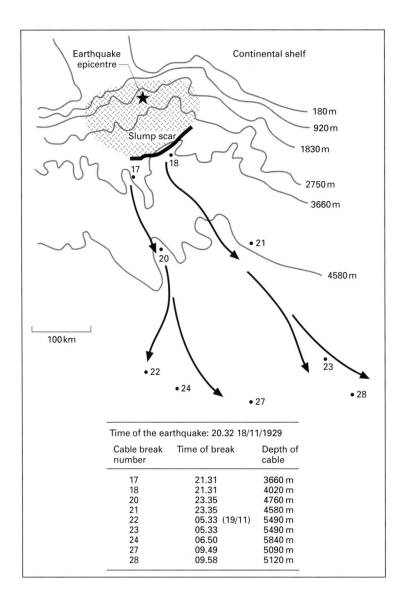

Time of the earthquake: 20.32 18/11/1929

Cable break number	Time of break	Depth of cable
17	21.31	3660 m
18	21.31	4020 m
20	23.35	4760 m
21	23.35	4580 m
22	05.33 (19/11)	5490 m
23	05.33	5490 m
24	06.50	5840 m
27	09.49	5090 m
28	09.58	5120 m

Fig. 15.3 Pathways of turbidity currents and the times of break recorded for a number of cables on the sea floor, Grand Banks, Newfoundland. (Data from Heezen & Drake 1964.)

15.2.4 Causes of mass flows in ocean basins

Turbidity currents and debris flows are usually considered by geologists to be instantaneous events and require some form of trigger to initiate the flow. Mass flows result from the mobilization of masses of unconsolidated or semi-consolidated sediment which mixes with water and flows as a dense mixture down a gradient. In order to set the material in motion some form of pulse of energy is probably needed. This may be provided by an earthquake: the shaking generated by a seismic shock can temporarily liquefy loose sediment and reduce its rigidity such that it starts to move. The impact of an ocean storm on shelf sediments may also act as a trigger for turbidites as the energy from waves is higher in shallow water and reaches deeper parts of the shelf. Initiation of a turbidity flow from a delta front may also follow a period of heavy rain in the drainage area of the feeder river: the increased supply of sediment to the delta caused by the rainstorm may result in an instability of the material accumulating at the delta front. Large rivers with high suspended sediment loads may supply turbidity flows which flow almost continuously for long periods of time. A situation may also be envisaged where the pile of sediment on a shelf gradually builds up until it reaches a critical unstable mass and volume. If very unstable the trigger required may be very small yet result in the movement of a large amount of material. Mass flows may also be triggered by falls in sea level.

15.3 Submarine fans

Canyons in the shelf edge localize the flow of water and sediment to the edges of ocean basins, creating a series of point sources from which flows spread out on to the ocean floor. This pattern is analogous to valleys on land which feed discrete alluvial fans at a mountain front *(8.4)*. In the submarine setting, a fan morphology is also recognized, radiating away from the feeder canyon to form a broad cone building out into the ocean basin. These morphological features are known as *submarine fans* (Fig. 15.4) (Normark 1978). Submarine fans are very variable in size, ranging from a few kilometres to over 2000 km across (Stow 1985). Although many show a true fan shape, others have a more elongate, lobate form.

Sedimentation on a submarine fan occurs mainly from turbidity currents carrying sediment from the shelf through the feeder canyon and depositing and decelerating across the fan surface. This creates a general trend from coarser material deposited near the apex to progressively finer sediment as the turbidity currents decelerate further out in the more distal parts of the fan. In most instances, this general pattern is substantially modified by localization of transport and deposition in channels and on discrete lobes. Modern submarine fans and ancient turbidites can be divided into proximal, medial and distal portions (Fig. 15.4), although the divisions between these three portions are not sharp.

It is worth emphasizing the distinction between submarine fans and other depositional geomorphological units also referred to as 'fans'. Fan deltas *(12.3)* may be largely submarine, but the term 'submarine fan' is normally restricted to deposits in deep marine environments where turbidity currents build up a fan-shaped body of sediment tens to hundreds of kilometres in radius. They are therefore large-scale features whose morphology can be discerned in modern oceans, but in the stratigraphic record it is usually the arrangement of turbidites in lateral and vertical patterns that leads to their interpretation as the deposits of a submarine fan.

15.3.1 Proximal (upper) fan characteristics

Turbidity currents issuing from the mouth of the feeder canyon retain their confined flow characteristics by remaining within a channel form in the upper, proximal part of the submarine fan. Upper fan channels are typically metres to tens of metres deep and hundreds of metres wide, although the deposits of larger submarine fan channels have been reported with thicknesses of up to 170 m and 20 km across (Macdonald & Butterworth 1990). At relatively high velocities maintained during the main part of the flow, turbidity currents deposit the very coarse sands and gravel in the channel, to form thick, structureless or crudely graded beds (Fig. 15.5).

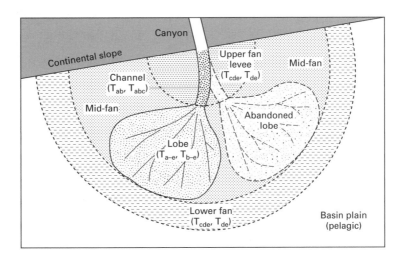

Fig. 15.4 Depositional environments on a submarine fan. (After Normark 1978, AAPG ©1978, reprinted by permission of the American Association of Petroleum Geologists, modified from Normark 1970.)

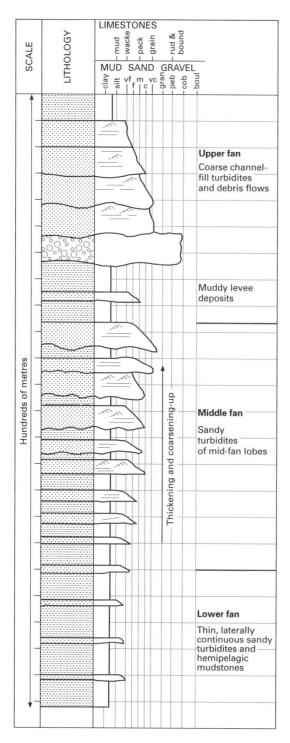

Fig. 15.5 Schematic graphic sedimentary log through submarine fan deposits: proximal, mid-fan lobe deposits and lower fan deposits.

These are mostly Bouma sequence 'a' division deposits (T_a). Finer-grained deposits may be deposited but subsequently eroded by later turbidity currents. The lateral extent of these coarse beds is limited by the width of the channel.

Most of an individual turbidity flow is confined to the channel in the proximal fan, but the upper, more dilute part of the flow may spill out of the channel laterally. This is analogous to a channel and overbank setting familiar from fluvial environments. The similarities to a river system are even more striking when some modern submarine fan channels are viewed in plan as they are seen to follow a strongly sinuous course which looks like a meandering river pattern. The overbank flow in the proximal part of a submarine fan contains fine sand, silt and mud from the upper part of the main channel flow and, as it is not confined, it is free to spread out as a fine-grained turbidity current away from the channel. The deposit will be characterized by the upper part of a Bouma sequence (T_{cde} or T_{de}). Palaeoflow indicators may show directions perpendicular to that of the coarse, channelized turbidite deposit.

An analogy can also be drawn with river behaviour in the way submarine fan channel deposits are preserved. The channel may either migrate laterally in a similar way to a meandering river or avulse to leave its former position abandoned. Typical upper fan deposits will comprise lenticular bodies made up of stacked coarse-grained turbidites which represent flow confined to the channel. The channel may be incised into other channel-fill units or into fine-grained turbidites which are also found as the laterally equivalent sediments. It should be noted that, although these are the most proximal deposits of a submarine fan system, the sediments are not all coarse and a substantial proportion of the upper fan comprises fine-grained 'overbank' sediments.

15.3.2 Medial (mid) fan characteristics

In the mid-fan area turbidity currents spread out from the confines of the upper fan channel. A *lobe* of turbidite deposits is built up which occupies a portion of the fan surface (Fig. 15.4). An individual lobe is constructed by a succession of turbidity currents which tend to deposit further and further out on the lobe through time. A simple progradational geometry results if the fan deposition is very ordered, with each turbidity current event of approximately the same magnitude and each depositing progressively further from the mouth of the channel. Under most circumstances turbidity currents are of varying magnitude and tend to be influenced by the topography created by the preceding deposits, leading to a more complex pattern of lobe deposits. As the lobe builds out the flow in the part closest to the upper fan tends to become channelized. Lobe progradation continues until the turbidity current pathway avulses to another part of the mid-fan surface. Avulsion occurs because an individual lobe will start to build up above the surrounding fan surface, and eventually flows start to follow the slightly steeper gradient on to a lower area of the fan surface.

The stratigraphy built up by submarine fan lobe progradation is ideally a coarsening-up succession capped by a channelized unit (Fig. 15.5). Individual turbidites will commonly be normally graded but as the lobe progrades currents will carry coarser sediment further out on the fan surface. Successive deposits therefore should contain coarser sediment and hence generate an overall coarsening-upward pattern. An upward thickening of the beds should accompany the coarsening-upward pattern (Fig. 15.6). Commonly this overall upward coarsening and upward thickening is not seen because of the complex, often random, pattern of deposition on mid-fan lobes (Anderton 1995). Therefore there may not be any consistent vertical pattern of beds deposited on a submarine fan lobe. The channel at the top of the lobe will have an erosive base and contain turbidites which may follow a fining-upward pattern as the lobe is abandoned in favour of another lobe building up elsewhere.

Mid-fan lobe deposits often contain the most complete Bouma sequences (T_{a-e} and T_{b-e}) and the whole lobe succession may be tens of metres thick. An individual lobe may be kilometres or tens of kilometres across. Other areas of the mid-fan will receive only fine-grained turbidites and hemipelagic sedimentation whilst an individual lobe is active. Lobes will be stacked both vertically and laterally against each other, although the lateral limits of a lobe are likely to be identified only by mapping out the submarine fan succession on the surface or in the subsurface over large areas. Such mapping has also revealed the existence of individual sheets deposited by turbidity currents which have not been restricted to deposition on a lobe but have spread out over a larger area of the fan.

15.3.3 Distal (lower) fan characteristics

Turbidity currents which flow along the upper fan channel and across a mid-fan lobe may eventually reach the lower fan (Fig. 15.4). By the time they have reached this outer area they have deposited most of their coarser sediment and only fine sand, silt and clay will be deposited on the lower fan. Thin, fine-grained turbidites characterized by Bouma divisions T_{cde} and T_{de} typify the distal fan succession with little or no organization into patterns or trends in grain size and bed thickness (Fig. 15.5). Hemipelagic sediments *(15.5)* are proportionally more significant in the lower fan area as the volume of turbidites decreases distally.

15.3.4 Muddy and gravelly submarine fans

The submarine fans first described by Normark (1978)

Fig. 15.6 A succession of sandy and muddy turbidite beds deposited on the middle part of a submarine fan complex, Oligocene, western Greece.

and discussed above are characterized by deposition of sand, or mixtures of sand and mud. However, not all submarine fans show these same characteristics due to differences in the grain size of the sediment supplied to the fan. The largest submarine fan systems in modern oceans are mud-rich (Stow *et al.* 1996), and are fed by very large rivers. These large mud-rich fans include the Bengal Fan fed by the Ganges and Brahmaputra rivers and the large submarine fan beyond the mouth of the Mississippi. Channels are well developed on the upper and medial parts of these fans, often with distinctly sinuous courses and high levees: fan lobes are dominated by muddy turbidites with low proportions of sand. At the other end of the grain size spectrum gravel-rich submarine fans typically lie at the toe of coarse-grained deltas *(12.3)*. Submarine fans in these settings are relatively steep, normally less than 10 km in radius, and in addition to coarse turbidites, debris flows and slump deposits commonly form part of the fan deposits (Stow *et al.* 1996).

15.4 Contourites

The importance of ocean bottom or geostrophic currents *(11.4)* in the transport and deposition of sediment was not recognized until relatively recently (Heezen *et al.* 1966). These currents flow along the sea floor parallel to, or nearly parallel to, the contours of the continental margin, giving rise to the general name *contourites* for the deposits of bottom currents. Contourites were first recognized from cores drilled into the floors of modern oceans. They consist of material of clay, silt and fine sand grade in centimetre-scale beds which show parallel, wavy and some cross lamination (Fig. 15.7) (Stow 1979; Stow & Lovell 1979). The origin of the sediment is mainly turbidites deposited elsewhere in the ocean which have been winnowed by the bottom currents. The turbidite sands are modified by removal and redeposition of the finer-grained material. Large amounts of sands may be reworked and redeposited by contour currents. Despite the common occurrence of contourites in modern oceans, ancient contourites in the stratigraphic record are relatively poorly known (Bouma 1972). It is quite likely that many units interpreted as fine- to medium-grained turbidites might be better interpreted as contourites and that modification to turbidity current deposits by bottom currents has been overlooked.

Criteria which may be useful in distinguishing turbidites and contourites in deep water sediments

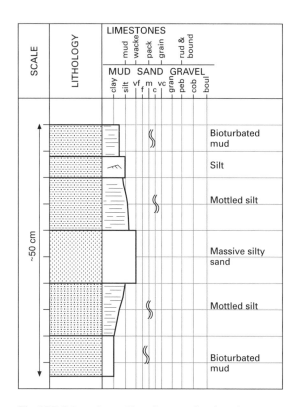

Fig. 15.7 Schematic graphic sedimentary log through a contourite.

include palaeocurrent data and provenance characteristics. Unfortunately, it is often difficult to demonstrate the direction of flow in fine-grained deposits, but differences in composition between units clearly deposited as turbidites and other fine-grained deposits may be a reasonable indicator of two different sources and therefore two different currents acting in the basin.

15.5 Pelagic and hemipelagic sedimentation

Fine-grained terrigenous and volcanic material which is temporarily held in suspension in the ocean waters is referred to as *hemipelagic* (cf. pelagic sediments which are formed from biogenic material which floats but sinks when the organism dies). This material is mainly clay and fine silt mixed with pelagic carbonate or siliceous sediment (Stow *et al.* 1996). It is brought into the ocean waters from terrestrial and volcanic sources by turbidity currents, storm currents and wind: aeolian dust makes up 50% of hemipelagic and pelagic sediments. It settles out to form mud which may be lithified into a shale or mudstone. Black shales *(11.6.3)* form in deep marine environments where there is an abundant supply of organic material and/or anoxic (poorly oxygenated) bottom water (Demaison & Moore 1980). The rate of hemipelagic mud deposition is normally slow, a few millimetres per thousand years, although the rate can be elevated in regions of high sediment input, such as offshore of major river deltas. Rates of hemipelagic sedimentation are lowest in open ocean areas, distant from any coastlines.

Pelagic sediments are composed of material which has at some stage been floating in the water (Jenkyns & Hsü 1974). Most material in this category is biogenic in origin being the skeletons and shells of organisms which had a free-swimming (*nektonic*) or free-floating (*planktonic*) lifestyle. Pelagic sediments are therefore mostly made up of biogenic calcareous or siliceous material and are mainly fine-grained. Not all pelagic material need be small particles because the remains of larger marine organisms may also find their way into open ocean settings: this could include anything up to the size of a whale. Coarse terrigenous detritus can be deposited in deep, distal ocean settings if carried by icebergs floating in the ocean *(7.3.5)*. Such dropstones are most common in high-latitude oceans near to ice caps but large icebergs can travel thousands of kilometres into more temperate regions before finally melting and depositing any sediment entrained in the ice.

15.5.1 Pelagic carbonate sediments

The surface waters of the oceans are rich in life. At the bottom of the food chain are algae on which both invertebrate and vertebrate animals feed which are in turn food for carnivorous animals. In open ocean areas all organisms either float (planktonic) or are free-swimming (nektonic). Although the larger animals are more obvious it is the small planktonic organisms that are the most abundant. Algae belonging to the group Chrysophyta include coccoliths which have spherical bodies of calcium carbonate a few tens of micrometres across *(3.1.2)*; organisms this size are commonly referred to as nannoplankton. Slightly larger are Foraminifera, a group of single-celled animals which includes a planktonic form with a calcareous shell about a millimetre or a fraction of a millimetre across. Vast quantities of these organisms have lived in warm seas since the Mesozoic and their remains contribute to soft calcareous muds on

the sea floor, *nannoplankton ooze* or *foraminiferal ooze*. A lithified calcareous ooze can be classified in a number of ways as a biomicrite, a calcareous mudstone or wackestone but is commonly called a pelagic limestone if it is known to have formed in an open ocean environment (Tucker & Wright 1990).

The solubility of calcium carbonate is partly dependent on pressure as well as temperature. At higher pressures and lower temperatures the amount of calcium carbonate which can be dissolved in a given mass of water increases. In oceans the pressure becomes greater with depth of water and the temperature drops so the solubility of calcium carbonate also increases. Near the surface most ocean waters are near to saturation with respect to calcium carbonate: animals and plants are able to extract it from sea water and precipitate either aragonite or calcite in shells and skeletons *(3.1.2)*.

When planktonic or nektonic organisms die their hard parts start to settle through the water column. Calcium carbonate in the form of calcite starts to dissolve at depths of around 3000 m and in most modern ocean areas will have been completely dissolved once depths of around 4000 m are reached (Fig. 15.8). This is the *calcite compensation depth* (CCD). It should be noted, however, that there are a number of variables which influence the dissolution of carbonates. Aragonite is more soluble than calcite and an aragonite compensation depth can be defined at a higher level in the water column than a calcite compensation depth (Scholle *et al.* 1983). The size and organic structure of the calcium carbonate particles also influence the solubility, with dissolution occurring more readily if there is a higher surface area: small pieces dissolve more quickly.

The CCD is not at a constant level throughout the world's oceans today and there is evidence that it varied in the past. The capacity for sea water to dissolve calcium carbonate depends on the amount that is already in solution. If a large amount of carbonate is introduced into the water body it becomes saturated with calcium carbonate to greater depths as higher pressures are required to put the excess of ions into solution. More carbonate is supplied from the upper parts of the water column in areas of higher biogenic productivity. The level of the CCD is therefore lower, deeper in the ocean, in tropical areas where organisms flourish and their skeletons provide a denser rain of particles down through the water column. The depth of the CCD is also known to vary with the temperature of the water and the degree of deep water circulation that is present.

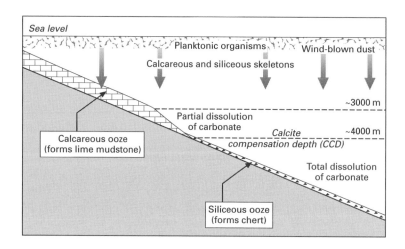

Fig. 15.8 Sources of material and controls on sedimentation in deep oceans.

Carbonate can be deposited at depths below the CCD if it is introduced by a mechanism other than settling through the water column. Carbonate brought into deep water by mass flows (turbidity currents and debris flows) will pass through the CCD quickly and will be deposited rapidly. The top of a calcareous turbidite may later start to dissolve at the sea floor, but the waters close to the sea floor will soon become saturated with the mineral and little dissolution of a calcareous turbidite deposit occurs.

15.5.2 Siliceous pelagic sediments

Below the CCD the absence of organic calcareous remains radically changes the composition of pelagic sediments. Two main groups of organisms, *radiolarians* and *diatoms*, have a planktonic lifestyle and hard parts of opaline silica. Both are relatively common in the oceans but are much less abundant than calcareous plankton such as foraminifera and coccoliths. Above the CCD the remains of siliceous organisms are swamped out by the carbonate material; below the CCD the skeletons of radiolaria can form the main biogenic component of a pelagic sediment (Stow *et al.* 1996). Individual organisms are typically a few tens to hundreds of micrometres across and under an electron microscope they can be seen to have an exquisite structure. High concentrations of siliceous organisms need not always indicate deep waters. The cold waters of polar regions may favour diatoms over calcareous plankton and in pre-Mesozoic strata foraminifera and calcareous nannoplankton are not present.

At water depths of around 6 km the opaline silica which makes up radiolarians and diatoms is subject to

dissolution because of the pressure: an *opal compensation depth* (or silica compensation depth) therefore can be recognized at the depth below which biogenic silica is not deposited. Particles of biogenic silica start to undergo dissolution as this depth is approached.

Accumulations of unconsolidated siliceous remains are called *siliceous ooze*. As they consolidate a hard siliceous rock (*chert*) is formed. Deep sea cherts are thin-bedded in layers a few centimetres thick (Fig. 15.9). They are dark grey, brown or red depending on the fine-grained clastic material they contain. White spots smaller than a pin-head on the surface of the rock may be visible and are due to the presence of radiolaria. In thin section these white spots are seen to be patches of fibres of chalcedony formed by the recrystallization of the

Fig. 15.9 Thin-bedded chert beds, Triassic, Nizwa, Oman: the deep water origin of these beds is indicated by the abundance of radiolaria in the chert.

silica which originally made up the radiolarian test. Diatoms are less commonly preserved.

A distinction can be drawn between chert beds formed from the lithification of a siliceous ooze deposited in deep water (*primary chert*) and a chert formed as nodules due to a diagenetic silicification of a rock (*secondary chert: 17.3.1*). Primary cherts have distinct thin beds, and white spots due to radiolaria may be seen; they may be associated with pillow basalts (*15.1.1*). Secondary cherts are developed in a host sediment (usually limestone) and have an irregular nodular shape. Nodular cherts do not provide information about the depositional environment but may be important indicators of the diagenetic history (Chapter 17).

Primary cherts always have small amounts of clay present in them as an impurity. With increasing amounts of clay in the original siliceous ooze it becomes a mud which is a *siliceous mudstone* upon lithification. Cherts and siliceous (cherty) mudstones may be interbedded, reflecting variations in the amount of clay washed out into the ocean basin. The colour of a chert or siliceous mudstone is controlled by the type and amount of fine-grained material present: a fine dust of iron oxides can make a deep water mudstone a strong red-brown colour, for example.

15.6 Other features of deep ocean sediments

Some areas of modern ocean floors are scattered with *manganese nodules*, concentrations of ferromanganese oxide which are forming in the present day. Similar nodules, ranging from a millimetre to centimetres across, are found in ancient deep water deposits. They form as precipitates of iron and manganese from sea water and the pore waters of sea floor muds, sometimes developed around a core of a rock fragment.

The volcanic activity at spreading centres and isolated seamounts is responsible for *hydrothermal deposits* precipitated from water heated by the magmas close to the surface (Oberhänsli & Stoffers 1988). Sea water circulates through the upper layers of the crust and at elevated temperatures it dissolves anions and cations from the igneous rocks. Upon reaching the sea floor, the water cools and precipitates minerals to form deposits localized around the hydrothermal vents. A wide variety of minerals may be present, but the commonest are base metal sulphides. These metalliferous deposits are potentially valuable ores, more readily accessible when

preserved in rocks, such as ophiolite suites (*23.2.6*), which originated as sea floor deposits now exposed on land, than in their original sites of formation on the ocean floor.

Pelagic limestones, cherts and siliceous mudstones can occur intimately associated with pillow lavas. Eruption of lava at the sea floor involves hot molten rock coming up through and moving sideways over sea floor oozes and muds. Sediment is trapped between the individual pillows where it is slightly metamorphosed by the heat of the lava and silicified by the hot, silica-rich water associated with the lava. *Inter-pillow sediment* occurs as cherts, cherty mudstones and limestones in small triangular areas between the rounded edges of the pillows (Fig. 15.10).

15.7 Fossils in deep ocean sediments

The most abundant fossils in Mesozoic and Tertiary deep ocean deposits are the skeletons of planktonic microscopic organisms such as foraminifera, coccoliths and radiolaria, but other organic remains may also be present. These include the shells of large free-swimming organisms such as cephalopods, bones and teeth of fish or aquatic reptiles and mammals. Sharks' teeth are found in the stratigraphic record in rocks as old as late Palaeozoic. Life in the open oceans in the early Palaeozoic was apparently dominated by graptolites, a hemichordate colonial organism with a free-swimming or floating lifestyle which had a 'backbone'. The compressed remains of graptolites are found in large quantities in Lower Palaeozoic mudrocks and are important in biostratigraphic correlation (Chapter 19).

Fig. 15.10 Chert between pillows of basaltic lava, Lower Palaeozoic, North Wales.

15.8 Ancient deep ocean deposits

Despite being such large areas of sedimentation, ocean basin deposits make up a small fraction of the material preserved in the stratigraphic record. This is because the likelihood of sediments laid down kilometres below sea level on ocean crust finding their way up on to a continent to be preserved on land is not very high. Ocean crust is normally subducted at destructive plate boundaries and may carry the sedimentary veneer down with it. However, there are two principal scenarios in which deep ocean sediments may become incorporated into strata exposed on land: as the components of an accretionary prism *(23.3.2)* or as part of an ophiolite complex *(23.2.6)*. Deep marine facies occur also in basins floored by continental crust — for example, foreland basins where a deep basin may be created by flexural loading of the lithosphere *(23.4)*.

Establishing what the water depth was at the time of deposition is problematic beyond certain upper and lower limits. The effects of waves, tides and storm currents can usually be recognized in sediments deposited on the shelf and are absent below about 200 m depth. There are almost no reliable palaeowater depth indicators between that point and the depths at which carbonate dissolution becomes a recognizable process at several kilometres water depth. Even then, establishing that deposition took place below the CCD is not always straightforward. Relative depth zones can sometimes be established from ichnofacies information *(11.7.2)* and the ratios of planktonic to benthic foraminifera (deeper water sediments tend to contain a higher proportion of planktonic forms of these organisms), but these do not provide absolute depths and at best only relative depth ranges. When describing a facies as 'deep water' it should be remembered that the actual palaeowater depth of deposition might have been anything between, say, 300 m and 3000 m or more.

There are problems also in relating ancient deep marine facies to the deposits of modern deep water environments. Knowledge of the deep water processes and products comes mainly from surveys and drilling in the world's major oceans as part of research programmes such as the Deep Sea Drilling Project and the Ocean Drilling Program. Many of the deep marine facies known in the stratigraphic record were probably the deposits of smaller basins (e.g. the Tertiary foreland basins of Europe). These smaller basins are not subject to thermo-haline bottom currents in the way that major ocean basins are and hence there is little evidence of reworking and redeposition as contourites, a process now known to be very important in modern oceans (Mutti & Normark 1987). Direct comparison between modern and ancient deep water facies is therefore not appropriate in many cases.

15.9 Recognition of deep ocean deposits: summary

Evidence in sedimentary rocks for deposition in deep seas is as much based on the absence of signs of shallow water as on positive indicators of deep water. Sedimentary structures, such as trough cross bedding, formed by strong currents, are normally absent from sediments deposited in depths greater than 100 m or so, as are wave ripples and any evidence of tidal activity. The main sedimentary structures in deep water deposits are likely to be parallel and cross lamination formed by deposition from turbidity currents and contour currents. Some authigenic minerals can provide some clues: glauconite does not form anywhere other than shelf environments, but is by no means ubiquitous there, and manganese nodules are characteristically formed at abyssal depths, but are not widespread. Absence of pelagic carbonate deposits may indicate deposition below the calcite compensation depth, although care must be taken not to mistake fine-grained redeposited limestones for pelagic sediments.

Some of the most reliable indicators of water depth are to be found from an analysis of body fossils and trace fossils. Many benthic organisms can only exist in shelf environments, but it must be remembered that shallow water deposits, including fossils, can be redeposited in deep water by turbidity currents. The ratio of benthic and planktonic fossils of organisms such as foraminifera is sometimes used as a relative depth indicator, as planktonic forms are likely to be more abundant in deeper water. Trace fossils provide some of the most reliable evidence of deep water sedimentation as there is a consistent relationship between ichnofacies and relative water depth.

All depth indicators in the deposits of deep oceans are very crude and only usually provide clues as to the relative depth of deposition within a stratigraphic succession. Absolute depth indicators between the edge of the continental shelf at 200 m and the abyssal plain at 4000 m or more are effectively non-existent.

Further reading

Hartley, A.J. & Prosser, D.J. (eds) (1995) *Characterization of Deep Marine Clastic Systems*, 247 pp. Geological Society of London Special Publication **94**.

Hsü, K.J. & Jenkyns, H.C. (eds) (1974) *Pelagic Sediments: On Land and Under the Sea*, 447 pp. International Association of Sedimentologists Special Publication **1**.

Nelson, C.H. & Nilson, T.H. (1984) *Modern and Ancient Deep-sea Fan Sedimentation*, 404 pp. Society for Economic Paleontologists and Mineralogists Short Course Notes **14**.

Pickering, K.T., Hiscott, R.N. & Hein, F.J. (1989) *Deep Marine Environments; Clastic Sedimentation and Tectonics*, 416 pp. Unwin Hyman, London.

Stow, D.A.V. (1985) Deep-sea clastics: where are we and where are we going? In: *Sedimentology, Recent Developments and Applied Aspects* (eds P.J. Brenchley & B.P.J. Williams), pp. 67–94. Blackwell Scientific Publications, Oxford.

Stow, D.A.V., Reading, H.G. & Collinson, J.D. (1996) Deep seas. In: *Sedimentary Environments: Processes, Facies and Stratigraphy* (ed. H.G. Reading), pp. 395–453. Blackwell Science, Oxford.

Walker, R.G. (1992) Turbidites and submarine fans. In: *Facies Models: Response to Sea Level Change* (eds R.G. Walker & N.P. James), pp. 239–264. Geological Association of Canada, St Johns, Newfoundland.

16   Volcanic environments

The contribution of volcanism to sedimentation and the importance of volcanic material to stratigraphic successions are often understated, except, of course, by geologists who live or work in volcanically active regions. The study of volcanism and its products is often considered to be within the remit of igneous geology, but volcanic and volcaniclastic rocks occur bedded with terrigenous clastic or carbonate sediments in stratigraphic successions. Volcanic rocks can occur in all the environments discussed in earlier chapters and many of the processes of transport and deposition are similar to those discussed for terrigenous clastic sediments. However, there are also important differences because some processes are unique to volcanic settings and these processes result in distinct volcanic environments in both continental and marine realms. Only a brief introduction to volcanigenic sedimentation can be provided here: a more comprehensive treatment of the topic is provided in specialist texts such as Cas and Wright (1987).

16.1 Volcanic and volcaniclastic rocks

The products of volcanism can be divided into material erupted as molten material which cools at the surface (*lavas*) and products which were solid fragments at the time of eruption or were subsequently fragmented (*volcaniclastic* material). Both lavas and volcaniclastic deposits can occur in any continental or marine environment.

16.2 Lavas

Lavas are molten rock extruded directly from volcanic vents and fissures. The mineralogical composition of the solidified lava flow reflects the chemistry of the magma from which it was formed. This chemistry also controls the physical properties of the lava as it forms, such as the temperature of the flow, its viscosity and dissolved gas content (Cas & Wright 1987). On the land surface very fluid lavas commonly have a *pahoehoe* texture, which is a ropy pattern on the surface, whilst viscous flows have a blocky texture known as *aa*. Lavas which erupt under water form a pillow structure (see Figs 15.1 & 15.10) due to the rapid cooling of the magma upon contact with water.

During and after formation lavas are subject to alteration upon contact with the surrounding environment. Submarine lavas react chemically with the sea water to form a sodium-enriched volcanic rock called *spilite*. Weathering processes attack subaerial lavas very quickly, particularly if the rock is of basaltic composition and made up of minerals which readily oxidize and hydrolyse on contact with air and water. In humid tropical regions a lateritic soil profile may start to develop on top of lavas in a matter of years.

16.3 Volcaniclastic rocks

Sediments composed of volcanic particles may be broadly classified on the basis of the processes by which they formed (Walker 1973). There are some important differences between the way primary volcaniclastic material behaves during transport and deposition and the terrigenous clastic detritus considered in earlier chapters. An important physical control on sedimentation is that the settling velocity is proportional to fragment size, shape and density *(4.2.5)*. Unlike terrigenous clastic material, the density of pyroclastic particles is very variable. In particular, pumice pyroclasts may have a very low density and can float until they become waterlogged. Pyroclastic deposits may show both normal and reverse grading of different components in the same bed. Lithic fragments and crystals will be normally graded with the coarsest material at the base. Pumice pyroclasts deposited in water may be reverse-graded because the larger fragments will take longer to become waterlogged and hence will be the last to be deposited.

16.3.1 Air fall deposits

When a volcanic eruption sends a cloud of debris into the air the pyroclastic fragments may return to the ground under gravity as a shower of *air fall* deposits. Volcanic blocks and bombs travel only a matter of hundreds of metres to kilometres up from the vent, depending on the force with which they were ejected. Finer lapilli and ash may be sent kilometres into the atmosphere and be distributed by wind. Large eruptions can result in ash distributed thousands of kilometres from the volcano. A distinctive feature of air fall deposits is that they mantle the topography, forming an even layer over all but the steepest ground surface (Fig. 16.1). The deposits become thinner and are composed of finer-grained material with increasing distance from the volcanic vent.

Pyroclastic falls range in size from small cinder cones to large volumes mantling topography over large areas. Small falls, referred to as *Hawaiian* or *Strombolian*, are typically basaltic in composition and form cones of *scoria* (glassy, vesicular fragments of basaltic composition) which dip gently away from and into the cone (Cas & Wright 1987; Francis *et al.* 1990). They typically have volumes of less than 1 km^3 and tend to be very localized in area. The fabric of the deposits of these small falls is typically crudely bedded, clast-supported and variably sorted due to particles being of differing vesicularity and hence density.

Plinian falls result from very large eruptions which progress from falls to flows during the course of the eruption. The deposits are typically clast-supported, angular, fragmented, pumice or scoria clasts with subordinate crystals and lithic fragments. The fabric may be massive or stratified, the former resulting from sustained eruptions, whereas stratification may result from fluctuations of eruption intensity or wind direction. Stratification can also result from a reworking of the material during which

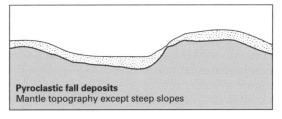

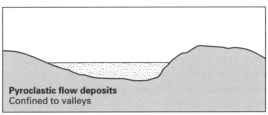

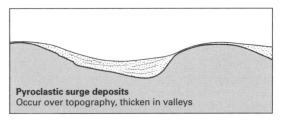

Fig. 16.1 Distribution of ash over topography from pyroclastic falls, pyroclastic flows and pyroclastic surges. (After Suthren 1985.)

process the pumice clasts will tend to become rounded. If normal grading is present in a single fall deposit it suggests that the eruption was most intense initially, whereas reverse grading results from an increase in eruption intensity with time. The bedding of Plinian falls tends to mantle topography except where it is reworked into depressions by secondary processes. The distribution of material from a Plinian eruption is very strongly influenced by the strength and direction of the prevailing wind.

16.3.2 Pyroclastic flows and surges

In addition to falls of ash, some eruptive events result in *ash flows*. These occur either when the column of pyroclastic debris forced into the air by the eruption collapses or when there is a laterally directed blast from a vent which sends a gas and debris mixture horizontally. These flows are a type of sediment gravity flow *(4.6)*. Flows produced by lateral blasts can be particularly devastating because they can travel at hundreds of kilometres per hour. They are usually low-density, turbulent mixtures of gas and debris. They generally produce a normally graded deposit analogous to a turbidite. Denser flows, which are a high-concentration mixture of air or gas and tephra, require a slope to maintain momentum and are hence more limited in extent to within a few kilometres of the volcano. They deposit a poorly sorted blanket of debris. Hot mixtures of gas and tephra form a *nuée ardente*, a 'glowing cloud' which includes fragments which are hot enough to fuse together when deposited and form a *welded tuff*. An *ignimbrite* is a pyroclastic flow composed of pumiceous material which is commonly, but not always, welded.

Ash flows can be divided into two types. The products of high-concentration mixes which move as laminar flows of fragments in gas are referred to as *pyroclastic flow* deposits. Deposits from low-concentration mixes which are turbulent suspensions are referred to as *pyroclastic surge* deposits.

PYROCLASTIC FLOWS

Three types of pyroclastic flows are recognized (Lajoie & Stix 1992) (Figs 16.2 & 16.3).

Pelée type events are the result of very strong lateral thrusts due to high gas pressure in the volcano. This results in very high velocities (about $130\,\mathrm{m\,s^{-1}}$ or $500\,\mathrm{km\,h^{-1}}$) in low-concentration, turbulent density currents. The deposits are characterized by a coarse, massive or reverse-graded basal unit (a traction carpet) which is overlain by a normally graded, stratified middle unit produced by deposition from a turbulent mix of gas and ash. Inclined stratification is interpreted as either antidune (dip upflow) *(4.3.5)* or dune (dip downflow)

Fig. 16.2 Merapi, St Vincent and Pelée types of pyroclastic flow. (After Lajoie & Stix 1992.)

formation. The top unit is a thin, fine-grained unit with accretionary lapilli.

Merapi type flows are primarily controlled by gravity, and in the absence of a lateral thrust the velocities are slower ($15–50\,\mathrm{m\,s^{-1}}$). The deposit is typically a single bed of reverse-graded, very poorly sorted lapilli tuff which lacks stratification. The mechanism of transport is a grain flow type of mass flow *(4.6.3)* which can only be maintained if there is a slope downflow.

A third type of pyroclastic flow is the *St Vincent type* which is not so well described. A moderate gas pressure drives a highly concentrated mass which descends radially from the crater at low velocities. The resulting deposit is massive, not graded or stratified.

PYROCLASTIC SURGES

Explosive magmatism may produce ejecta plumes with a horizontal component called a *base surge* or *pyroclastic surge* (Sparks & Walker 1973). These behave like turbidity currents as they are gravity-controlled density currents with superheated steam as the medium. Surges are referred to as 'dry' or 'wet' depending on the proportion of water present. *Dry surges* have a complex flow behaviour, evolving from turbulent flow in the proximal area becoming laminar flow as gas is lost and concentration is increased. Proximal parts of flows may therefore be cross bedded but pass distally to massive deposits. The behaviour and deposits of *wet surges* are very similar to those of a normal turbidity current.

16.3.3 Brecciated volcaniclastic rocks

The surface of a subaqueous or subaerial lava flow cools and solidifies to form a crust which can become broken up as the lava continues to move (Orton 1996). This broken lava rock forms an *autoclastic* breccia which may

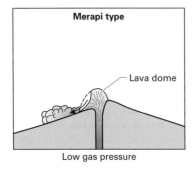

Low gas pressure

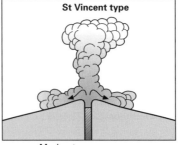

Moderate gas pressure

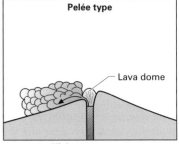

High gas pressure

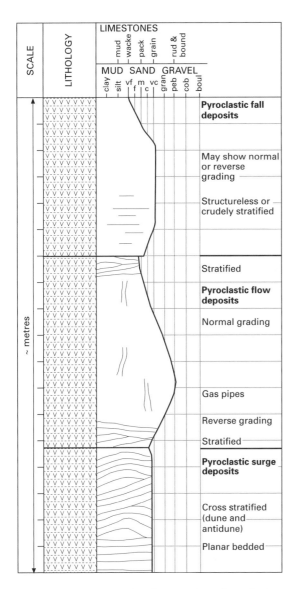

Fig. 16.3 Schematic graphic sedimentary logs of pyroclastic deposits.

Fig. 16.4 Brecciated andesitic lava, an autoclastic volcanic rock formed by breakup of lava erupting from a subaqueous vent, Miocene, south Spain.

16.3.4 Epiclastic volcaniclastic rocks

Epiclastic volcaniclastic rocks are formed by the reworking of volcanic and volcaniclastic material (Fig. 16.5). In the case of air fall or pyroclastic flow deposits this may occur soon after the initial deposition. Unconsolidated pyroclastic sediments are very susceptible to reworking especially if deposited on slopes. Rainfall can trigger dense flows of volcaniclastic material and water as a form of debris flow known as a *lahar*. These flows have the characteristics of terrigenous clastic debris flows but are distinctive because of the wholly volcanic composition of the sediment.

A further source of volcaniclastic detritus is the breakdown and erosion of lithified volcanic or volcaniclastic material—for example, a rock formed from a lava flow

be recognized by a fabric which is typically coarse, angular blocks of volcanic rock, all of the same composition (Fig. 16.4). They are normally localized deposits which may occasionally be reworked by water currents. *Hyaloclastites* are also poorly sorted breccias, but are made up of fragments of volcanic glass formed by the rapid quenching of a molten lava. They often occur associated with pillow lavas, filling in the gaps between the pillows.

Fig. 16.5 Water-worn epiclastic volcanic rock, Tertiary, Bacan Island, Moluccas, Indonesia.

being eroded by a river or on a beach. Such reworked pyroclastic material can become incorporated into any depositional environment and may be mixed with terrigenous clastic or carbonate deposits. This type of volcanic detritus is referred to as *epiclastic*, having been broken up at the surface. Epiclastic volcanic deposits will often be mixed with non-volcanic sediment, whereas primary, pyroclastic deposits will be purely volcanigenic. They may also show primary sedimentary structures which reflect the processes of reworking of the material.

16.4 Environments of deposition

Volcanism can create distinctive volcanic environments both on land and in the sea. Volcanic material can also contribute to the sediment in almost all other depositional environments in amounts which largely depend on the proximity of the volcanic centre and the volume of material erupted.

16.4.1 Environments created by volcanism

Environments created by volcanism can be classified according to their geomorphic form which is in turn controlled by the nature of the material erupted (Cas & Wright 1987).

The classic volcanoes forming steep conical mountains with a vent in a crater near the summit are *stratavolcanoes*. These volcanic landforms are composite bodies resulting from repeated eruptions of pyroclastic falls, pyroclastic flows and relatively short lava flows. They typically result from the eruption of intermediate to acidic magmas. Volcanoes formed by the repeated eruption of basaltic magmas tend to have lower slopes built up mainly of basaltic lava: these are *shield volcanoes*. In continental areas with intense thermal activity over a mantle hot-spot there may be eruption of large amounts of lava and pyroclastic material from multiple vents and fissures, forming a *flood basalt* province. Valleys become filled and pre-existing landforms completely enveloped when flood basalts cover many thousands of square kilometres. A build-up of volcanic rocks several kilometres thick may occur. *Rhyolitic volcanic centres* are commonly complex with multiple eruption points and may lack the topographic expression of a volcanic cone. They are very variable in form and may be quite diffuse.

Subaerial eruptions of basaltic pyroclastic material generate limited areas of scoria which form *scoria cones*, circular landforms which may be only a few hundred metres across but with steep sides. Scoria cones are very common in areas of basaltic magmatism but are very susceptible to weathering and erosion. *Maars*, *tuff rings* and *tuff cones* are all morphological types of volcanic crater composed of volcaniclastic material. Maars have steep-sided craters and gentle outer slopes, tuff rings have roughly equal slopes either side of the rim and tuff cones have steep outer cones and small craters.

Huge amounts of submarine volcanism occur along ocean *spreading ridges* where pillow lavas and hyaloclastites are the main rock types. Isolated volcanic centres associated with hot-spots in the oceans form *seamounts*, basaltic lava piles which form steep-sided cones. Once emergent, volcaniclastic material may form a cone of scoria. Seamounts may be important sites of carbonate sedimentation as isolated platforms *(14.8.2)* especially once the volcanic activity has ceased.

16.4.2 Volcanic material in continental environments

Volcaniclastic clasts have little preservation potential in aeolian environments because the processes of abrasion and attrition during wind transport are too severe for the relatively fragile lithic and crystal grains to survive. Preservation of ashes is favoured by low-energy continental environments where there is active sediment accumulation. Lakes provide ideal conditions for the preservation of an ash fall deposit in a stratigraphic succession. Floodplain environments may also be suitable but pedogenic and weathering processes will rapidly alter the volcanic material. Ash bands also occur bedded with coals in swamps and mires where organic detritus is rapidly accumulating.

At the interface between continental and marine environments lie deltas. Volcanic deposits which can be classified as deltas are rare, but have been documented as either cones of lava and hyaloclastite advancing into the sea (Porebski & Gradzinski 1990) or aprons of volcaniclastic material (Nemec 1990).

16.4.3 Volcanic material in marine environments

The black sands of a beach on the south coast of the main island of Hawaii were famous until they were engulfed by lava flows in the early 1990s. These beach sands composed of grains of basalt were not unique and can be found along the coasts of many volcanic islands. Such sands exhibit a high degree of textural maturity and classify as mature or even supermature, but their compositional maturity is very low, consisting as they do of unstable lithic fragments. The ease with which volcanic sediments break down limits their distribution and preservation potential in high-energy shallow marine environments. An association with carbonate sediments is quite common. The fringes of volcanic islands in tropical seas are ideal locations for carbonate sedimentation because of the absence of terrigenous clastic detritus. In the periods between volcanic eruptions faunal communities are able to develop and provide a source of carbonate as bioclastic sands or reef build-ups.

In sedimentary basins bordered by island arcs the volcanoes may be the primary source of clastic detritus. Some forearc and backarc basins *(23.3.4)* may contain volcanigenic sediments almost to the exclusion of terrigenous clastic detritus, especially in cases where the arc is intra-oceanic and the basin is distant from any continental land mass (Nichols & Hall 1991). Primary air falls and pyroclastic flows will be important contributors to the basin fill adjacent to the arc, but further into the basin reworked epiclastic material dominates. The main process of transport and deposition is by turbidity currents which can carry volcanic detritus from the volcanic edifices many hundreds of kilometres out into the basin. In these circumstances the only other source of sediment may be pelagic material.

16.4.4 Global distribution of volcanic environments

Most volcanicity around the world is associated with plate margins. The great chains of volcanic islands in the western Pacific are related to subduction of ocean plates, as are the volcanoes in the Caribbean, the Andes and the Indonesian archipelago islands of Sumatra, Java and others further east. Volcanism also occurs in extensional tectonic regimes along all the oceanic spreading ridges and in intra-continental rifts such as the East African rift system. Strike-slip plate boundaries may also be sites of volcanism. Exceptions to this pattern of association with plate boundaries are volcanoes situated above 'hot-spots', sites around the surface of the Earth where *mantle plumes* provide exceptional amounts of heat to the crust. The Hawaiian island chain is the best present-day example of a volcanic centre located at a hot-spot and distant from any plate boundary. Areas of flood basalts form in continental areas located over hot-spots. Geochemical work has shown distinct chemical signatures for the volcanic rocks associated with each of these different tectonic settings.

Important clues to the plate tectonics of the past are provided by the recognition of volcanigenic deposits in the stratigraphic record and analysis of their chemistry, making it possible to recognize ancient plate boundaries a long way back through Earth history.

16.5 Recognition of volcanigenic deposits: summary

Recognition of volcanic-related sediments in modern environments is normally very straightforward when the volcanic centre is nearby. However, the distances that volcaniclastic material can travel from the vent can be enormous. Dust from major eruptions may be deposited literally all over the surface of the globe and incorporated as fine-grained material in all depositional environments. The darkening of skies after events such as the 19th-century eruption of Krakatau is a testament to the volumes of detritus ejected into the atmosphere and circulated around the world. Even larger volcanic events such as the prehistoric eruptions of Crater Lake and Yellowstone in North America and the Toba Volcano in Sumatra shed volcaniclastic sediment over millions of square kilometres (Smith & Braile 1994).

These widespread volcaniclastic deposits can be recognized in Recent sediments as material which mantles the topography and is petrographically distinct from other contemporaneous deposits. In the stratigraphic record the distribution of a widespread volcaniclastic deposit as a blanket of detritus has usually been obscured by superficial reworking in marine and continental settings. The original thickness distribution of the material may only be preserved in low-energy environments such as lakes and deep oceans.

Closer to the source of the volcanism, the mantling effect of the erupted material is more marked because anything from a few tens of centimetres to tens of metres of material may be deposited in a single episode. In these circumstances the layer of volcaniclastic detritus may additionally be preserved on alluvial fans, river floodplains and shallow marine environments to become part of the stratigraphic succession.

The single most important criterion for the recognition of volcanigenic deposits is the composition of the material. Lavas and primary volcaniclastic detritus rarely contain any material other than the products of the eruption. The nature of that material depends on the chemical composition of the magma and the nature of the eruption (see above). Recognition of the volcaniclastic origin of rocks in the stratigraphic record becomes more difficult if the material is fine-grained, altered or both. In hand specimen a fine-grained volcaniclastic rock could be confused with a terrigenous clastic rock of similar grain size. Microscopic examination of a thin section usually resolves the problem by making it possible to distinguish the crystalline forms within the volcaniclastic deposit from the eroded, detrital grains of terrigenous clastic material. Alteration can destroy the original volcanic fabric of the rock principally by breakdown of feldspars and other minerals to clays. Rocks of basaltic composition are particularly susceptible to alteration at or near the Earth's surface. Complete alteration may mean that the original nature of the material can be determined only from relic fabrics such as the outlines of the shapes of feldspar crystals remaining despite total alteration to clay minerals and the chemistry of the clays as determined by X-ray diffractometer analysis *(2.5.4)*.

16.6 Volcanic rocks in stratigraphy: flows, dykes and sills

Distinguishing between intrusive sills and lava flows in the field is important for a number of reasons. First, the presence of a lava is evidence of volcanism during the period when the beds above and below were deposited. This is useful information when attempting to develop a picture of the palaeoenvironment. Second, volcanic rocks may be directly dated by isotopic analysis *(20.1)*. The date for a lava provides a date for that part of the sedimentary succession, whereas the date for a sill is some time after the sediments were deposited. Dates from flows, dykes and sills are important for developing a chronostratigraphy in stratigraphic successions *(18.3.1)*.

Sills may be identified by features which are not seen in lava flows such as a *baked margin*, where there is evidence of heating at the contact with both the beds below and the beds above. Also, when tracing the sill laterally it may be found locally to cut through beds up or down stratigraphy, behaving as a dyke at these points. Lava flows, on the other hand, may display characteristics such as a pillow structure if the eruption was under water, or a weathered top surface of the flow in the case of subaerial eruptions.

Further reading

Cas, R.A.F. & Wright, J.V. (1987) *Volcanic Successions: Modern and Ancient*, 528 pp. Unwin Hyman, London.

Fisher, R.V. & Schmincke, H.-U. (1994) Volcaniclastic sediment transport and deposition. In: *Sediment Transport and Depositional Processes* (ed. K. Pye), pp. 351–388. Blackwell Scientific Publications, Oxford.

Fisher, R.V. & Smith, G.A. (1991) *Sedimentation in Volcanic Settings*, 257 pp. Society of Economic Paleontologists and Mineralogists Special Publication **45**.

Lajoie, J. & Stix, J. (1992) Volcaniclastic rocks. In: *Facies Models: Response to Sea Level Change* (eds R.G. Walker & N.P. James), pp. 101–118. Geological Association of Canada, St Johns, Newfoundland.

Orton, G.J. (1995) Facies models in volcanic terrains: time's arrow vs. time's cycle. In: *Sedimentary Facies Analysis: A Tribute to the Teaching and Research of Harold G. Reading* (ed. A.G. Plint), pp. 157–193. International Association of Sedimentologists Special Publication **22**.

Orton, G.J. (1996) Volcanic environments. In: *Sedimentary Environments: Processes, Facies and Stratigraphy* (ed. H.G. Reading), pp. 485–567. Blackwell Science, Oxford.

17   Sediments into rocks: post-depositional processes

The preceding chapters have dealt with the processes which lead to the accumulation of sediments in different depositional environments. The characteristics of the deposits can then be used to interpret sedimentary environments of the past. There are, however, a number of physical and chemical processes which occur after deposition of the material during formation of a sedimentary rock. Some of these processes occur very soon after deposition and are seen as a mechanical deformation of the sedimentary layering. Other post-depositional (diagenetic) processes occur over a longer period of time and principally result in the sediment becoming lithified as a sedimentary rock. The nature of the bedding, the colour, the texture and the chemistry of the sediment can also be changed during diagenesis. An understanding of the causes and effects of diagenetic processes is needed in order to help interpret the rocks in terms of their original depositional setting. It also leads to a more complete knowledge of the history of the sediments and the basin in which they have accumulated. The changes to the physical properties of the sediment are particularly important for economic reasons as they affect the porosity and permeability of the sediment and determine its attributes as a hydrocarbon reservoir or aquifer.

17.1 Post-depositional modification of sedimentary layers

Sediment deposited as bedload from a current or out of suspension will initially display sedimentary structures and layering formed during the process of deposition. These primary sedimentary structures include ripple cross lamination, cross bedding, horizontal lamination and so on *(4.3)*. They may be preserved intact within the sediment and be recognizable in a sedimentary rock. However, these primary features are subject to modification by fluid movement and gravitational effects if the sediment remains soft and deformable. Disruption of the sedimentary layers may occur within minutes of deposition or may happen at any time up to the point where the material becomes lithified *(17.2)*.

17.1.1 Early post-depositional structures

Soft sediment deformation is the general term for changes to the fabric and layering of a bed of recently deposited sediment (Leeder 1982). Sand and mud are deposited in rivers, lakes and seas along with water which is gradually expelled from between the grains as the sediment settles and compacts. If the sediment is disturbed in any way before much water has been lost it is easily disrupted to form deformation structures. Disruption may be prompted by the addition of more sediment on top of the layer which traps water in the sediment or by shocks such as an earthquake which may temporarily liquefy the sediment. Deformed layered beds are generally referred to as *convolute bedding* or *convolute lamination* depending on the scale, but these structures may have a variety of origins (Fig. 17.1).

Dewatering structures seen in bedded sediment result from the expulsion of water due to loading or a shock, and liquefied flow occurs *(4.6.4)*. The water moves upwards, disrupting overlying layers to form *dish and pillar* structures which consist of concave layers of sediment and vertical pipes through layers of sand and mud (Fig. 17.2). Large dewatering pipes may extend to the bed surface where sand, carried up by the water,

Slump structures

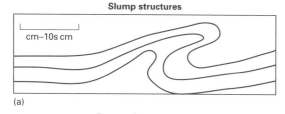

(a)

Dewatering structures

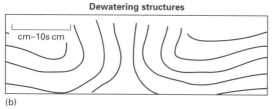

(b)

Load structures

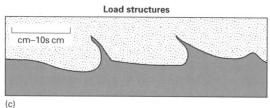

(c)

Fig. 17.1 Soft sediment deformation structures: (a) convolute bedding produced by slumping; (b) convolute bedding formed by dewatering; (c) load structures at the base of a sandstone bed.

Fig. 17.2 Dewatering structure in laminated sandstone and siltstone.

forms a cone a few tens of centimetres across known as a *sand volcano* (Collinson & Thompson 1982; Leeder 1982).

Disturbance of beds which involves a lateral movement of material occurs when the soft sediment has been deposited on a slight slope or there is a shear stress on the layers due to a flow over the top. *Slumped beds* deformed into layers which are folded in one direction (downslope or downflow) are common in the silty 'c' and 'd' divisions of turbidites, for example (Fig. 17.3). Larger-scale slumping can leave surfaces from which sediment has been removed by slumping which are distinctive as slump scars *(4.8.3)*. The shear stress caused by a strong current over a set of cross beds in sands can lead to the formation of *overturned cross stratification* (Fig. 17.4).

The deposition of a bed of sand on a layer of wet mud can result in localized disruption of the sediment. *Load structures* where the sand has partially sunk into the underlying mud are common at the bases of sandy

Fig. 17.3 Convolute lamination in thinly bedded sandstone and mudstone formed as a result of slumping.

Fig. 17.4 Overturned cross stratification in 60 cm thick Cambrian sandstone beds, Sinai Peninsula, Egypt; these would have been deposited originally as simple cross beds by the migration of a subaqueous dune bedform and subsequently the upper part of the cross bed set was deformed by the shear stress of a flow over the top.

turbidite beds and other situations where sand is deposited directly on wet muds. The mud may also be forced up into the overlying sand bed to form a *flame structure* (Collinson & Thompson 1982). Flame structures formed by injection of material upwards are distinct from *neptunian dykes* which are vertical sheets of sediment formed by the infilling of a fissure from above.

17.1.2 Post-depositional structures due to density contrasts

When clay-rich sediments are deposited, they initially contain a high proportion of water, up to 80–90%. Under most circumstances this water is gradually expelled from between the clay particles as more sediment accumulates on top. If, however, sandy sediment is added on top of the clay at a higher rate than the water can be expelled from the relatively impermeable clay layer, a density inversion can occur. The clay-rich layer with a high water content is trapped below a sandy layer of higher density. In this unstable situation the lower-density mixture of clay and water will attempt to move up through the overlying sand.

Shale or *mud diapirs* are bodies of mud a few metres to hundreds of metres across which move up through layers of overlying strata. They may remain attached to the mud layer from which they originated, or may become detached. Diapirs are common at delta fronts where fine-grained sediments are trapped beneath coarser strata as the delta progrades. Shale diapirs reaching the sea floor have been recognized in the offshore areas of the Mississippi and Niger deltas and are seen in ancient, muddy delta facies (Morgan *et al.* 1968; Elliott 1986a). On a smaller scale of a few centimetres, injection of mud up into overlying sand occurs when the sandy layer is rapidly deposited. These *ball-and-pillow structures* are soft sediment deformation structures which are commonly seen in turbidite successions.

17.2 Post-depositional physical and chemical processes

As sediments accumulate physical and chemical changes occur to alter the characteristics of the material deposited. These changes are collectively referred to as the processes of *diagenesis*. Diagenesis can be considered to be anything that happens to a sediment after it has been deposited. Diagenetic changes are those which occur in sediments at relatively low pressures and temperatures

(Fig. 17.5). There is a continuum between diagenesis and metamorphism, the latter being considered to be those processes which occur at higher temperatures (typically above 200°C: McLane 1995) and pressures. A number of different diagenetic processes may be occurring at the same time, commonly dependent on each other.

17.2.1 Lithification

Most sediments are soft, unconsolidated material at the time of deposition and are in the form of loose sand or gravel, soft mud or accumulations of the body parts of dead organisms. In order for sediment to be transformed into sedimentary rock, processes of *lithification* take place after deposition. These processes are both chemical and physical and they may take place at any time after initial deposition. The time between sedimentation and lithification may be effectively an instant or it may be thousands or millions of years. Some sediments never become consolidated, remaining as loose material millions of years after deposition.

In some circumstances the processes of lithification are contemporaneous with deposition: limestones, evaporite deposits and volcaniclastic sediments may all form rocks at the time of deposition. Some limestones are formed from the frameworks of organisms which build up solid masses of calcium carbonate as bioherms—for example, coral reefs; loose material between the coral mass may be subsequently lithified but the main framework of the rock is formed *in situ*. Chemical precipitation out of water in lakes or seas results in solid crystalline evaporite minerals in the first instance which build into beds of evaporites. A further example is that of pyroclastic deposits deposited from hot clouds of ash and gases

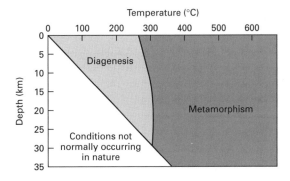

Fig. 17.5 Depth and temperature ranges of diagenetic processes.

(*nuées ardentes*): the temperatures may be high enough for the ash particles to fuse together on deposition as a welded tuff.

17.2.2 Burial diagenesis: compaction

The accumulation of sediment at a point results in the earlier deposits being overlain by younger material which exerts a pressure on it. This *overburden pressure* acts vertically on a body of sediment and increases as more sediment, and hence more mass, is added on top. Loose aggregates initially respond to overburden pressure by changing the packing of the particles; clasts move past each other into positions which take up less volume for the sediment body as a whole (Fig. 17.6). This is one of the processes of *compaction* which increases the density of the sediment. Compaction occurs in all loose aggregates as the clasts rearrange themselves under moderate pressure: this may reduce the volume of a sand by around 10%. When muds are deposited they may contain up to 80% water by volume (Perrier & Quiblier 1974): this is reduced to around 30% under burial of 1000 m, representing a considerable compaction of the material. Certain sediments (e.g. boundstones formed as a coral reef) may not compact at all under initial burial. Compaction has little effect on horizontal layers of sediment except to reduce the thickness. Internal sedimentary structures such as cross stratification may be slightly modified by compaction and the angle of the cross strata with respect to the horizontal may be decreased slightly.

Where there is a lateral change in sediment type *differential compaction* occurs. Possible examples are lime muds deposited around an isolated patch reef, a sand bar surrounded by mud and a submarine channel cut into muds and filled with sand. In each case the degree to which the finer material will compact under

overburden pressure will be greater than for the sand body or reef. A 'draping' of the finer sediments around the isolated body will occur under compaction (Fig. 17.7). This can occur on all scales from bodies a few metres across to masses hundreds of metres wide. Note that the differential compaction effect is less marked in fluvial successions where sand-filled channel bodies are surrounded by overbank mudstones. This is because the fine sediment on the floodplain dries out between flood events and loses most of its pore waters at that stage. As a consequence, the effect of overburden pressure on overbank muds and channel sands may be the same. Differential compaction effects can also be seen on the scale of millimetres and centimetres where there are contrasts in sediment type. Mud layers may become draped around lenses of sand formed by ripple and dune bedforms. Local compaction effects also occur around nodules and concretions where there is early cementation. In all cases where the sediment is deposited in an aqueous environment the *pore* spaces between the grains (the *porosity*) will be occupied by water (*pore water*). As compaction occurs this water is expelled from the sediment—that is, it begins to *dewater*.

Under thick piles of sediment the overburden pressure may be high enough to result in more extreme physical changes in the sediment. Weaker grains may be deformed or broken by the pressures exerted by stronger grains around them: mica flakes are commonly bent, crystal and lithic fragments may be crushed between quartz grains. At grain contacts pressures are concentrated at the points where the grains touch. At these high-pressure points the mineral may locally go into solution

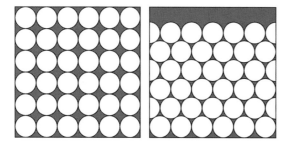

Fig. 17.6 Changes to the packing of spheres can lead to a reduction in porosity and an overall reduction in volume.

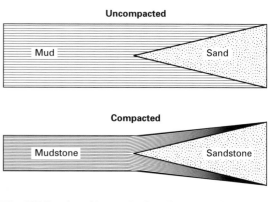

Fig. 17.7 Draping of layers of soft mudstone around an isolated sandstone body due to the differential compaction of mud and sand.

in the encompassing pore waters in a process called *pressure solution* or *pressure dissolution.* Pressure dissolution results in interlocking grains which provide a rigidity to the sediment, that is, it becomes lithified. Beds of limestone may show extensive effects of pressure dissolution *(17.5.1).*

The degree of compaction in an aggregate can be determined by looking at the nature of the grain contacts (Fig. 17.8) (Tucker 1991). If the sediment has been subjected to very little overburden pressure the clasts will be in contact mainly at the point where they touch *(point contacts).* Reduction in porosity by changes in the packing will bring the edges of more grains together as *long contacts.* Pressure solution between grains results in *concavo-convex contacts* where one grain has dissolved at the point of contact with another (Fig. 17.9). Under very high overburden pressures the boundaries between grains become complex *sutured contacts*, a pattern more commonly seen under the more extreme conditions of metamorphism.

17.2.3 Chemical processes of diagenesis: cementation

Chemical processes are important in diagenesis because most sediments are saturated with pore waters with which the mineral grains can react. The processes of pressure solution result in ions being dissolved from mineral grains and held in the pore waters. Other lithic, biogenic and crystal grains may simply dissolve without enhanced pressure conditions. In particular, calcium carbonate is readily dissolved in slightly acid pore waters and it is common for shelly debris within a sediment to be dissolved in this way.

Ions dissolved by pore waters in one place may be carried by fluid flow through the body of sediment to places where the chemical conditions may be favourable for the precipitation of the ions as new minerals. For example, a reduction in the acidity may favour the precipitation of calcium carbonate from solution in the pore waters. Where localized pressure solution occurs, precipitation may occur in adjacent pore spaces, but in other cases ions may be carried hundreds of metres or kilometres laterally or vertically through the sediment body before being precipitated.

The nucleation and growth of crystals within pore spaces in sediments is the process of *cementation.* Minerals precipitated within pore spaces during diagenesis are *cements* and must be distinguished from the fine material which is deposited with the coarser grains as a matrix

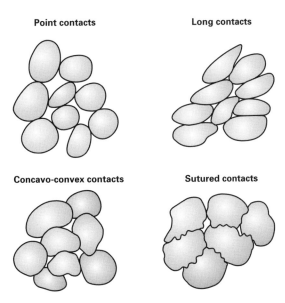

Point contacts   **Long contacts**

Concavo-convex contacts   **Sutured contacts**

Fig. 17.8 Types of grain contact: there is generally a progressive amount of compaction from point to long contacts (involving a reorientation of grains), to concavo-convex and to sutured contacts (which both involve a degree of pressure dissolution).

Fig. 17.9 The contacts between the pebbles in this conglomerate show evidence of pressure solution occurring: the edges of some pebbles are indented into others, a 'pitted pebble' fabric which indicates that the conglomerate has undergone burial. Lower Palaeozoic, Connemara, Ireland (2 cm coin, bottom centre, for scale).

(2.4.7). The most common cement in sandy and gravelly sedimentary rocks deposited in marine environments is calcium carbonate. Calcite and aragonite in shells and skeletons of organisms are relatively easily dissolved and pore waters may become saturated with calcium carbonate, leading to reprecipitation under appropriate

chemical conditions. The carbonate cement may be in the form of *sparry* calcite, large crystals which grow out into the pore spaces, or may completely envelop individual clasts (Fig. 17.10) (Tucker & Wright 1990), or may occur as a fine-grained *micritic* precipitate between the grains. Silica is also abundant in sediments but is much less soluble than calcium carbonate. Silica cements are only commonly found in circumstances where there is an absence of calcium carbonate (e.g. in quartz-rich sands deposited in a continental environment). The breakdown of volcanic and other lithic fragments in sands leads to the formation of clay minerals which precipitate as cements between sand grains. This effect increases with burial depth as more feldspar grains break down. In practice it is often difficult to distinguish clay mineral cements from clays deposited with the sand as a matrix.

A feature of cementation is that if the cement and the adjacent grain are of the same chemical composition the cement may form an *overgrowth* extending the mineral grain into the pore space. Overgrowths are commonly seen in silica-cemented quartz sands; thin section examination reveals the shape of a quartz crystal formed around a detrital quartz grain. In carbonate rocks overgrowths of sparry calcite form over biogenic fragments of organisms such as crinoids and echinoids which are made up of single calcite crystals. These are referred to as *syntaxial overgrowths*.

Cementation lithifies the sediment into a rock and as it does so it reduces both the porosity and the *permeability* (the ease with which a fluid can pass through the rock).

Fig. 17.10 The voids between pebbles in a conglomerate have been partly filled by calcite which has formed in layers on the edges of the voids. Miocene fan delta deposits, near Mukalla, Yemen (3 cm coin, bottom right, for scale).

Cements form around the edges of grains and grow out into the pore spaces. This cement growth tends to block up the gaps between the grains and can result in a situation where a rock still has a moderate porosity but a very low permeability. Pore spaces can be completely filled by cement, resulting in a complete lithification of the sediment and a reduction of the porosity and permeability to zero.

17.3 Nodules and concretions

Cementation of a sediment rarely occurs evenly throughout the body of material. The heterogeneity of sediment influences the passage of pore waters which carry cementing minerals, and local chemical differences also affect cementation processes. In some cases the distinction between well-indurated patches of sediment and the surrounding body of material is very marked. Irregular cemented patches are normally referred to as *nodules*; more symmetrical, round or discoid features are called *concretions*.

Nodules and concretions can form in any sediment which is porous and permeable. They are commonly seen in sand beds (where large nodules are sometimes referred to as *doggers*), mudrock and limestone. Sometimes they may be seen to have nucleated around a specific feature, such as the body of a dead animal or plant debris, but in other cases there is no obvious reason for the localized cementation. Concretions formed at particular levels within a succession may coalesce to form bands of well-cemented rock (Fig. 17.11). A variety of different minerals can be the cementing medium, including calcite, siderite, pyrite and silica. There is evidence that some concretions in mudrocks form very soon after deposition from the pattern of compaction of mud around the body.

17.3.1 Flints and other secondary cherts

It was noted in Section 3.4 that cherts may form either from the direct precipitation of siliceous material on the sea floor to form sedimentary layers or as a result of the concentration of silica into nodules during diagenesis. In both cases the source of the silica is biogenic material. The tests of diatoms and radiolarians are an amorphous form of opal which crystallizes as microcrystalline opal as the material is buried. The opal further changes to microquartz and chalcedonic quartz during diagenesis. The radiating fabric of chalcedonic quartz replacing the

Fig. 17.11 Beds of the Jurassic Bridport Sandstone on the Dorset coast, southern England. The more prominent sandstone beds show a greater degree of cementation, making them more resistant to erosion.

skeleton of a radiolarian is a characteristic feature of deep marine cherts.

Secondary, diagenetically formed cherts are common in sedimentary rocks, particularly limestones. They are generally in the form of nodules sometimes coalesced to form layers. The clue to the diagenetic origin to these cherts is that they have a *replacement fabric*. Within the nodules the structures of organisms which originally had carbonate hard parts can be seen and the edges of a chert nodule may cut across sedimentary layering. They form by the dissolution of the original material and precipitation of silica, often on such a fine scale that original biogenic structures can still be seen.

The source of the silica is almost certainly the remains of siliceous organisms deposited with the calcareous sediment (Calvert 1974; Tucker 1991). These organisms are sponges, diatoms and radiolarians which originally have silica in a hydrated, opaline form. In shelf sediments sponge spicules are the most important sources of silica. The *opaline silica* is relatively soluble and it is transported through pore waters to places where it precipitates, usually around fossils or burrows as microcrystalline or chalcedonic quartz in the form of a nodule. *Flint* is the specific name given to nodules of chert formed in the Cretaceous Chalk.

17.4 Clastic diagenesis

The processes of compaction *(17.2.2)* are important in all terrigenous clastic rocks which are subject to burial. Overburden pressure expels water from between grains,

reducing the porosity, and the contact points between sand grains are subject to pressure solution. Cementation occurs as a result of the growth of authigenic minerals within beds of sand and mud. Minerals formed as cements within terrigenous clastic rocks during diagenesis include feldspar, calcite and clay minerals, as well as zeolite minerals and iron oxides which also may be important in colouring the rock.

17.4.1 Colour changes during diagenesis

The colour of a sedimentary rock can be very misleading when interpretation of the depositional environment is being attempted. It is very tempting to assume, for example, that all strongly reddened sandstone beds have been deposited in a strongly oxidizing environment such as a desert. Whilst an arid continental setting will result in oxidation of iron oxides in the sediment, changes in the oxidation state of iron minerals, the main contributors to sediment colour, can occur during diagenesis (Turner 1980). A sediment may be deposited in a reducing environment but if the pore waters passing through the rock long after deposition are oxidizing then any iron minerals are likely to be altered to iron oxides. Conversely, reducing pore waters may change the colour of a sediment from red to green.

Diagenetic colour changes are obvious where the boundaries between the areas of different colour are not related to primary bedding structures. In fine-grained sediments *reduction spots* may form around particles of organic matter: the breakdown of the organic matter draws oxygen ions from the surrounding material and results in a localized reduction of oxides from red or purple to green. Bands of colour formed by concentrations of iron oxides in irregular layers within a rock are called *liesegangen bands*. The bands are millimetre-scale and can look very much like sedimentary laminae. They can be distinguished from primary structures as they can cut across bedding planes or cross strata, and there is no grain size variation between the layers of liesegangen bands. They form by precipitation of iron oxides out of pore waters. Other colour changes may result from the formation of minerals such as zeolites (Fig. 17.12) which are much paler than the dark volcanic rocks within which they form.

17.4.2 Clay mineral diagenesis

When a mud is deposited in water it has a high water content which is lost as the sediment compacts and cements into a mudrock. In addition to the water between clay particles, clay minerals have water molecules within the lattice structure of the phyllosilicate. At temperatures over 100°C this water may also be driven off and the clay mineral dehydrates, further compacting the mudrock. Mineralogical changes also occur in clay minerals during diagenesis. Kaolinite and smectite tend to be replaced by illite and chlorite with increasing pressure and temperature; illite clay minerals show increasing crystallinity, a feature which can be recognized using an X-ray diffractometer *(2.5.4)* (Leeder 1982). Once metamorphic conditions are reached most clay minerals deposited as mudrock and in the matrix of sandstone are

transformed into the phyllosilicate minerals chlorite and sericite.

17.4.3 Diagenesis of organic matter in marine muds

Shallow marine environments are regions of high biogenic productivity *(11.7)* and the mud deposited on the sea floor of the shelf is rich in the remains of organisms. The diagenesis of this organic material takes place in a series of depth-defined zones (Curtis 1977; Burley *et al.* 1985). Organic material at the surface is subject to bacterial oxidation, a process which dominates the upper few centimetres of the sediment which is oxygenated by diffusion and bioturbation. Below this surface layer *sulphate reduction* takes place down to about 10 m (Fig. 17.13). Bacteria are involved in reducing sulphate ions to sulphide ions and in the same region ferric iron is reduced to ferrous iron. Under these reducing conditions calcite is precipitated and the ferrous iron reacts with the sulphide ions to form *pyrite* (iron sulphide). At deeper levels within the sediment pile no sulphate ions remain and the dominant reactions are bacterial fermentation processes which break down organic material into carbon dioxide and methane. Carbonate minerals such as calcite and siderite are precipitated in this zone which extends down to about 1000 m. At deeper levels any remaining organic matter is broken down inorganically.

17.5 Carbonate diagenesis

Cements in carbonate rocks are not surprisingly mainly

Fig. 17.12 Cretaceous rocks in Citadel Bastion cliff, Alexander Island, Antarctic Peninsula. This cliff face shows strong colour variations between darker and lighter layers. However, the pale colour is entirely due to the formation of laumontite, a zeolite mineral, during diagenesis and is not related to grain size or textural variations in the dark volcaniclastic sediments.

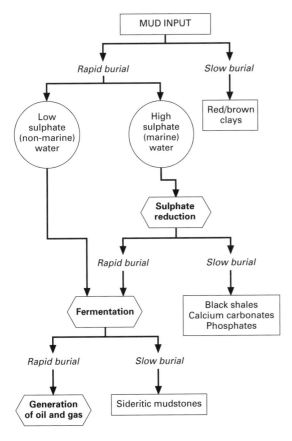

Fig. 17.13 Diagenesis of mudrocks containing organic matter. (After Curtis 1977.)

made up of calcium carbonate derived from the host sediment. Lithification of aggregates of carbonate material can occur contemporaneously with deposition—for example, in beachrock *(13.4.1)* where the carbonate debris washed up on the beach is cemented by calcium carbonate from water in the intertidal to supratidal zone. In warm tropical shallow marine environments the sea water is often saturated with respect to calcium carbonate and cementation takes place on the sea floor, sometimes involving microbial activity. In contrast, colder sea water is undersaturated with calcium carbonate and dissolution of carbonate material occurs.

If a carbonate sediment is buried cementation can occur by precipitation from *formation water* (water in the pores of the rock) during burial diagenesis. Alternatively, *meteoric diagenesis* occurs where carbonate rock is exposed at the surface. Rainwater is undersaturated with respect to calcium carbonate so dissolution of the

sediment occurs. This forms a *karst* surface (Fig. 17.14), a zone of solution of limestone which may be recognized in the stratigraphic record as a *palaeokarst* (Hird & Tucker 1988). Precipitation of calcium carbonate as cements in voids is also known from within the *vadose zone*, the region between the surface and the water table. A carbonate sediment may go through a number of stages of diagenetic dissolution and cementation if the groundwater conditions change through time due to changes in relative sea level or climate.

17.5.1 Compaction effects in limestones: bedding planes and stylolites

Limestone undergoes pressure dissolution under a few hundred metres of overburden. The pressure dissolution is often found to be distributed not evenly through the

Fig. 17.14 A present-day karst surface formed on Carboniferous limestone, County Clare, Ireland.

rock but at certain horizons (Bathurst 1987). Surfaces parallel to the horizontal become sites of pressure dissolution and are seen as distinct partings in the succession which are often thought of as bedding planes (Fig. 17.15). In fact they may not be bedding planes at all and can be seen to cut through the middle of graded beds or other sedimentary structures. The surfaces are accentuated in outcrop by slight concentrations of clay or other insoluble material which is left as a residue when the calcium carbonate is dissolved. Bedding in conglomerates, sandstones and mudstones is picked out by changes in lithology due to changes which occur at the time of deposition. In limestones, bedding planes may not represent primary depositional surfaces at all: they can be diagenetic in origin and may even cut across depositional bedding.

The pressure solution surfaces within a limestone are rarely flat, horizontal planes. Under greater overburden pressures the irregular nature of the surface becomes more apparent as *stylolites* form (Tucker & Wright 1990). These are surfaces within the rock which are seen as complex, highly irregular fine lines in vertical section. They are picked out by concentrations of iron oxide or clay left as a residue when the calcium carbonate dissolves. Extreme examples of pressure solution and *stylolitization* result in loss of most of the calcium carbonate, leaving only isolated nodules of limestone in a wavy-bedded mudstone. *Nodular limestones* of this type are likely to have contained a high proportion of insoluble clay either disseminated throughout the rock or, more commonly, concentrated into muddier layers. Pressure solution tends to accentuate irregular distributions of clay and limestone.

17.5.2 Dolomitization and dedolomitization

Dolomite is a calcium magnesium carbonate mineral ($CaMg(CO_3)_2$) which is found in carbonate sedimentary rocks of all ages, but its origin has long been something of an enigma. The few modern environments where dolomite is actively forming are places where there are hypersaline waters in lakes, lagoons and tidal flats (Shinn *et al.* 1965; Tucker & Wright 1990); it is not found to be forming in other modern carbonate environments. However, dolomite rocks in the stratigraphic record occur in successions in which fossils indicate normal marine conditions. Most of the dolomite in ancient rocks is therefore considered to be secondary in origin, and this hypothesis is supported by evidence from *replacement fabrics* within the carbonate rocks. Material which was clearly originally made up of calcite or aragonite has been wholly or partially replaced by dolomite. This replacement process is called *dolomitization*, and a number of different models for the conditions under which it may occur have been proposed (Land 1985; Tucker & Wright 1990), of which four main models are considered here (Fig. 17.16).

The precipitation of gypsum from marine waters in sabkha and hypersaline lagoon environments results in a decrease in the concentration of calcium ions and a relative increase in magnesium. Dense residual groundwaters with high magnesium concentrations sink through the limestone beneath the sabkha or lagoon where

Fig. 17.15 Cliff of bedded Jurassic limestone, Portland Bill, southern England: the bedding surfaces in a thick, carbonate succession like this are often planes along which pressure solution has taken place.

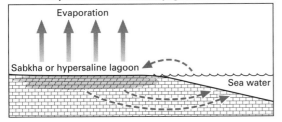

Evaporite brine residue/seepage reflux model

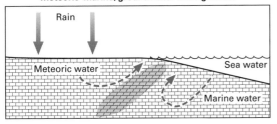

Meteoric–marine/groundwater mixing model

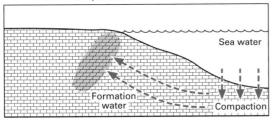

Burial compaction/formation water model

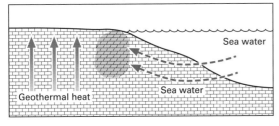

Sea water/convection model

Fig. 17.16 Four of the models proposed for the processes of dolomitization. (After Tucker & Wright 1990.)

dolomitization occurs. This is referred to as the *evaporite brine residue* or *seepage reflux* model for dolomitization. An alternative model is applicable under conditions where there are fresh, meteoric groundwaters mixing with sea water. The theory of this *meteoric–marine* or *groundwater mixing* model is based on the effect of mixing decreasing the saturation of calcite but leaving the waters saturated with respect to dolomite. Dolomitization would

occur in the subsurface within the zone of mixing of the marine and meteoric waters. Diagenesis of clay minerals at depth can lead to groundwaters enriched in magnesium passing up through overlying beds of limestone. The dolomitization in this *burial* or *formation water* model is enhanced by the raised temperatures at depth. *Sea water dolomitization* is also thought to occur when large volumes of cold sea water are driven through a carbonate platform by ocean currents or a geothermal gradient through the platform. The cold water is saturated with respect to dolomite but the saturation of calcite is reduced at low temperatures.

A reversal of the process which causes dolomitization in association with evaporites can cause dolomite to be replaced by calcite. This *dedolomitization* occurs when beds of evaporite rich in gypsum are dissolved, enriching groundwaters in calcium sulphate. Sulphate-rich waters passing through dolomitized limestone result in the solution of dolomite and the precipitation of calcite.

17.6 Post-depositional changes to evaporites

Evaporite minerals may be either dissolved out from beds in the subsurface or replaced by other, less soluble minerals such as calcite and silica. Dissolution by pore waters passing through the beds leaves vugs and caverns which collapse under the weight of the overburden, forming a *dissolution breccia*. Breccias formed in this way consist of angular pieces of the strata bedded with or immediately overlying the evaporite units. There is no sign of transport of the clasts, and the composition of the clasts will show a stratigraphy which reflects the order of the beds prior to collapse. Gaps between the clasts will be filled by a crystalline cement.

Initial burial and heating of gypsum leads to its replacement by anhydrite. However, if anhydrite beds are uplifted to a near-surface environment a change to gypsum may occur. Volume changes associated with these transitions may result in local deformation and disruption of the bedding. Replacement of halite, gypsum and anhydrite by calcite and silica may occur at any stage in diagenesis (McLane 1995). The original crystal form of halite as a cubic shape or the selenite form of gypsum may be preserved by the replacement and provide the evidence for the original mineralogy. These replacement forms are called *pseudomorphs*. Siliceous replacement of anhydrite minerals is in the form of microcrystalline or chalcedonic quartz.

17.7 Diagenesis of volcaniclastic sediments

All the crystal, lithic and vitric particulate material in a volcaniclastic deposit is relatively susceptible to diagenetic alteration, especially if the chemical composition of the material is intermediate or mafic. Lithic fragments of basalt and crystals of minerals such as hornblende, pyroxene and plagioclase feldspar are likely to react with pore waters to form clay minerals. Volcanic glass changes form in the absence of any other medium because it is metastable and *devitrifies* (changes from glass to mineral form) to form very finely crystalline minerals (Cas & Wright 1987).

Volcanic ashes in sedimentary succession may therefore be recognized only by the products of the diagenesis of volcaniclastic material. Two specific examples of this are *tonsteins* which are kaolinite-rich mudrocks formed from volcanic ashes (Spears & Kanaris-Sotirious 1979), and *bentonite*, formed mainly of smectite clays as alteration products of basaltic rocks. The interaction of volcaniclastic material and alkaline waters results in the formation of members of the zeolite group of minerals. Where the original volcanic material has been largely altered during diagenesis the only clues to the origin of the sediment may be the composition of the clays in a mudrock, especially if they are mainly smectite, and the relics of glass shards and mineral crystals preserved in the sediment.

17.8 Formation of coal and hydrocarbons

The branch of geology which has the greatest economic importance world-wide is the study of *fossil fuels* (coal, oil and natural gas). These compounds form by diagenetic processes which alter material made up of the remains of organisms. The places where the original organic material forms can be understood by studying depositional processes, but the formation of coal from plant material and the migration of volatile hydrocarbons as oil and gas require an understanding of the diagenetic history of the sedimentary rocks where they are found.

17.8.1 Coal formation

The process of *coalification* of peat to form brown and black coals *(3.7)* is a complex series of chemical and physical processes, including compaction (Fig. 17.17). The initial degradation of the humic compounds in peat to form coal is accomplished in the process of burying the plant material in the absence of air over a period of time. The *lignite* (soft brown coal) which forms first will increase in rank to sub-bituminous after a longer period of burial with elevated temperatures and pressures. Water is eliminated as the temperature rises to between 100°C and 200°C during the formation of *bituminous coals*. The ultimate rank reached by a coal will depend on the temperature and the period of time over which the

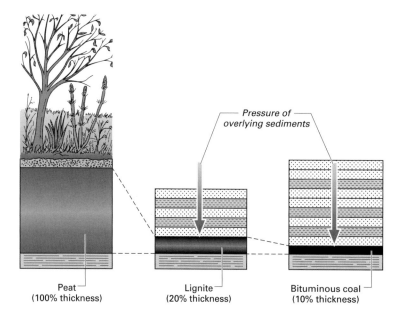

Fig. 17.17 The formation of coal from peat involves a considerable amount of compaction. (After Skinner & Porter 1987; copyright ©1987. Reprinted by permission of John Wiley & Sons, Inc.)

temperature is maintained. The temperature at any point in a sedimentary succession is determined by the *geothermal gradient*, the rate of increase of temperature with depth. The longer a coal is kept at a high temperature the higher the rank achieved. *Anthracite* coals with volatile (water, carbon dioxide, hydrogen and methane) contents amounting to less than 10% form as a result of long periods of burial at high temperatures. Coals of this rank will have undergone considerable compaction and may be only 5–10% of the thickness of the original peat layer (Fig. 17.17). The rank of a coal is determined in detail by assessing the reflectance of plant fragments preserved as vitrinite *(3.7.2)*. *Vitrinite reflectance* is an analytical technique which can be used to determine the burial and temperature history of organic material within a sedimentary succession; it is an important tool in working out the *geothermal history* of a sedimentary succession and hence its hydrocarbon prospectivity.

17.8.2 Hydrocarbon formation

The organic matter which ultimately forms petroleum starts off as planktonic algae which accumulate in anaerobic conditions in anoxic marine environments and in lakes. As the remains of these organisms break down during diagenesis they form a material called *kerogen* which is made up of long-chain hydrocarbons. This process is called *maturation*. Kerogen starts to generate shorter-chain hydrocarbons once temperatures of 70–100°C are reached which, with a geothermal gradient of around 30°C km^{-1}, involves burial to a depth of 2–3 km. This depth zone for the generation of liquid hydrocarbons is called the *oil window*, and it can be shallower or deeper depending on the geothermal gradient. Oil is not generated instantly and many millions of years of burial within the oil window are required for kerogen to release significant quantities of oil (Tissot & Welte 1984; Tucker 1991). With increased temperature (e.g. due to greater burial), the kerogen generates gas, and at over 150°C only methane is produced. Once 300°C is exceeded all hydrocarbons are broken down completely.

Oil is formed by the maturation of kerogen derived largely from planktonic algal matter. The breakdown of higher plants forms a type of kerogen which tends to yield gas and very little oil when it is buried. Similarly, burial of coal generates natural gas (principally methane) and no oil. Methane generated from coal may be stored in fractures in the coal seam as *coal bed methane* which is exploited economically in some areas.

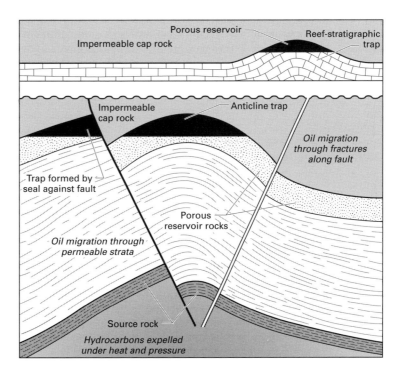

Fig. 17.18 Cartoon of the relationships between the source rock, migration pathway, reservoir, trap and cap rocks required for the accumulation of oil and gas in the subsurface.

17.8.3 Hydrocarbon accumulation

Hydrocarbons generated from kerogen are compounds which have a lower density than the formation water which is present in most sedimentary successions. They are also immiscible with water and droplets of oil or bubbles of gas tend to move upwards through the pile of sedimentary rocks due to their buoyancy. This *hydrocarbon migration* proceeds through any rock which is permeable until the hydrocarbons reach an impermeable barrier. These barriers are formed by impermeable rocks such as well-cemented lithologies, mudrock and evaporite beds known as *cap rocks*. The hydrocarbons will find their way around the barrier unless there is some form of *trap* which prevents further upward migration (Fig. 17.18). Traps are either structural configurations of strata (such as anticlines) or stratigraphic relationships (such as reefs or sand bars). In the absence of traps and caps the hydrocarbon reaches the surface and leaks to the atmosphere. Partial release of hydrocarbons from the subsurface as *oil seeps* and *gas seeps* can be important indicators of the presence of hydrocarbons. Accumulations are economic if the rock below the cap and trap is sufficiently porous and permeable to contain significant quantities of hydrocarbons in pore spaces, that is, it is a *reservoir rock*. Porous sandstone, oolitic limestone and reef boundstone are typical reservoir lithologies.

Exploration for economic reserves of hydrocarbons requires a knowledge of the depositional history of an area to determine whether suitable source rocks are likely to have formed and if there are any suitable reservoir and cap lithologies in the overlying succession. This analysis of the sedimentology is an essential part of oil and gas exploration. Knowledge of post-depositional events is also important to provide an assessment of the thermal and burial history which controls the generation of hydrocarbons. A further discussion of petroleum geology is to be found in specialist texts such as North (1985).

17.9 Diagenetic processes: summary

The geologist is often concerned with the study of sedimentary rocks rather than unconsolidated sediments. The processes which occur after the deposition of a sediment may produce subtle changes or fundamentally change the physical structure and chemical composition of the material. If sedimentary rocks are to be interpreted in terms of depositional processes and environments it is essential to have an appreciation of how diagenetic processes may have modified the original sediment. Post-depositional processes are also important economically in controlling the formation of resources such as coal, oil and gas.

Further reading

Burley, S.D., Kantorowicz, J.D. & Waugh, B. (1985) Clastic diagenesis. In: *Sedimentology, Recent Developments and Applied Aspects* (eds P.J. Brenchley & B.P.J. Williams), pp. 189–228. Blackwell Scientific Publications, Oxford.

Dickson, J.A.D. (1985) Diagenesis of shallow marine carbonates. In: *Sedimentology, Recent Developments and Applied Aspects* (eds P.J. Brenchley & B.P.J. Williams), pp. 173–188. Blackwell Scientific Publications, Oxford.

Leeder, M.R. (1982) *Sedimentology, Process and Product*, 344 pp. Unwin Hyman, London.

McLane, M. (1995) *Sedimentology*, 423 pp. Oxford University Press, Oxford.

North, F.K. (1985) *Petroleum Geology*, 607 pp. Allen & Unwin, Boston, MA.

Tucker, M.E. & Wright, V.P. (1990) *Carbonate Sedimentology*, 482 pp. Blackwell Scientific Publications, Oxford.

18 Stratigraphy: concepts and lithostratigraphy

The history of the Earth is recorded in a very incomplete manner in the rocks formed at different times and places. Evidence for changes in the form and shape of the oceans and continents, in global and local climates and in the environments in which different forms of life have evolved is to be found mainly in sedimentary strata. Stratigraphy provides the framework to unravel the events of the past few billion years. The first stage in a stratigraphic analysis is to determine the order in which rocks were formed. Beds of rock are grouped into units which have similar characteristics which are then considered in relation to other units. This process of grouping rock units by their physical characteristics is called lithostratigraphy; as well as being the fundamental technique in stratigraphic analysis, it provides the basis for the construction of geological maps.

18.1 Stratigraphy and geological time

There are two distinct issues in stratigraphy which are easily confused. The first is geological time, the millions of years which have passed since the formation of the Earth. The second is the material evidence from rocks, minerals and fossils for events that have occurred during Earth history. The events are often described in terms of time and the basis for determining the passage of time comes from geological materials, so the two issues are inextricably related. However, from an objective scientific view there are separate concepts to be considered.

18.1.1 Geological time

The passage of time since the formation of the Earth is divided into *geochronologic units*. These are divisions of time which may be measured in years or referred to by name. The Cretaceous Period, for example, was the time between 144 and 65 million years before present. This is analogous to historians referring to the time between 1837 and 1901 as the Victorian period. Geological time is normally expressed in millions of years or thousands of years before present (and to be precise, 'present' is defined as 1950; although this distinction is not necessary on a scale of millions of years!). Geological time units are abstract concepts, they do not exist in any physical sense.

The abbreviations used for dates are 'Ma' for millions of years before present and 'ka' for thousands of years before present. The time thousands of millions of years before present is abbreviated to 'Ga' (gigayears). The North American Stratigraphic Code (North American Commission on Stratigraphic Nomenclature 1983) suggests that to express an interval of time of millions of years in the past abbreviations such as 'my', 'm.y.' or 'm.yr' could be used. This convention has the advantage of distinguishing dates (e.g. 45 Ma) from intervals of time (e.g. 45 my), but it is not universally applied.

A hierarchical set of terms for the geological time units has been established (Table 18.1).

Eons. These are the longest periods of time, with the history of the Earth now commonly divided into three eons: the Archaeozoic Eon up to 2.5 Ga, the Proterozoic Eon from 2.5 Ga to 570 Ma, and the Phanerozoic Eon up to the present.

Eras. There are three time divisions of the Phanerozoic: the Palaeozoic Era up to 248 Ma, the Mesozoic Era from then until 65 Ma and the Cenozoic Era up to the present. Some Precambrian eras have also been defined.

Table 18.1 Geological time units and chronostratigraphic units.

Geological time units (Intervals of time measured in years)	Chronostratigraphic units (Material units defined by the ages of the rocks in them)
Eons (Phanerozoic Eon)	—
Eras (Mesozoic Era)	—
Periods (Jurassic Period)	Systems (Jurassic System)
Epochs (Lias Epoch)	Series (Lias Series)
Ages	Stages (Toarcian Stage)
Divisions into 'Early', 'Middle' and 'Late'	Divisions into 'Lower', 'Middle' and 'Upper'

Periods. The basic unit of geological time is the period, and this is the division most commonly used when referring to Earth history. The Mesozoic Era, for example, is divided into three periods: the Triassic Period, the Jurassic Period and the Cretaceous Period.

Epochs. These are the major divisions of periods. Some have names (e.g. the Llandovery, Wenlock, Ludlow and Pridoli in the Silurian). Others are simply Early, Middle and Late divisions of the period (e.g. Early, Middle and Late Devonian).

Ages. The smallest commonly used divisions of geological time are ages. These are typically a few million years in duration. The Oligocene Epoch is divided into the Rupelian and Chattian Ages, for example.

Chrons. Short periods of time are sometimes determined from palaeomagnetic information, but these units do not have widespread usage outside of magnetostratigraphy *(20.2)*.

18.1.2 Differences in geological time terminology and usage

1 The Carboniferous Period is recognized as a single period in Europe, but in North America it is considered to be two periods, the Mississippian and the Pennsylvanian Periods.
2 The hierarchy of divisions of the Cenozoic has been revised and different schemes are in common use. The Tertiary was formerly considered to be an era and the subdivisions (Palaeocene through to Pliocene) were considered to be periods. It is now more common to consider the Cenozoic Era to be made up of two periods, the Tertiary Period (divided into two sub-periods, the Palaeogene and the Neogene) and the Quaternary Period. In this scheme Palaeocene, Eocene, Oligocene, Miocene and Pliocene are epochs.

18.1.3 Material stratigraphic units

In contrast to geological time, stratigraphic units are based on material entities. Five principal types of material stratigraphic unit are recognized.

Lithostratigraphic unit: a body of rock which is distinguished and defined by its lithological characteristics and its stratigraphic position relative to other bodies of rock. Lithostratigraphic units are generally layered sedimentary rocks. Bodies of metamorphic or igneous rocks which are not layered may be referred to as *lithodemic* units *(18.4)*.

Biostratigraphic unit: a body of rock which is defined and characterized by its fossil content. The fossil content is used to place the rock in an ordered succession relative to other rocks which contain fossils of older or later lineages (Chapter 19).

Allostratigraphic unit: a body of rock defined by its position relative to unconformities or other correlatable surfaces which reflect changes in base level during the deposition of the sedimentary succession. 'Sequence stratigraphy' is an allostratigraphic approach to the definition of material stratigraphic units (Chapter 21).

Magnetostratigraphic unit: a body of rock which exhibits magnetic properties which are different from those of adjacent bodies of rock in the stratigraphic succession. The main property used is the polarity of the remnant magnetism relative to the present-day magnetic field *(20.2)*.

Chronostratigraphic unit: a body of rock which has lower and upper boundaries which are each isochronous surfaces. An *isochronous* surface is one which formed at the same time everywhere. A defined chronostratigraphic unit is a reference for all other rocks formed within the same span of time and provides a means for organizing rocks on the basis of their age relations.

18.2 Chronostratigraphy

At first the distinction between chronostratigraphic units and units of geological time is not obvious. A chronostratigraphic unit is defined by obtaining ages for the rocks within it either directly from isotopic ages *(20.1)* or by calibration of biostratigraphic information (Chapter 19). A chronostratigraphic unit is a physical entity rather than an abstract concept. Chronostratigraphic units have direct equivalents in geological time units (Table 18.1).

System: the equivalent of a period, and the name given to the rocks which were formed during the span of time represented by the period. The names given are the same: for example, the Jurassic System rocks are those which were formed during the Jurassic Period.

Series: rocks which constitute a system are divided into series, which are the material equivalents of the spans of time called epochs. Again, the names are the same, with Miocene Series rocks formed in the Miocene Epoch.

Stage: the material equivalent of the time represented by an age. An outcrop of rocks may be called Tortonian Stage strata if they were deposited during the Tortonian Age of the Miocene Epoch.

There are also material equivalents to the longer time periods (eonothem and erathem) but these terms are rarely used. The rock unit equivalent of a chron is a *chronozone*.

18.2.1 Use of terminology: time and rock, early and lower

It is easy to confuse chronostratigraphic and geochronologic terminology. The difference in usage is between the rock itself and the time when the rock was formed. The chalk rocks in north-west Europe form part of the Cretaceous System and were deposited in shallow seas which existed over the area during the Cretaceous Period. Whenever the rock itself is being referred to, chronostratigraphic terminology is used (system, series, stage) whereas when a period of time is being considered, the geochronologic time units are used (period, epoch, age).

The adjectives 'early', 'middle' and 'late' are used in conjunction with geochronologic time units. The equivalent material rock units are 'lower', 'middle' and 'upper'. Hence we can more specifically state that the chalk rocks are the main lithologies seen in the Upper Cretaceous System, as the appropriate shallow marine conditions for their formation existed predominantly during the Late Cretaceous Period. These adjectives are capitalized when a span of time has been formally subdivided (e.g. the Early Permian and Late Permian Epochs). Where the subdivision of time is informal a lower-case initial letter is used (e.g. the early and late Silurian, because this period is formally divided into four named epochs). The same rules apply to the equivalent chronostratigraphic units: Lower and Upper Permian are formal divisions of the rock units, lower and upper Silurian are informal chronostratigraphic divisions.

18.2.2 Applications of chronostratigraphy

One of the objectives of sedimentological and stratigraphic analysis of rocks is the reconstruction of palaeoenvironments. The facility to determine the spatial and temporal distribution of environments of deposition is of academic interest in helping us understand how the surface of the planet has evolved. It is also of economic interest as a means of determining the position and extent of rock types which contain mineral or hydrocarbon reserves.

In order to reconstruct the palaeoenvironment at any time in the past it is necessary to know what depositional environments existed and where they were to be found at that time. This requires a knowledge of chronostratigraphy. In order to be able to compare rocks formed in different depositional settings at the same time a *time-line* through the succession of rocks across the sedimentary basin is required. The environment of deposition is determined from the sedimentary facies of the rocks that the time-line intersects and used to reconstruct a *palaeogeography*. With a series of chronostratigraphic surfaces in the body of rock it is possible to make a series of reconstructions and to develop a model for the evolution of palaeoenvironments.

Chronostratigraphy is a key part of the analysis of sedimentary rocks, but it is not a simple matter of looking to obtain a numerical age of formation from all of the rocks. A numerical age can only be determined from a limited range of rock types *(20.1)* and more indirect lines of evidence must be used in order to establish chronostratigraphic units. These involve all the other aspects of stratigraphy: lithostratigraphy, allostratigraphy, biostratigraphy and magnetostratigraphy, as well as isotopic age determinations.

18.3 Physical stratigraphy

The study of stratigraphy started with a few basic principles which are just as important today: they form the starting point of any stratigraphic analysis. These relationships should be considered before any other, more detailed, techniques are used.

18.3.1 Stratigraphic relationships

SUPERPOSITION

Provided the rocks are the right way up *(18.3.2)* the beds higher in the stratigraphic sequence of deposits will be younger than the lower beds. This rule can be simply applied to a layer-cake stratigraphy but must be applied

with care in circumstances where there is a significant depositional topography (e.g. forereef deposits may be lower than reef crest rocks: Fig. 18.1).

UNCONFORMITIES

A break in sedimentation with erosion of the underlying strata provides a clear relationship in which the beds below the unconformity are clearly older than those above it (Figs 18.2 & 18.3). All rocks which lie above the unconformity or a surface which can be correlated with it must be younger than those below. This is one of the principles on which the concept of sequence stratigraphy has been built (Chapter 21).

CROSS-CUTTING RELATIONSHIPS

Any unit which has boundaries which cut across other strata must be younger than the rocks it cuts. This is most commonly seen with intrusive bodies such as batholiths on a larger scale and dykes on a smaller scale (Fig. 18.4).

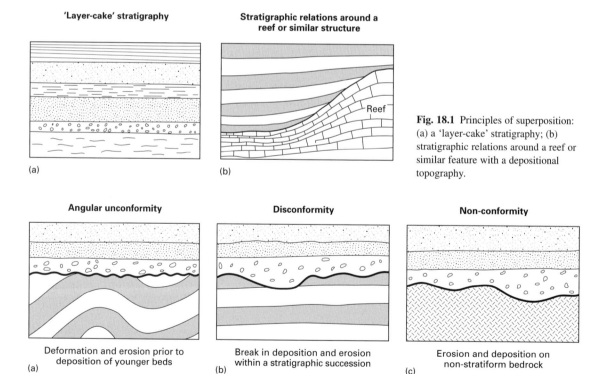

'Layer-cake' stratigraphy

(a)

Stratigraphic relations around a reef or similar structure

Reef

(b)

Fig. 18.1 Principles of superposition: (a) a 'layer-cake' stratigraphy; (b) stratigraphic relations around a reef or similar feature with a depositional topography.

Angular unconformity

Deformation and erosion prior to deposition of younger beds

(a)

Disconformity

Break in deposition and erosion within a stratigraphic succession

(b)

Non-conformity

Erosion and deposition on non-stratiform bedrock

(c)

Fig. 18.2 Types of stratal boundary: (a) angular unconformity; (b) disconformity; (c) non-conformity.

Fig. 18.3 An angular unconformity between dipping Eocene limestone beds and horizontal Oligocene conglomerate beds, near Graus, northern Spain.

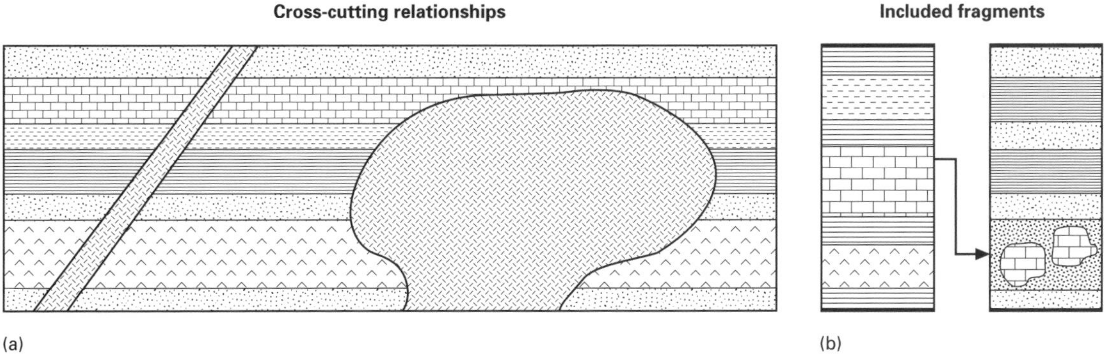

Cross-cutting relationships

Included fragments

(a)

(b)

Fig. 18.4 Stratigraphic relationships: (a) cross-cutting relations; (b) included fragments.

This relationship is also seen in sedimentary or *neptunian dykes* which form by younger sediments filling a crack or chasm in older rocks. These are most often seen as cracks in limestone formed by karstic solution.

INCLUDED FRAGMENTS

This is a common-sense rule which dictates that the clasts in a clastic rock must be made up of a rock which is older than the strata in which they are found (Fig. 18.4). The same relationship holds true for igneous rocks which contain clasts of the surrounding country rock as *xenoliths* (literally 'foreign rocks'). This relationship can be useful in determining the age relationship between rock units which are some distance apart. Pebbles of a characteristic lithology can provide conclusive evidence that the source rock type was being eroded by the time a later unit was being deposited tens or hundreds of kilometres away.

18.3.2 Way-up indicators in sedimentary rocks

The folding and faulting of strata during mountain building can rotate whole successions of beds (formed as horizontal or sub-horizontal layers) through any angle, resulting in beds which may be vertical or completely overturned. In any analysis of deformed strata, it is essential to know the direction of *younging*—that is, the direction through the layers towards younger rocks. The direction of younging can sometimes be worked out from evidence on a broad scale—for instance, by dating the rocks using fossils or radioactive ages, or by recognizing unconformities which must always place younger rocks on older. It is also useful to have information from a more local scale. The features used are mostly sedimentary structures which are consistently asymmetric and can indicate the *way up* of the beds (Fig. 18.5).

18.4 Lithostratigraphy

In *lithostratigraphy* rock units are considered in terms of the lithology of the strata and their relative stratigraphic positions. Biostratigraphic and isotopic ages form no part of the definitions of lithostratigraphic units. The basic unit of lithostratigraphic division of rocks is the

formation. Formations may be divided into members and assembled into groups.

18.4.1 Definition of a formation

The definitions used by the North American Stratigraphic Code (North American Commission on Stratigraphic Nomenclature 1983) and the British Stratigraphic Code (Whittaker *et al.* 1991) require that a *formation* should be a 'body of material which can be identified by its lithological characteristics and by its stratigraphic position'. It must be mappable at the surface or traceable

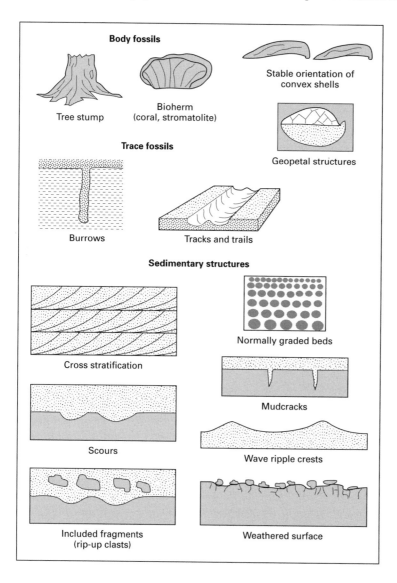

Fig. 18.5 Way-up indicators in sedimentary rocks.

in the subsurface. A formation should have some degree of lithological homogeneity and its defining characteristics may include chemical and mineralogical composition, texture, primary sedimentary structures and fossil content, in addition to essential lithological composition. Note that the material does not necessarily have to be lithified and that all the discussion of terminology and stratigraphic relationships considered below applies equally to unconsolidated sediment. However, as geologists are most commonly dealing with lithified material most of the following sections refer to rock units for simplicity.

A formation is not defined either in terms of its age by isotopic dating or in terms of biostratigraphy. Information about the fossil content of a mapping unit is useful in the description of a formation but the detailed taxonomy of the fossils which may define the relative age in biostratigraphic terms does not form part of the definition of the formation or other lithostratigraphic unit. The 'stratigraphic position' refers to the relationship of the formation to others which lie above, below or lateral to it. A formation may be, and often is, a *diachronous* unit, that is, a deposit with the same lithological properties which was formed at different times in different places *(18.6)*.

18.4.2 Description of a formation

A geologist making a map of an area will normally describe a formation in terms of the following.
• *Lithology*: the field characteristics of the rock (e.g. an oolitic grainstone, interbedded coarse siltstone and claystone, a basaltic lithic tuff, etc.).
• *Sedimentary structures*: ripple cross laminations, normal grading, etc.
• *Petrography*: the mineralogical composition and textural characteristics as determined in thin section.
• *Definition of top and base*: the criteria which are used to distinguish beds of this unit from those of underlying and overlying units; this is most commonly a change in lithology from, say, calcareous mudstone to coral boundstone.
• *Thickness*: as measured in the type section, and the range of thicknesses of the unit if it is not constant.
• *Faunal/floral content*: body and trace fossils, used for environmental analysis and in determining the age of the unit.
• *Depositional environment*: determined from the characteristics noted above.

• *Age*: as determined by fossil content, radiometric dating or relationships with other rock units, but remember that this does not form part of the definition of the formation.
• *Geographical extent*: the geographical area over which the unit is recognized as a formation.
• *Type section*: normally the place where the unit is best exposed and has been measured and described in detail; the name of the formation is normally taken from the type section.
• *Former definitions*: terminology and definitions used by previous workers and how they differ from the usage proposed.

It must be emphasized that only the lithological and textural characteristics (particularly those recognizable in the field) and the boundaries with other units form part of the definition of the formation. These are features which are purely descriptive and do not rely on interpretation of data. A formation would not be defined as, for example, 'rocks of Burdigalian age', because an interpretation of the fossil content or isotopic dating information is required to determine the age. Similarly, a formation should not be defined in terms of depositional environment, for example, 'lagoonal deposits', as this is an interpretation of the lithological characteristics.

18.4.3 Subdivision of formations: members and beds

A formation may be divided into smaller units in order to provide more detail on the distribution of lithologies. The term *member* is used for rock units which have limited lateral extent and are consistently related to a particular formation (or, rarely, more than one formation). An example would be a formation composed mainly of sandstone but which included beds of conglomerate in one or more parts of the area of outcrop. A number of members may be defined within a formation (or none at all) and the formation does not have to be completely subdivided in this way: some parts of a formation may not have a member status.

Occasionally an even smaller scale of lithostratigraphic division is applied. Individual *beds* or sets of beds which are very distinctive by virtue of their lithology, fossil content or chemistry may be named. These beds may have economic significance or be useful in correlation because of their easily recognizable characteristics across an area. They may be only decimetres or metres thick. Named beds may sometimes be included on geological maps although they are commonly too thin to be represented as anything more than a single line. Definitions of members and beds are made purely on the basis of lithologies and stratal relationships in the same way as formations are defined.

18.4.4 Sets of formations: groups and supergroups

Where two or more formations are found associated with each other and share certain characteristics they are considered to form a *group*. Groups are commonly bound by unconformities which can be traced over some distance. The constituent formations of a group can change laterally if formations are not laterally continuous. Not all formations are included in groups and some may exist as separate entities. A formation cannot cross group boundaries. In some circumstances, a group may change laterally into a formation of the same name if the characteristics of the individual formations are not present. Unconformities which can be identified as divisions in the stratigraphy over the area of a continent are sometimes considered to be the bounding surfaces of an association of two or more groups known as a *supergroup*.

18.4.5 Scales of lithostratigraphic units

There are no formal upper or lower limits to the thickness and extent of rock units defined as a member, formation or group. The variability of rock types within an area will be the main constraint on the number and thickness of lithostratigraphic units which can be described and defined. Quality and quantity of exposure will also play a role as finer subdivision is possible in areas of good exposure. The thickness of a lithostratigraphic unit is also often a function of the degree of detail to which an area has been studied. In southern and eastern Britain the study of Jurassic strata over several centuries has resulted in a very fine subdivision, including some formations which are only a few metres thick (Sellwood & Jenkyns 1975). In the eastern islands of Indonesia, where little geological fieldwork has been carried out, a single, defined formation may be over a thousand metres thick (Nichols & Hall 1991).

18.4.6 Type sections and type areas

When a formation is defined reference is always made to a *type section*. This is the location where the lithological characteristics are clear and, if possible, where the lower and upper boundaries of the formation can be seen. Sometimes it is necessary for a type section to be a composite within a *type area* with different sections described from different parts of the area. The type section will normally be described and presented as a graphic sedimentary log with the base and often the top clearly marked. It should be possible for any other geologist to visit the type section and see the boundaries and the lithological characteristics described. The boundaries of a formation or member should be distinct changes in lithology. In cases where there is a gradation from one lithology to another an arbitrary point has to be chosen and defined. An example is a succession which is all mudrock of one formation at the base and limestone of another at the top but with interbedded lithologies occurring in between: the boundary may be chosen at the first limestone bed.

The information from the type section provides the basis for the *stratotype* of a lithostratigraphic unit. A common convention is for only the base of a unit to be defined at the type section. The top is taken as the defined position of the base of the overlying unit. This convention is used because at another location there may be beds at the top of the lower unit which are not present at the type locality: these can be simply added to the top with no need to redefine the formation boundaries. It is important for stratigraphic correlation that boundaries between units should be unambiguous such that it can be established whether any comparable succession lies above or below a defined lithostratigraphic boundary.

18.5 Lithostratigraphic nomenclature

To minimize confusion and standardize the use of terminology, international commissions have collectively spent years in committee sessions establishing codes of practice in stratigraphic nomenclature. The minutiae of these deliberations are set out in publications such as 'A guide to stratigraphical procedure' published by the Geological Society of London (Whittaker *et al.* 1991) and the North American Stratigraphic Code published by the American Association of Petroleum Geologists (North American Commission on Stratigraphic Nomenclature 1983). The rules governing definitions and nomenclature must be followed if a rock unit is to be defined in the literature as a formation or other stratigraphic unit.

Establishing a name for a formation does not offer as

much scope as the geologist might wish: the Dead Duck Formation would not be allowed (even though it might be memorable for the person who mapped the area), unless there was a place called Dead Duck and this was the site of the type section. The name of the formation, group or member is taken from a distinct and permanent geographical feature as close as possible to the type section. The lithology is often added to give a complete name such as the Kingston Limestone Formation, but this is not essential, or necessarily desirable, if the lithological characteristics are varied. Sometimes the word 'formation' is omitted from the formal name—for example, Greenville Shale.

The choice of geographical name should be a feature or place marked on topographic maps such as a river, hill, town or village. Names which are liable to change should be avoided, although in remote areas the stratigrapher may have little scope in the choice of names. It is not permissible to use a name which is already in use or to use the same name for two different ranks of lithostratigraphic unit.

Members always have the word 'member' in the name and the lithological term is optional: Redham Shale Member or Redham Member would both be acceptable. Groups and supergroups are named from a geographical locality, with a lithological term optional. There are, of course, some exceptions to these rules of nomenclature which are largely due to historical precedent. It is simpler and less confusing to leave a well-established name as it is rather than dogmatically to revise it.

18.5.1 Informal stratigraphic nomenclature

The rules of stratigraphic nomenclature are enforced in published literature to ensure a conformity of usage. Informal use of stratigraphic terminology is widespread and necessary for individual workers or groups of workers during the course of developing the stratigraphic framework for an area. When an area is being mapped it is often useful to take an objective view and start from a blank stratigraphic column. The lithostratigraphy developed objectively can then be compared with published schemes from the same or adjacent areas. New information can necessitate changes in nomenclature.

18.5.2 Changing lithostratigraphic nomenclature

Once a formation has been defined it should not be necessary to change the definition: the lithology at the type section will always be the same and if the boundaries have been chosen to be at distinct changes in lithological characteristics it should not be necessary to move them. Nevertheless modifications are sometimes required, although there is a sensible resistance to making radical changes because constant revision of definitions causes confusion in scientific literature.

Revision of stratigraphic nomenclature becomes necessary when more detailed work is carried out or more information becomes available. A new exposure of rock due to a new road cut or quarry may clarify stratigraphic relations and lithological descriptions. New, detailed work in an area may allow a formation to be subdivided more effectively. In these cases a formation may be elevated to the rank of group and members may become formations in their own right. For the sake of consistency the geographical name is retained when the rank of the unit is changed.

Even if a new, more complete section through a formation is found the type section is not moved from its original locality: information from this new reference section may be used to supplement the descriptions at the type section.

18.5.3 Rock units which are not stratiform

The concepts of division into stratigraphic units were developed for rock units which are essentially layers. These layers may vary in thickness and be laterally discontinuous but sedimentary rocks and some volcanic rocks are largely stratiform in nature. However, some volcanic rocks are not layered, such as the plugs formed in the vents of volcanoes and in many cases plutonic and metamorphic rocks occur in non-stratiform units. These units do not necessarily follow the rules of superposition and a slightly modified lithostratigraphic terminology is used.

Non-stratiform bodies of rock are called *lithodemic* units. The basic unit is the *lithodeme*, equivalent in rank to a formation. They are defined on lithological criteria and should have established boundaries within a type area. The word 'lithodeme' is itself rarely used in the name: the body of rock is normally referred to by its geographical name and lithology, such as the White River Granite or Black Hill Schist. An association of lithodemes which share lithological properties such as a similar metamorphic grade is referred to as a *suite*, although the term *complex* is also used as the equivalent to a group for volcanic rocks.

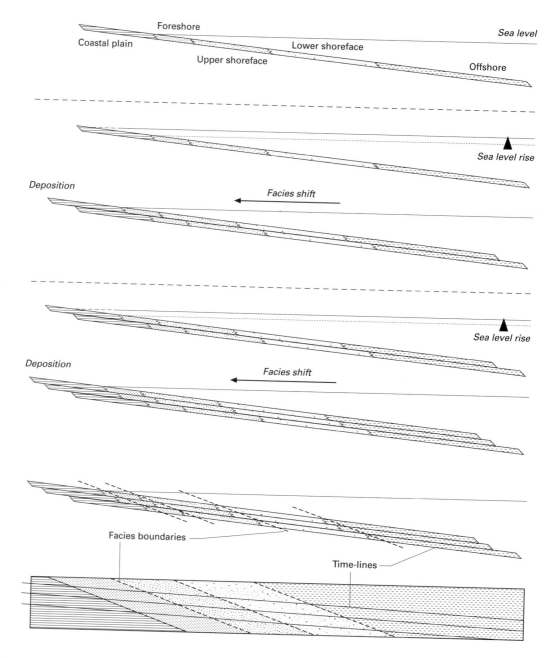

Fig. 18.6 Relationships between the boundaries of
lithostratigraphic units (defined by lithological characteristics
resulting from the depositional environment) and time-lines in
a succession of strata formed during gradual sea level rise
(transgression).

18.6 Lithostratigraphy and environments

It is clear from the earlier chapters on the processes and
products of sedimentation that the environment of deposition has a fundamental control on the characteristics of
a rock unit. A formation is therefore likely to be composed of strata deposited in a particular sedimentary

environment. This has two important consequences for any consideration of formations in any chronostratigraphic (time) framework.

First, in any modern environment it is obvious that fluvial sedimentation can be occurring on land at the same time as deposition is proceeding on a beach, on a shelf and in deeper water. In each environment the characteristics of the sediments will be different and hence they would be considered to be different formations if they are preserved as sedimentary rocks. It inevitably follows that formations have a limited lateral extent, determined by the area of the depositional environment in which they formed and that two or more different formations can be deposited at the same time.

Second, depositional environments do not remain fixed in position through time. Consider a coastline (Fig. 18.6) where a sandy beach (foreshore) lies between a vegetated, muddy coastal plain and a shoreface succession of mudstones coarsening upward to sandstones (see Fig. 13.9a). The foreshore is a spatially restricted depositional environment: it may extend for long distances along a coast, but seawards it passes into the shallow marine, shoreface environment and landwards into continental conditions. The width of deposit produced in a beach and foreshore environment may therefore be only a few tens of metres. However, a foreshore deposit covering a much larger area can be generated by the gradual rise or fall of sea level relative to the land. If sea level is slowly rising the shoreline will move landwards and after an interval of time the place where sands are being deposited on a beach could have moved several hundreds of metres (Fig. 18.6). The main depositional environments, the coastal plain, the sandy foreshore and the shoreface, will each have distinct lithological characteristics which would allow them to be distinguished as mappable formations. The foreshore deposits could therefore constitute a formation, but it is also clear that the beach deposits were formed earlier in one place (at the seaward extent) than another (at the landward extent). The same would be true of formations representing the deposits of the coastal plain and shoreface environments: through time the positions of the depositional environments migrate in space. From this example, it is evident that the body of rock which constitutes a formation can be diachronous *(18.4.1)*. Most importantly, the upper and lower boundaries of the formation are time-transgressive, diachronous surfaces.

A formation is typically composed of the sediments formed in a particular depositional environment, such as a playa lake in a desert setting. Other deposits also found in a desert might be aeolian sand dunes and alluvial fan conglomerates: each of these could constitute a formation, but as they are closely associated in both time and space these formations would be likely to constitute a group. A change of lithofacies to a different depositional setting (e.g. due to marine flooding of the area) would mark the boundary of the group. An example of a member in this context could be local screes of debris found within the aeolian sands at the margins of the basin. These would constitute a conglomerate or breccia member of the formation which is predominantly sandstone.

18.7 Lithostratigraphy and geological maps

Geological maps are familiar methods of displaying the distribution of rocks in an area. Some show the distribution of rocks on the basis of their age. This is the commonest practice for maps which cover a large area, such as a whole country or a continent. A colour code is used for each age using the stratigraphic column for the area as the basis of division. Other maps at this scale may show tectonic features, such as the areas affected by orogenies at different stages in the history of the area.

The chronostratigraphic subdivision of the rocks is logical at these scales, but at larger scales, or in areas where rocks are all more or less the same age, this method is no longer appropriate. A method of subdividing the rock units of an area principally on the basis of their lithology and stratigraphic position is required. A geological map at a scale of 1 : 50 000 or 1 : 100 000 will normally show areas of outcrop of rock units which have been defined in terms of their lithofacies (lithology and sedimentary structures: *5.2*) and their position relative to other bodies of rock in a stratigraphic column (normally presented along the side of the map sheet). This basic unit of mapping is therefore the formation, defined as a rock unit which has consistent characteristics which can be recognized and traced across an area.

Formations, members and groups used in making a geological map represent the distribution of lithofacies. Some sort of age control is usually used to supplement the lithofacies data, normally from biostratigraphic information or occasionally radiometric dating of certain rock units. The age of strata which are barren of biostratigraphically useful fossils and cannot be dated by other means can only be determined by comparison with rock units above or below which can be dated. In most cases, the lithostratigraphic units used in making a geological map can be ascribed with an age, although the precision may vary considerably. However, the resolution of the dating is commonly insufficient to distinguish different formations clearly on the basis of their age, or to demonstrate the diachroneity of a formation.

18.8 Lithostratigraphy and correlation

In most cases, correlation is attempted within some sort of temporal framework—that is, we are trying to determine which rock units were formed at the same time, which are older and which are younger. Correlation on the basis of lithostratigraphy is therefore fraught with problems because lithostratigraphic units are likely to be diachronous. In the example of the lithofacies deposited in a beach environment during a period of rising sea level (Fig. 18.6) the lithofacies is of different ages in different places. Therefore the upper and lower boundaries of this lithofacies will cross *time-lines* (imaginary lines drawn across and between bodies of rock which represent a moment in time).

If we are to attempt to reconstruct the palaeogeography of an area we need to know what was happening at the same time in different places. Chronostratigraphy (Chapter 20) provides an absolute time-scale but is not easy to apply because only certain rock types can be usefully dated. Biostratigraphy provides the most widely used time framework, a relative dating technique which can be related to an absolute time-scale, but it often lacks the precision required for reconstructing environments

and in some depositional settings appropriate fossils may be partly or totally absent (e.g. in deserts). Palaeomagnetic reversal stratigraphy provides time-lines, events when the Earth's magnetism changed polarity, but is difficult to apply in many circumstances. The concept of sequence stratigraphy provides an approach to analysing succession of sedimentary rocks in a temporal framework. In practice a number of different correlation techniques (Chapters 19, 20 and 21) are used in developing a temporal framework for rock units.

Further reading

Blatt, H., Berry, W.B. & Brande, S. (1991) *Principles of Stratigraphic Analysis*, 512 pp. Blackwell Scientific Publications, Oxford.

Boggs, S. (1995) *Principles of Sedimentology and Stratigraphy*, 2nd edn, 774 pp. Prentice Hall, Upper Saddle River, NJ.

Doyle, P., Bennett, M.R. & Baxter, A.N. (1994) *The Key to Earth History: An Introduction to Stratigraphy*, 231 pp. Wiley, Chichester.

Friedman, G.M., Sanders, J.E. & Kopaska-Merkel, D.C. (1992) *Principles of Sedimentary Deposits: Stratigraphy and Sedimentology*, 717 pp. Macmillan, New York.

North American Commission on Stratigraphic Nomenclature (1983) North American Stratigraphic Code. *American Association of Petroleum Geologists Bulletin* **67**, 841–875.

Prothero, D.R. & Schwab, F. (1996) *Sedimentary Geology: An Introduction to Sedimentary Rocks and Stratigraphy*, 575 pp. W.H. Freeman, New York.

Whittaker, A., Cope, J.C.W., Cowie, J.W., Gibbons, W., Hailwood, E.A., House, M.R., Jenkins, D.G., Rawson, P.F., Rushton, A.W.A., Smith, D.G., Thomas, A.T. & Wimbledon, W.A. (1991) A guide to stratigraphical procedure. *Journal of the Geological Society of London* **148**, 813–824.

19 Biostratigraphy

The value of fossils in the interpretation of sedimentary environments has been discussed in earlier chapters. Whilst they have proved to be useful in sedimentology as indicators of depositional environment, fossils can be considered to be essential to stratigraphy. The occurrence of fossils in strata led to the early detailed division of the stratigraphic column, and biostratigraphy now provides a high-resolution technique in stratigraphic analysis. The evolution of organisms into new species which can be identified in the rocks provides a framework of events which can be considered to have occurred at specific points in geological time. These events in evolution cannot be directly dated, but the age of the rocks in which the fossils occur can be constrained using combinations of physical stratigraphic relationships and radiometric dating techniques. Correlation between biostratigraphic units and the geological time-scale provides the temporal framework for the analysis of successions of sedimentary rocks.

19.1 Strata and fossils

The study of fossils has provided a wealth of information about the history of life on Earth. Skeletons and shells of animals or pieces of plant found as fossils which have no living counterparts are clear evidence that life on Earth has changed through geological time. The more specta-cular of these fossils tend to capture the imagination with visions of times in the past when, for example, dinosaurs occupied ecological niches on land, in the sea and even in the air. Even casual fossil hunting reveals the remains of aquatic animals such as ammonites and fragments of plants which are unlike anything we see living around us today.

Cataloguing the fossils found in sedimentary rocks carried out in the 18th and 19th centuries provided the first clues about the passage of geological time (Conkin & Conkin 1984; Fritz & Moore 1988). Early scientists and naturalists observed that different rock units contained either similar fossil remains or assemblages of fossils which were quite different from one unit to another. Moreover, the units which contained the same fossils could sometimes be traced laterally and shown to be part of the same layer. Those with different fossils could be shown by general stratigraphic principles (18.3.1) to be either younger or older. The rocks which contained a particular fossil type were often the same lithology, but, crucially for the development of stratigraphy, sometimes the same fossil type was found in a different rock type.

As the study of fossils, the science of palaeontology, developed it became evident that there were patterns in their distribution. Certain types of organism were found to be dominant in particular groups of strata. This led to the erection of the scheme of systems which were ini-tially grouped into deposits formed in three eras of geological time: 'ancient life', the Palaeozoic; 'middle life', the Mesozoic; and 'recent life', the Cenozoic. The actual time periods that these represented were pure speculation when these concepts were first introduced in the 19th century. The occurrence of certain types of fossil in particular stratigraphic units was simply an obser-vation at this stage. An explanation for the distribution of the fossils of organisms in the stratigraphic record came once ideas of the evolution of life on Earth were developed.

19.1.1 Evolutionary trends

There is a general trend of increasing complexity and sophistication of life forms starting from those which occur in older rocks and finishing with the biosphere around us today. Along with this trend, there is evidence of the emergence and diversification of organisms as well as signs of decline of some groups from abundance to insignificance or even extinction. Looking at particular groups, we see that many of them display trends in form through the stratigraphic record. The morphologies change in ways which are attributed to the evolutionary development of that group of organisms. It is one of the fundamental precepts of evolutionary theory that these changes are a one-way process: after a particular type of organism has developed a new feature to become more 'advanced' later changes do not result in a return to the more 'primitive' form. The concept of evolutionary trends therefore provides us with a way of interpreting the fossil content of rocks in terms of biological changes through time. This provides a means of correlating rocks and determining their relative ages by the fossils that they contain.

19.1.2 Biological classification

All living organisms are now classified according to the *Linnaean system*, a scheme of nomenclature which is based on the idea that organisms can be related to each other in a hierarchical manner. This hierarchical system is shown in Table 19.1. The basic unit is the species, and kingdoms are the largest groups. The Linnaean system was developed for living organisms, but the same nomenclature and classification scheme are also used in palaeontology. The general term for any one of the taxonomic ranks defined by the Linnaean system is a *taxon* (plural *taxa*) and the fundamental rank of the taxon is the species.

A *species* can be considered to encompass organisms which share genetic characteristics which may allow individuals to interbreed with each other and produce

Table 19.1 The Linnaean hierarchical system for the taxonomy of organisms.

Kingdom Phylum Class Order Family Genus Species	Many of these categories may be grouped or subdivided (e.g. 'superfamily', 'suborder', 'subspecies')

fertile offspring. This definition is straightforward and practicable for living species because the capacity of individuals to interbreed and the genetic make-up of an animal or plant can be readily determined. Applying these definitions of a species to fossils is problematic because it is not usually possible to demonstrate a capacity to interbreed. Recent advances have been made in extracting material from fossils which has allowed their genetic code to be established, but in the vast majority of cases the DNA material required is degraded in fossils. Palaeontologists therefore have to work on similarities or differences of form to define *morphospecies*, which is the strictly correct term for fossil 'species', although it is not often used.

Morphospecies have to be defined mainly on external characteristics, and as the soft parts of organisms are only preserved in extraordinary circumstances, it is the hard parts which are principally used. There is always an element of doubt about whether a similarity of skeletal form is a sufficient basis for assuming membership of the same species: one only has to think of different species of birds with strikingly contrasting plumages but with essentially identical skeletons. The converse is also true: breeds of dog are very diverse in physical form, but they can all theoretically, if not practically, interbreed because they belong to the same species. Fossils of these different breeds would almost certainly be assigned to separate species if intermediate forms between extremes were not available.

Subspecies and *races* are distinct sets which show common characteristics which set them apart from others but which can still be considered to be part of the same species. The variations are often due to geographical separation of the sets, leading to the development of different characteristics. The concept of the subspecies is used in palaeontology in cases where a high-resolution biostratigraphy is developed.

Genus (plural *genera*) is the rank between species and family. When an organism is named it is given a genus as well as a species—for example, *Homo sapiens* is the Linnaean classification name for the human species. Closely related species are considered to belong to the same genera. In palaeontology species-level identification is normally only required for biostratigraphic purposes, otherwise it is common to identify and classify a fossil only to generic level. For example, if fossil oysters are found in a limestone, they may be simply referred to as *Ostraea*, as identification to this level provides sufficient palaeoenvironmental information without the need to identify the species of *Ostraea*. Note the conventions used in referring to species and genera: the names of genera always begin with a capital letter, while those of species do not, and *italics* are used in printed text.

The higher ranks in the hierarchy are family, order, class, phylum and kingdom in order of scale (Table 19.1). The major phyla (Mollusca, Arthropoda, etc.: Fig. 19.1) have existed throughout the Phanerozoic, and it is possible to compare fossils to modern representatives of these subsets of the main kingdoms (animal and plant). However, some classes, many orders and a large number of families which have been identified as fossils have no modern equivalents. The ammonites, for example, formed a very large and diverse order from Ordovician to Cretaceous times, but there are no modern equivalents, only organisms such as nautiloids which belong to the same class, the Cephalopoda, in the phylum Mollusca. The graptolites, which are commonly found in Palaeozoic rocks, form a class of which there are no modern representatives. As the similarities to modern organisms become fewer, the problems of classification become greater as the significance of morphological differences is less apparent. The classification of fossils in the Linnaean hierarchy is therefore in a constant state of flux as new fossil discoveries are made which shed light on the probable relationships between fossil organisms.

19.1.3 Evolutionary lineages and patterns of evolution

From the point of view of application to stratigraphy there are a number of aspects of the processes of evolution which are important. An *evolutionary lineage* or *phylogeny* is a line of descent of organisms and its pattern will display two components. First, it may show numerous branches as new species form when a subset of one species (a race) develops characteristics which are sufficient to distinguish it as a separate species. By this process the number of species is increased. Second, a single species may change through time to become a different species which is genetically and morphologically distinct from its antecedent. This latter process is called *phyletic evolution* and does not increase the number of taxonomic groups. Both of these patterns of evolution are useful in stratigraphy as they involve the appearance of new taxonomic groups through time. The extinction of a species can either be due to phyletic evolution into a

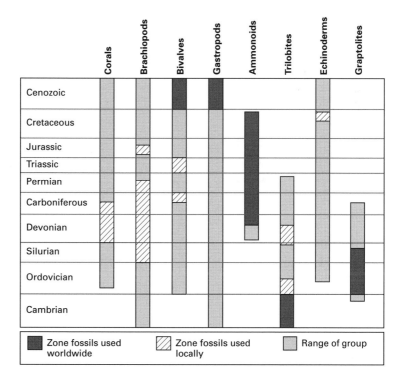

Fig. 19.1 Major groups of organisms preserved as macrofossils in the stratigraphic record and their age ranges.

new form or involve a complete end to that branch of a particular lineage.

In its simplest form, the theory of evolution implies that genetic changes occur in organisms in response to environmental factors and that these changes eventually result in an organism sufficiently different to be considered a separate species. This would suggest that evolution is a gradual process, a *phyletic gradualism*. If this were the case the stratigraphic record should provide us with a succession of fossils which exhibit a gradual transition of forms. What we actually find are morphologically distinct species, with intermediate forms rarely seen. This observation may be explained in two ways. One hypothesis is that evolution does proceed by phyletic gradualism and the stratigraphic record is far too incomplete to reveal the intermediate forms. It is the case that the stratigraphic record is very incomplete in a general sense, but in places there are successions of fossiliferous rocks which apparently contain a continuous record for millions of years and the intermediate forms (missing links) should be present in these strata. A second theory is that a lineage does not evolve gradually but in an episodic way (Fig. 19.2). It is envisaged that there are periods of *stasis* when a species does not change genetic make-up or form, followed by short periods of rapid change when new species develop. This stop–start process is called the *punctuated equilibria* theory of evolution (Eldredge & Gould 1972). It is likely that elements of both theories actually occur, with evolutionary change occurring slowly most of the time, punctuated by periods of rapid change in response to environmental stresses.

19.2 Fossils in stratigraphy

Fossils are the most powerful and widespread tool avail-

able for establishing a detailed stratigraphy and correlation scheme in sedimentary rocks. The strata deposited during nearly 600 million years of Earth history have been subdivided and correlated on the basis of the fossils that they contain. The level of detail that has been achieved is determined by the nature of depositional environments and the ways in which different taxa lived and evolved.

The ideal fossil for stratigraphic purposes would be of an organism which lived in all depositional environments all over the world and was abundant; it would have easily preserved hard parts and would be part of an evolutionary lineage which frequently developed new, distinct species. Not surprisingly, no such fossil taxon has ever existed and a variety of schemes are used in biostratigraphy, including assemblages of taxa *(19.5.1)*.

19.2.1 Depositional environments

The conditions vary so much between different depositional environments that no single species, genus or family can be expected to live in all of them. The adaptations required to live in a desert compared to a swamp, or a sandy coastline compared to a deep ocean, demand that the organisms which live in these environments are different. There is a strong environmental control on the distribution of taxa today and it is reasonable to assume that the nature of the environment strongly influenced the distribution of fossil groups as well. Some environments are more favourable to the preservation of body fossils than others: for example, preservation potential is lower on a high-energy beach than in a low-energy lagoon. There is a fundamental problem with correlation between continental and marine environments because very few animals or plants are found in both settings. In the marine environment the most widespread organisms

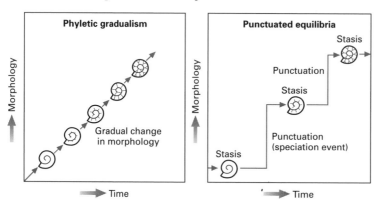

Fig. 19.2 Patterns of evolution: phyletic gradualism and punctuated equilibria.

are those which are *planktonic* (free floating) or animals which are *nektonic* (free swimming lifestyle). Those which live on the sea bed, the *benthonic* or *benthic* creatures and plants, are normally found only in a certain water depth range and are hence not quite so useful.

19.2.2 Geographical distribution

Two environments may be almost identical in terms of physical conditions but if they are on opposite sides of the world they may be inhabited by quite different sets of animals and plants. The contrasts are greatest in continental environments where geographical isolation of communities due to continental drift has resulted in quite different families and orders. The mammal fauna of Australia are a striking example of geographical isolation resulting in the evolution of a group of animals which are quite distinct from animals living in similar environments in Europe or Asia. This geographical isolation of groups of organisms is called *provincialism*, and it also occurs in marine organisms, particularly benthic forms which cannot easily travel across oceans. Present or past oceans have been sufficiently separate to develop localized communities even though the depositional environments may have been similar. This faunal provincialism makes it necessary to develop different biostratigraphic schemes in different parts of the world.

19.2.3 Abundance and size

To be useful in biostratigraphy a species must be sufficiently abundant to be found readily in sedimentary rocks. It must be possible for the geologist to be able to find representatives of the appropriate taxon without having to spend an inordinate amount of time looking. There is also a play-off between size and abundance. In general, smaller organisms are more numerous so the fossils of small organisms tend to be the most abundant. The problem with very small fossils is that they may be difficult to find and identify. The need for biostratigraphic schemes to be applicable to subsurface data from boreholes has led to an increased use of *microfossils*, fossils which are too small to be recognized in hand specimen, but which may be abundant and readily identified under the microscope (or electron microscope in some cases). Schemes based on microfossils have been developed in parallel to macrofossil schemes. Although a scheme based on ammonites may work very well in the field, the chances of finding a whole ammonite in the

core of a borehole are remote. Microfossils are the only viable material for use in biostratigraphy where drilling does not recover core but only brings up pieces of the lithologies in the drilling mud *(22.6)*.

19.2.4 Preservation potential

It is impossible to determine how many species or individuals have lived on Earth through geological time because very few are ever preserved as fossils. The fossil record represents a very small fraction of the biological history of the planet for a variety of reasons. First, some organisms do not possess the hard parts which can survive burial in sediments: we therefore have no idea how many types of worm may have existed in the past. Sites where there is exceptional preservation of the soft parts of fossils (*lagerstätten*) provide tantalizing clues to the diversity of life forms about which we know next to nothing (Whittington & Conway-Morris 1985; Clarkson 1993). Second, the depositional environment may not be favourable to the preservation of remains: only the most resistant pieces of bone survive in the dry, oxidizing setting of deserts and almost all other material is destroyed. All organisms are part of a food chain which means that their bodies are normally consumed by either a predator or a scavenger. Preservation is therefore the exception for most animals and plants. Finally, the stratigraphic record is very incomplete *(24.9)* with only a fraction of the environmental niches which have existed preserved in sedimentary rocks.

The low preservation potential severely limits the material available for biostratigraphic purposes, restricting it to those taxa which had hard parts and which existed in appropriate depositional environments. The incomplete record also hampers interpretation of lineages of taxa and assemblages of fossils because the clues available are only a very small part of the whole story.

19.2.5 Rate of speciation

The frequency with which new species evolve and replace former species in the same lineage determines the precision which can be applied in biostratigraphy. Some organisms seem to have hardly evolved at all: the brachiopod *Lingula* seems to look exactly the same today as the fossils found in Lower Palaeozoic rocks. The biostratigraphic value of *Lingula* is limited to indicating that the rocks are Phanerozoic. Other groups appear to have developed new forms regularly and at frequent intervals: new species of ammonites appear to have evolved every million years or so during the Jurassic and Cretaceous. A biostratigraphy for these periods can be developed which has an equivalent geological time resolution of about a million years. A *high-resolution biostratigraphy* with subdivisions of a few hundred thousand years has been achieved in Mesozoic and Cenozoic marine strata in some areas using stages and sub-stages defined by planktonic marine microfossils such as foraminifera *(19.3.2)*.

19.2.6 Mobility of organisms

The lifestyle of an organism not only determines its distribution in depositional environments, but also affects the rate at which an organism migrates from one area to another. If a new species evolves in one geographical location its use as a stratigraphic marker in a regional or world-wide sense will depend on how quickly it migrates to occupy ecological niches elsewhere. Again, planktonic and nektonic organisms tend to be most useful in biostratigraphy because they move around relatively quickly. Some benthic organisms have a larval stage which is nektonic and may therefore be spread around oceans relatively quickly. Organisms which do not move much (a *sessile* lifestyle) generally make poor fossils for biostratigraphic purposes.

19.3 Taxa used in biostratigraphy

No single group of organisms fulfils all the criteria for the ideal zone fossil and a number of different groups of taxa have been used for defining biozones through the stratigraphic record (Clarkson 1993). Some, such as the graptolites in the Ordovician and Silurian, are used for world-wide correlation; others are restricted in use to certain facies in a particular succession (e.g. corals in the Carboniferous of north-west Europe). Some examples of taxonomic groups used in biostratigraphy are listed below.

19.3.1 Marine macrofossils

The hard parts of invertebrates are common in sedimentary rocks deposited in marine environments throughout the Phanerozoic (Fig. 19.1) (Clarkson 1993). These fossils formed the basis for the divisions of the stratigraphic column into Systems, Series and Stages (see Table 18.1) in the 18th and 19th centuries. The fossils of organisms

such as molluscs, arthropods, echinoderms, etc., are relatively easy to identify in hand specimen and provide the field geologist with a means for establishing the age of rocks to the right Period or possibly Epoch. Expert palaeontological analysis of marine macrofossils provides a division of the rocks into Stages based on these fossils.

TRILOBITES

These Palaeozoic arthropods are the main group used in the zonation of the Cambrian. Most trilobites are thought to have been benthic forms living on and in the sediment of shallow marine waters. They show a wide variety of morphologies and appear to have evolved quite rapidly into taxa with distinct and recognizable characteristics. They are rarely abundant as fossils.

GRAPTOLITES

These exotic and somewhat enigmatic organisms are interpreted as being colonial groups of individuals connected by a skeletal structure. They appear to have had a planktonic habit and are widespread in Ordovician and Silurian mudrocks. Preservation is normally as a thin film of flattened organic material on the bedding planes of fine-grained sedimentary rocks. The shapes of the skeletons and the 'teeth' where individuals in the colony were located are distinctive when examined with a hand lens or under a microscope. Lineages have been traced which indicate rapid evolution and have allowed a high-resolution biostratigraphy to be developed for the Ordovician and Silurian systems. The only real drawback in the use of graptolites is the poor preservation in coarser-grained rocks such as sandstones.

BRACHIOPODS

Shelly, sessile organisms like brachiopods generally make poor zone fossils but in shallow marine, high-energy environments where graptolites were not preserved, brachiopods are used for regional correlation purposes in Silurian rocks and occasionally in later Palaeozoic strata.

AMMONOIDS

This taxonomic group of cephalopods (phylum Mollusca) includes goniatites from Palaeozoic rocks as well as the more familiar ammonites of the Mesozoic. The nautiloids are the most closely related living group. The large size and nektonic habit of these cephalopods made them an excellent group for biostratigraphic purposes. Fossils are widespread, found in almost all fully marine environments, and they are relatively robust. Morphological changes through time were to the external shape of the organisms and to the 'suture line', the relict of the bounding walls between the chambers of the coiled cephalopod. Goniatites have been used in correlation of Devonian and Carboniferous rocks, whilst ammonites and other ammonoids are the main zone fossils in Mesozoic rocks. Ammonoids became extinct at the end of the Cretaceous.

ECHINODERMS

This phylum includes crinoids (sea lilies) and echinoids (sea urchins). Most crinoids probably lived attached to substrate, and echinoids are benthic, living on or in soft sediment. They are only used for regional and world-wide correlation in parts of the Cretaceous.

CORALS

The extensive outcrops of Devonian and Lower Carboniferous (Mississippian) shallow marine limestones in some parts of the world contain abundant corals. This group is therefore used for zonation and correlation within these strata, despite the fact that they are not really suitable for biostratigraphy because of the very restricted depositional environments they represent.

GASTROPODS

Marine 'snails' are abundant as fossils in Cenozoic rocks. Distinctive shapes and ornamentation on the calcareous shells make identification relatively straightforward of a wide variety of taxa within this group. They are very common in the deposits of almost all shallow marine environments.

19.3.2 Marine microfossils

Microfossils are any taxa which leave fossil remains which are too small to be seen clearly with the naked eye or hand lens. They are normally examined using an optical microscope although some forms can only be analysed in detail using a scanning electron microscope. Three main groups are used in biostratigraphy.

FORAMINIFERA

Forams (the common abbreviation for 'foraminifera') are marine protozoans (single-celled organisms) which have been found as fossils in strata as old as the Ordovician. The fossils of Mesozoic and Cenozoic planktonic forams are calcareous and are around a millimetre or less across although some larger benthic forms also occur. Planktonic forams make very good zone fossils as they are abundant, widespread in marine strata and appear to have evolved rapidly. Schemes using forams for correlation in the Mesozoic and Cenozoic are widely used in the hydrocarbon industry because microfossils are readily recovered from boreholes. Both regional and world-wide zonation schemes are used.

RADIOLARIA

This sub-class of planktonic protozoans has silica skeletons. They are roughly spherical, often spiny organisms a fraction of a millimetre across. They are important in the dating of deep marine deposits because the siliceous skeletons survive in siliceous oozes deposited at depths below the CCD *(15.5.1)* where calcareous fossils dissolve. Radiolaria cherts *(15.5.2)* are found in deep marine strata throughout the Phanerozoic.

NANNOFOSSILS

Fossils which cannot be seen with the naked eye and are only just discernible using a high-power optical microscope are referred to as nannofossils. They are micrometres to tens of micrometres across and are best examined using a scanning electron microscope *(2.5.4)*. The most common nannofossils are coccoliths, the spherical calcareous cysts of marine algae *(3.1.2)*. Coccoliths may occur in huge quantities in some sediments and are the main constituent of some fine-grained limestones such as the Chalk of the Upper Cretaceous in north-west Europe. They are found in fine-grained marine sediments deposited on the shelf or at any depths above the CCD, below which they are not normally preserved. They are used biostratigraphically in Mesozoic and Cenozoic strata.

19.3.3 Other taxonomic groups used in biostratigraphy

All the above are marine organisms. Correlation in the

deposits of continental environments is always more difficult because of the poorer preservation potential of most materials in a subaerial setting. Only the most resistant materials survive to be fossilized in most continental deposits, and these include the organophosphates that vertebrate teeth are made of and the coatings of pollen, spores and seeds of plants. Local stratigraphic schemes have been set up using the teeth of small mammals and reptiles which are sufficiently common and distinctive to be used in some circumstances (Woodburne 1977).

Pollen, spores and seeds (*palynomorphs*) are much more commonly used. The carbonaceous material they are made from is highly resistant to chemical attack and can be dissolved out of siliceous sedimentary rocks using hydrofluoric acid. Airborne particles such as pollen, spores and some seeds may be widely dispersed and they therefore have some good characteristics as zone fossils. Identification is carried out with an optical microscope or an electron microscope. One of the main drawbacks of zonation using palynomorphs is that they rarely provide such a high resolution as marine fossils.

19.4 Correlating different environments

If the rocks being studied do not contain representatives of the taxa used in the world-wide biostratigraphic zonation scheme used for that particular part of the stratigraphic record, a more roundabout method of correlation using fossils may be required. A local or regional zonation scheme may be set up using the taxa which are represented in the depositional facies of the strata. The strata containing the fauna or flora of the local scheme must then be correlated with the global scheme by finding a succession elsewhere in which both the locally used and globally used taxa are preserved. Such a process leads to errors and uncertainties but these often cannot be avoided.

Correlation using an intermediary stratigraphic method may also be used. For example, a continental succession and a marine succession can rarely be related to each other using fossils, but both may have a remnant magnetic signature preserved in the rocks. The two successions may therefore be correlated using the magnetic polarity reversal record *(20.2)*.

19.5 Biostratigraphic nomenclature

A biostratigraphic unit is a body of rock defined by its fossil content. It is therefore fundamentally different from a lithostratigraphic unit which is defined by the lithological properties of the rock. The fundamental unit of biostratigraphy is the *biozone*. Biozones are units of stratigraphy which are defined by the fossil taxa (usually species or subspecies) that they contain. In theory they are independent of lithology, although environmental factors often have to be taken into consideration in the definition and interpretation of biozones. In the same way that formations in lithostratigraphy must be defined from a type section, there must also be a type section designated as a stratotype and described for each biozone. They are named from the characteristic or common taxon (or occasionally taxa) which define the biozone.

19.5.1 Zonation schemes

There are several different ways in which biozones can be designated in terms of the *zone fossils* that they contain (Fig. 19.3) (Whittaker *et al.* 1991; North American Commission on Stratigraphic Nomenclature 1993).

Total range biozone: this is defined by the stratigraphic interval between the first appearance of a single taxon and the disappearance of that taxon. The lineage of the taxon is not considered.

Consecutive range biozone: where a taxon can be recognized as having followed another and preceding a third as part of a phyletic lineage, the biozone defined by this taxon is called a consecutive range biozone.

Partial range biozone: a biozone defined on this basis is characterized by a single taxon, but instead of the whole range of that taxon defining the biozone the base of it is marked by the extinction of a second taxon and the top by the appearance of a third.

Assemblage biozone: in this case the biozone is defined by a number of different taxa which may or may not be related. The presence and absence, appearance and disappearance of these taxa are all used to define a stratigraphic interval.

Acme biozone: the abundance of a particular taxon may vary through time, in which case an interval containing a certain statistically high proportion of this taxon may be used to define a biozone.

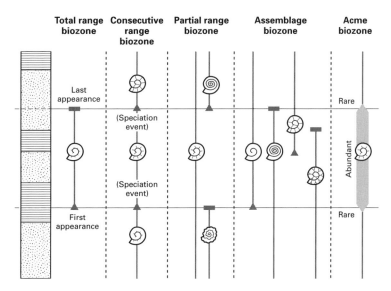

Total range biozone · Consecutive range biozone · Partial range biozone · Assemblage biozone · Acme biozone

Last appearance

(Speciation event)

(Speciation event)

First appearance

Rare

Abundant

Rare

Fig. 19.3 Zonation schemes used in biostratigraphic correlation. (After North American Commission on Stratigraphic Nomenclature 1983; AAPG ©1983, reprinted by permission of the American Association of Petroleum Geologists.)

19.5.2 Biozones and depositional environments

If the appearance and disappearance of particular taxa from a stratigraphic succession are to be used in correlation we need to consider the possible reasons for the presence and absence of fossils from sedimentary rocks (Eicher 1976). If the depositional environment has remained the same, the appearance of a taxon may be due to a speciation event and this will therefore have stratigraphic significance. However, an alternative explanation may be that the species had already existed for a period of time in a different geographical location before migrating to the area of the studied section. In this case the fossil taxon may only have a stratigraphic significance in a restricted area. The disappearance of a species from the stratigraphic succession is likely to represent an extinction event if the depositional environment has not changed: a taxon is unlikely to move away from a favourable setting.

The interpretation of the appearance and disappearance of fossil taxa in successions which contain evidence of changes in environment is more complicated. Variations in water depth, rate of sedimentation, proportion of mud, and so on may favour one taxon over another by changing the environmental conditions. Organisms which are tolerant of different conditions therefore have the widest application and most value as zone fossils. Taxa which are very sensitive to environmental conditions, such as corals, are only useful in circumstances where the environment of deposition has been constant.

19.5.3 Graphical correlation schemes

The thickness of a biostratigraphic unit at any place is determined by the rate of sediment accumulation during the time period represented by the biozone. A succession which is considered to have been a site of continuous, steady sedimentation is chosen as a reference section and the positions of biozone markers (appearance and disappearance of taxa) are noted on it. Another vertical succession of strata containing the same biozone markers can then be compared with this reference section. Tie-points are established using the biostratigraphic information and intermediate levels can be correlated graphically (Fig. 19.4). This approach is particularly effective at identifying changes in rates of sedimentation and hiatuses (periods of erosion or non-deposition) in a succession. Depositional hiatuses are important in stratigraphic analysis using sequence stratigraphic concepts (Chapter 21). This graphical correlation method was first developed by Shaw (1964) and has subsequently been used extensively in subsurface correlation (Chapter 22).

19.6 Biostratigraphy and chronostratigraphy

It can be argued that biozones are chronostratigraphic units. If a speciation event takes place rapidly enough to be considered to be an 'instant' in geological time and the new taxon is very quickly dispersed (again geologically instantaneously) then the base of a biozone can be

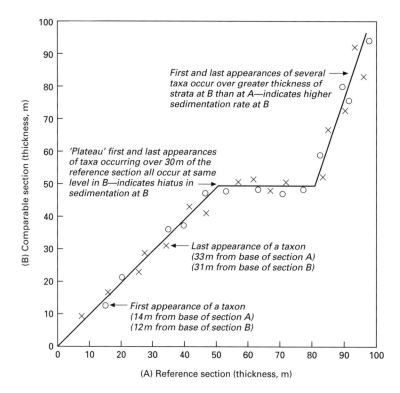

Fig. 19.4 Graphical correlation methods are used to identify changes in rates of sedimentation or a hiatus in deposition. (Adapted from Shaw 1964.)

regarded as an isochronous horizon. It is difficult to judge the rates of evolutionary change in lineages and the time taken for a new species to be distributed. However, if the concept of 'punctuated equilibria' *(19.1.3)* is applied to nektonic or planktonic organisms then the bases of biostratigraphically defined units are effectively chronostratigraphic surfaces within the resolution available. It may also be the case that certain taxa or groups of taxa become extinct in a geologically short period of time, so upper boundaries defined by these events can also be considered to approximate to isochronous surfaces. In other cases it must be assumed that biozone boundaries are to some extent diachronous.

Further reading

Blatt, H., Berry, W.B. & Brande, S. (1991) *Principles of Stratigraphic Analysis*, 512 pp. Blackwell Scientific Publications, Oxford.

Clarkson, E.N.K. (1993) *Invertebrate Palaeontology and Evolution*, 323 pp. Allen & Unwin, London.

Doyle, P., Bennett, M.R. & Baxter, A.N. (1994) *The Key to Earth History: An Introduction to Stratigraphy*, 231 pp. Wiley, Chichester.

Eicher, D.L. (1976) *Geologic Time*, 2nd edn, 150 pp. Prentice Hall, Englewood Cliffs, NJ.

Skelton, P.W. (1992) Fossils and evolutionary theory. In: *Understanding the Earth, A New Synthesis* (eds G.C. Brown, C.J. Hawkesworth & R.C.L. Wilson), pp. 458–482. Cambridge University Press, Cambridge.

It has always been a goal of geologists to determine the age of the Earth and give dates to events in Earth history. Rock relationships, lithostratigraphy and biostratigraphy provide essential information about the order in which things have happened in a relative sense, but give no idea of the absolute age of anything. A geological time-scale has become possible thanks to radiometric dating, which uses the decay of naturally occurring isotopes as a geological clock. A number of different techniques for absolute radiometric dating have been developed but they do not provide a universal means of determining the age of rocks. The age of formation of an igneous rock can be obtained in some circumstances but most sedimentary rocks cannot be dated in this way. Relative dating and correlation techniques are essential if the radiometric ages obtained from suitable lithologies are to be useful for the majority of rocks. Of these other techniques magnetostratigraphy is becoming increasingly important, although it can only be used in conjunction with other methods. In practice the whole process of dating and correlating rocks relies on the integration of information from a number of different sources and techniques.

20.1 Radiometric dating

The discovery of radioactivity and the radiogenic decay of isotopes in the early part of the 20th century opened the way for dating rocks by an absolute, rather than relative, method. Up to this time estimates of the age of the Earth had been based on assumptions about rates of evolution, rates of deposition, the thermal behaviour of the Earth and the Sun or interpretation of religious scriptures (Eicher 1976). *Radiometric dating* uses the principle that if the rate of the decay of isotopes of elements in minerals within rocks is known and the proportion of these isotopes is known when the rock was formed the age of the rock can be determined. This dating method can be used for determining the age of formation of igneous rocks and authigenic minerals in some sedimentary rocks: radiometric dating of minerals in metamorphic rocks usually indicates the age of the metamorphism.

20.1.1 Radioactive decay series

A number of elements have *isotopes* (forms of the element with different atomic mass) which are unstable and change by radioactive decay to the isotope of a different element. Each radioactive decay series (Fig. 20.1) takes a characteristic length of time known as the *radioactive half-life*, which is the time taken for half of the original (*parent, P*) isotope to decay to the new (*daughter, D*) isotope. The decay series of most interest to geologists are those with half-lives of tens, hundreds or thousands of millions of years. If the proportions of parent and daughter isotopes of these decay series can be measured, then periods of geological time in millions to thousands of millions of years can be calculated.

If the age of a rock is to be calculated it is necessary to know the half-life of the radioactive decay series used, the amount of the parent and daughter isotopes present in the rock when it formed and the present proportions of these isotopes. The relationship can be expressed in an equation as:

$$N = N_0 e^{-\lambda t}$$

in which N_0 is the number of parent atoms at the start, N the number of parent atoms after a period of time (t) and λ is the rate at which parent decays to daughter (the *decay constant*); 'e' is approximately 2.718, the base of

Alpha decay

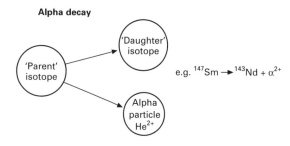

e.g. $^{147}Sm \rightarrow {}^{143}Nd + \alpha^{2+}$

Beta decay

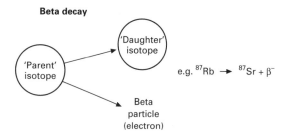

e.g. $^{87}Rb \rightarrow {}^{87}Sr + \beta^{-}$

Fig. 20.1 Radioactive decay results in the formation of a new 'daughter' isotope from the 'parent' isotope.

the natural logarithm (ln). The radiometric age can be determined using the relationship:

$$t = 1/\lambda \ln(D/P + 1)$$

It is normally assumed that when a mineral crystallizes out of a melt as part of an igneous rock, is precipitated chemically in a sediment or forms by solid-state recrystallization in a metamorphic rock there is only the parent isotope present. The radiometric 'clock' starts as the mineral is formed. It must also be assumed that it is possible to determine the isotopes found in the rock today which are a result of the decay of the parent (daughter isotopes) and that there has not been addition or loss of these isotopes during the period since the mineral formed (that is, it is a 'closed system'). Errors in the

results of radiometric dating arise when these conditions are not met—for example, when the minerals and rock have been partly metamorphosed or otherwise altered between formation and the present day.

The radiometric decay series commonly used in radiometric dating of rocks are detailed in the following sections. The choice of method of determination of the age of the rock is governed by its age and the abundance of the appropriate elements in minerals. Further details of these dating methods and their applications are to be found in texts such as Eicher (1976), Faure (1986) and Dalrymple (1991).

20.1.2 Rubidium–strontium dating

This is a widely used method for dating because the parent element, rubidium, is common as a trace element in many silicate minerals and rocks. The isotope ^{87}Rb decays by shedding an electron (*beta decay*) to ^{87}Sr with a half-life of 48 billion years (Table 20.1). The proportions of two of the isotopes of strontium, ^{87}Sr and ^{86}Sr, are measured by a mass spectrometer (*20.1.6*). The ratio of ^{87}Sr to ^{86}Sr will depend on two factors. First, it will depend on the proportions in the original magma: this ratio will be constant for a particular magma body but will vary between different bodies. Second, the amount of ^{87}Sr present will vary according to the amount produced by the decay of ^{87}Rb: this depends on the amount of rubidium present in the rock and the age. The rubidium and strontium concentrations in the rock can be measured by geochemical analytical techniques such as X-ray fluorescence. Two unknowns remain: the original $^{87}Sr/^{86}Sr$ ratio and the ^{87}Sr formed by decay of ^{87}Rb (which provides the information needed to determine the age). The principle of solving simultaneous equations can be used to resolve these two unknowns. If the determination of the ratios of $^{87}Sr/^{86}Sr$ and Rb/Sr are carried

Table 20.1 The main decay series used in radiometric dating of rocks.

Parent isotope	Daughter isotope	Half-life (10^9 years)	Parent's decay constant (year^{-1})	Practical dating range (Ma)
^{40}K	^{40}Ar	1.250	4.692×10^{-10}	1 to >4500
^{87}Rb	^{87}Sr	48.8	1.42×10^{-11}	10 to >4500
^{147}Sm	^{143}Nd	1.06	6.54×10^{-12}	>200
^{176}Lu	^{176}Hf	3.5	1.94×10^{-11}	>200
^{232}Th	^{208}Pb	14.01	4.95×10^{-11}	10 to >4500
^{235}U	^{207}Pb	0.704	9.85×10^{-10}	10 to >4500
^{238}U	^{206}Pb	4.468	1.55×10^{-10}	10 to >4500
^{14}C	^{14}N	5730 years	1.29×10^{-4}	<80 000 years

out for two different minerals (e.g. orthoclase and muscovite) each will start with different proportions of strontium and rubidium because they are chemically different. An alternative method is to take samples from different parts of the igneous body which will have crystallized at different times and hence have different amounts of rubidium and strontium present (e.g. crystallization of a mineral such as biotite depletes the amount of rubidium present). This latter approach is known as *whole-rock dating* and provides the best results if large numbers of samples from different parts of the igneous body are measured. It is more straightforward than dating individual minerals as it does not require the separation of these minerals.

20.1.3 Potassium–argon (and argon–argon) dating

One of the isotopes of potassium, ^{40}K, decays partly by electron capture (a proton becomes a neutron) to an isotope of the gaseous element argon, ^{40}Ar, the other product being an isotope of calcium, ^{40}Ca (Table 20.1). Potassium is a very common element in the Earth's crust. It occurs in many silicate minerals, including micas and feldspars, and its concentration is easily measured. However, the proportion of potassium present as ^{40}K is very small at only 0.012%, and most of this decays to ^{40}Ca, with only 11% forming ^{40}Ar. Argon is an inert rare gas and the isotopes of very small quantities of argon can readily be measured by a mass spectrometer by driving the gas out of the minerals. K–Ar dating has therefore been widely used in dating igneous rocks and can be used for dating the authigenic sedimentary mineral glauconite *(11.6.1)*. However, a significant problem with potassium–argon dating is that the daughter isotope can escape from the rock by diffusion because it is a gas. The amount of argon measured is therefore commonly less than the total amount produced by the radioactive decay of potassium. This results in an underestimate of the age of the rock.

The problems of argon loss have been overcome by the argon–argon method. The first step in this technique is the irradiation of the sample by neutron bombardment to form ^{39}Ar from ^{39}K occurring in the rock. The ratio of ^{39}K to ^{40}K is a known constant so if the amount of ^{39}Ar produced from ^{39}K can be measured, then an indirect method of calculating the ^{40}K present in the rock becomes available. Measurement of the ^{39}Ar produced by bombardment is made by mass spectrometer at the same time as measuring the amount of ^{40}Ar present. Before an

age can be calculated from the proportions of ^{39}Ar and ^{40}Ar present it is necessary to find out the proportion of ^{39}K which has been converted to ^{39}Ar by the neutron bombardment. This can be achieved by bombarding a sample of known age (a 'standard') along with the samples to be measured and comparing the results of the isotope analysis. The principle of the Ar–Ar method is therefore the use of ^{39}Ar as a proxy for ^{40}K.

Although a more difficult and expensive method, Ar–Ar is now preferred to K–Ar as a method for dating many igneous rocks. The effects of alteration can be eliminated by step-heating the sample during determination of the amounts of ^{39}Ar and ^{40}Ar present by mass spectrometer. Alteration (and hence ^{40}Ar loss) occurs at lower temperatures than the original crystallization so the isotope ratios measured at different temperatures will be different. The sample is heated until there is no change in ratio with increase in temperature (a 'plateau' is reached): this ratio is then used to calculate the age. If no 'plateau' is achieved and the ratio changes with each temperature step the sample is known to be too altered to provide a reliable date.

20.1.4 Uranium–lead dating

Isotopes of uranium are all unstable and decay to daughter elements which include thorium, radon and lead. Two decays are important in radiometric dating: ^{238}U to ^{206}Pb with a half-life of 4.47 billion years and ^{235}U to ^{207}Pb with a half-life of 704 million years (Table 20.1). The naturally occurring proportions of ^{238}U and ^{235}U are constant, with the former the most abundant at 99% and the latter 0.7%. By measuring the proportions of the parent and daughter isotopes in the two decay series it is possible to determine the amount of lead in a mineral produced by radioactive decay and hence calculate the age of the mineral. Trace amounts of uranium are to be found in minerals such as zircon, monazite, sphene and apatite. Of these zircon is the most useful because it contains the highest proportions of uranium and the lowest amounts of non-radiogenic lead. Dating of zircons has produced the most precise ages in older rocks and has been used to establish the age of the oldest rocks in the world.

20.1.5 Neodymium–samarium dating

These two rare earth elements in this decay series are normally only present in parts per million in rocks. The parent isotope is ^{147}Sm; it decays by alpha particle emission to ^{143}Nd, with a half-life of 106 billion years (Table 20.1). The slow generation of ^{143}Nd means that this technique is best suited to older rocks as the effects of analytical errors are less significant. The advantage of using this decay series is that the two elements behave almost identically in geochemical reactions and any alteration of the rock is likely to affect the two isotopes to equal degrees. This eliminates some of the problems encountered with Rb–Sr caused by the different reactivity and mobility of the two elements in the decay series. Measurements of the ratios of the two elements Sm and Nd and the proportions of the two isotopes of neodymium (^{143}Nd and ^{144}Nd) are required to determine the amount of radiogenically formed ^{143}Nd. Measurements from two different minerals are therefore required to solve two simultaneous equations using the same technique as in Rb–Sr dating.

20.1.6 Practical radiometric dating

As indicated above, any alteration of the rock to be dated may result in an error in the age calculated, so it is important that the material collected for analysis is fresh, unaltered and unweathered. A large sample (several kilograms) is usually collected to eliminate local inhomogeneities in the rock. The samples are crushed to sand and granule size, thoroughly mixed to homogenize the material and a smaller subsample selected.

In cases where particular minerals are to be dated, these are separated from the other minerals by using heavy liquids in which some minerals will float and others sink, or magnetic separation using the different magnetic properties of minerals. A pure separation of the mineral to be dated is required to reduce errors. The mineral concentrate may then be dissolved for isotopic or elemental analysis (except in the case of the argon

isotopes, in which case the mineral grains are heated in a vacuum and the composition of the argon gas driven off measured directly).

Measurement of the concentrations of different isotopes is carried out with a *mass spectrometer* (Fig. 20.2). In these instruments a small amount (micrograms) of the sample is heated in a vacuum to ionize the isotopes, and these charged particles are then accelerated along a tube in a vacuum by a potential difference. Partway along the tube a magnetic field across the tube induced by an electromagnet deflects the charged particles. The amount of deflection will depend upon the atomic mass of the particles so different isotopes are separated by their different masses. Detectors at the end of the tube record the number of charged particles of a particular atomic mass and provide a ratio of the isotopes present in a sample. A particular mass spectrometer is often designed and set up to measure a certain range of atomic masses. Mass spectrometers are used to measure the ratios of different isotopes present in a sample, rather than absolute amounts of a particular isotope. In the case of Rb–Sr dating this is the ratio of ^{86}Sr to ^{87}Sr, in argon–argon dating it is the ratio of ^{39}Ar to ^{40}Ar, and so on.

20.1.7 Applications of radiometric dating

Radiometric dating is the only technique which can provide absolute ages of rocks through the stratigraphic record, but it is limited in application by the types of rocks which can be dated. The age of formation of minerals is determined by this method, so if orthoclase feldspar grains in a sandstone are dated radiometrically, the date obtained would be that of the granite the grains were eroded from. It is therefore not possible to date the formation of rocks made up from detrital grains and this excludes virtually all sandstones, mudrocks and conglomerates. Limestones are formed largely from the remains

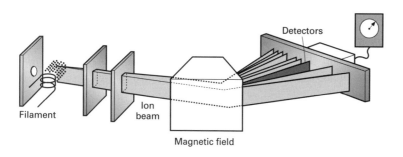

Fig. 20.2 In a mass spectrometer charged ions are accelerated through a magnetic field where they are deflected by an amount which is proportional to their mass: the number of ions which reach the detectors is recorded to provide a measure of the proportion of ions (isotopes) of different masses.

of organisms with calcium carbonate hard parts, and the minerals aragonite and calcite cannot be dated radiometrically in pre-Quaternary rocks. Hence almost all sedimentary rocks are excluded from this method of dating and correlation. An exception to this is the mineral glauconite, an authigenic mineral which forms in shallow marine environments *(11.6.1)*: glauconite contains potassium and may be dated by K–Ar or Ar–Ar methods, but the mineral is readily altered and limited in occurrence.

The formation of igneous rocks can usually be successfully dated, provided that they have not been severely altered or metamorphosed. Intrusive bodies, including dykes and sills, and the products of volcanic activity (lavas and tuff) may be dated and these dates used to constrain the ages of the rocks around them by the laws of stratigraphic relationships *(18.3.1)*. Dates from metamorphic rocks may provide the age of metamorphism, although complications can arise if the degree of metamorphism has not been high enough to reset the radiometric 'clock', or if there have been multiple phases of metamorphism.

20.1.8 Radiometric dating and the geological time-scale

The division of rocks into units which were thought to be chronostratigraphic had been carried out long before any method of determining the geological time periods involved had been developed. The main systems had been established and partly divided into series and stages by the beginning of the 20th century by using stratigraphic relations and biostratigraphic methods. Radiometric dating has provided a time-scale for the chronostratigraphic division of rocks. The published geological time-scales (Table 20.2: Harland *et al.* 1989) have been constructed by integrating information from biostratigraphy, magnetostratigraphy and data from radiometric dating to determine the chronostratigraphy of rock units throughout the Phanerozoic.

20.1.9 Use of the geological time-scale

It is rarely possible to obtain a radiometric age from a unit which contains the fauna used to define the stages of the stratigraphic column, so correlation between the biostratigraphically defined stages and chronostratigraphic units has been achieved indirectly from dated lavas, cross-cutting relationships, and so on. Increased knowl-

edge about these relationships and refinements in the techniques of radiometric dating have led to a number of modifications over the years, and there is every reason to believe that the time-scale will be amended again in the future with the boundaries between units becoming younger or older with these changes. A stratigraphic unit should therefore not be described as '150 million years old' unless it has been directly dated as such. The stratigraphic age given to it should be defined in terms of the method used to determine that age, so if it contains fossils which are considered to be Oxfordian or fossils used to define that stage, then the age given should be 'Oxfordian'. Although the Oxfordian stage is about 150 Ma, the correlation between the biostratigraphic and the chronostratigraphic scales may be revised in the future.

20.2 Magnetostratigraphy

Understanding the characteristics of the Earth's magnetic field and interpretation of magnetism preserved in rocks (*palaeomagnetism*) played a major role in the development of the theory of plate tectonics (Tarling 1983). Relative movements between the continents and the magnetic poles and between the continents themselves can be demonstrated by polar wander curves, constructed by plotting changes in geological time of the apparent position of the magnetic pole indicated by the magnetism in the rocks. The process by which continents move relative to each other was indicated by a second feature of the record of magnetism in the past. Using sea-borne magnetometers to look at the strength of the magnetic field on the sea floor, geophysicists discovered a regular pattern of stripes of increased field strength alternating with a slightly lower field strength parallel to the mid-ocean ridges. Two conclusions were drawn from this: first, that there were regular reversals in the polarity of the Earth's magnetic field; and second, that the oceans grew by addition of new material at the mid-ocean ridges. As basalt erupts and cools at the mid-ocean spreading centres, it becomes magnetized in the direction of the Earth's field at the time of cooling: when that field is in the same direction as today (*normal polarity*) the field strength recorded today will be greater than in the region of basalts cooled under an opposite field direction (*reversed polarity*) which tends to counteract the present-day field. This phenomenon of reversals in the Earth's magnetic field has now been more widely applied in stratigraphy.

Table 20.2 The geological time-scale. (From Harland *et al.* 1989.)

Era	Period	Epoch	Age (Ma)
Cenozoic	Quaternary	Holocene	0.01
		Pleistocene	1.64
	Tertiary — Neogene	Pliocene	5.2
		Miocene	23.3
	Tertiary — Palaeogene	Oligocene	35.4
		Eocene	56.5
		Palaeocene	65.0
Mesozoic	Cretaceous	Senonian	88.5
		Gallic	131.8
		Neocomian	145.6
	Jurassic	Malm	157.1
		Dogger	178.0
		Lias	208.0
	Triassic	Late	235.0
		Middle	241.1
		Scythian	245.0
Palaeozoic	Permian	Zechstein	256.1
		Rotliegend	290.0
	Carboniferous — Pennsylvanian	Stephanian	303.0
		Westphalian	318.3
		Namurian	332.9
	Carboniferous — Mississippian*	Visean	349.5
		Tournaisian	362.5
	Devonian	Late	377.4
		Middle	386.0
		Early	408.5
	Silurian	Pridoli	410.7
		Ludlow	424.0
		Wenlock	430.4
		Llandovery	439.0
	Ordovician	Ashgill	443.1
		Caradoc	463.9
		Llandeilo	468.6
		Llanvirn	476.1
		Arenig	493.0
		Tremadoc	510.0
	Cambrian	Merioneth	517.2
		St. David's	536.0
		Caerfai	570.0

Eon			Age (Ma)
Proterozoic			570.0
Archean			2500
Priscoan			4000
			4560

*The Pennsylvanian–Mississippian boundary is at 322.8 Ma.

20.2.1 The behaviour of the Earth's magnetic field

The Earth's magnetic field is generated by currents in the relatively fluid iron-rich material in the Earth's outer core. These currents appear to be unstable over short and long periods of geological time. Over short periods (years) the positions of the magnetic poles move with respect to the poles of rotation of the Earth: this difference is noted on topographic maps and corrections have to be made to compass readings according to the geographical location and the year. These rapid field variations are assumed to average to a position close to the poles of rotation over thousands of years.

The longer-period fluctuation is a periodic reversal of the polarity of the Earth's magnetic field. At times in the past, a compass needle would have indicated that the direction of magnetic north was towards the geographical south. These *magnetic polarity reversals* (Fig. 20.3) have occurred at regular intervals during geological time: short-term reversals have been as little as a few tens of thousands of years apart, whilst there was a period of nearly 30 million years in the Cretaceous when the magnetic field apparently remained the same. Through most of the Tertiary polarity reversals have occurred every few hundred thousand to a few million years. There is no evidence for any intermediate condition (that is, with the magnetic poles located somewhere other than close to the geographical poles) and the time taken for a reversal to occur appears to be 'instantaneous' in the context of geological time.

20.2.2 The magnetic record in rocks

A magnetic material acquires the polarity of the ambient magnetic field as it cools through the *Curie point*, a temperature above which the magnetic dipoles in the material are mobile and free to reorientate themselves. Once below the Curie point, the material retains the same field when it is moved or the magnetic field changes around it. A rock may contain a number of different magnetic minerals which each have their own Curie point temperature, and alteration of the rock may occur to create new minerals which will record the ambient magnetic field at the time of their formation. The preserved magnetization or *remnant magnetism* in a rock may therefore be a complex mixture of different field orientations resident in different minerals.

The objective of analysing the remnant magnetism is normally to determine the orientation of the Earth's

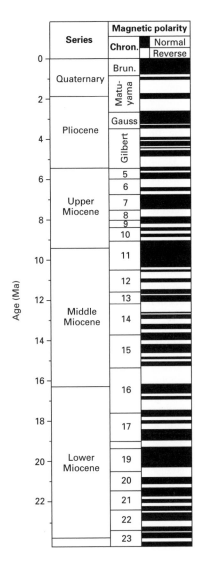

Fig. 20.3 Reversals in the polarity of the Earth's magnetic field through time. (After Haq *et al.* 1988; courtesy of the Society for Sedimentary Geology.)

magnetic field relative to the sample at the time of the formation of the rock (Hailwood 1989). In extrusive igneous rocks this will be recorded by the remnant magnetism in minerals such as magnetite and haematite as they cool below their Curie point. This strong signal is relatively easily detected by a magnetometer, but of more use to the stratigrapher is a much weaker remnant magnetism preserved in sedimentary rocks.

As fine magnetized particles settle out of water they tend to orientate themselves parallel to the Earth's magnetic field. Clearly not all of these particles will line up perfectly parallel to the ambient field, but there will be a statistically significant pattern in their orientation which will give the sediment a remnant polarity. The effect is strongest in fine-grained sediments deposited from suspension with a high proportion of iron minerals. In coarser-grained sediment the particles will be orientated by the flow which deposited them and the remnant magnetism in sediments with a low iron content may not be detectable. The measurement of the remnant magnetism will be focused on the detrital grains (usually haematite) which were initially magnetized at high temperatures when the mineral was formed and have acted as small magnets settling out in the sediment. Magnetization due to diagenesis and weathering would be ignored. The remnant magnetism in a rock will be reset when the minerals are heated above their Curie point during metamorphism or when the minerals are altered diagenetically or by weathering.

20.2.3 Practical aspects of magnetostratigraphy

An anagram of 'paleomagnetism' (the North American spelling) is 'not a simple game', and perhaps this should be remembered by a geologist tempted to use this technique for stratigraphic analysis. To determine the record of reversals in a succession of sedimentary rocks requires careful sampling of the rocks in the field (Fig. 20.4) and a series of measurements to be made in the laboratory (Hailwood 1989).

Field sampling is normally carried out by drilling out a core of rock from the outcrop about 20 mm across and 100 mm long. The orientation of this core in three dimensions and the attitude of the bedding must be measured. At least six cores are normally taken from a single bed in order to provide enough samples for a statistically significant analysis of the remnant magnetism at a single site. The vertical interval between sampling sites will depend on the rates of accumulation of the sediments and the time interval between field reversals during that period of Earth history. In successions deposited at slow rates, samples may need to be taken every few metres up the succession in order to be sure of detecting all the polarity reversals, whereas higher rates of accumulation allow a wider spacing of sample sites. Once a reversal is identified, the precise location in the succession may be determined by resampling at closer intervals between the sites which show opposite field directions.

The remnant magnetism in the samples is determined

Fig. 20.4 Drilling bedrock to obtain samples for a palaeomagnetic study on the island of Bacan, Indonesia.

in the laboratory by a *magnetometer*. Modern instruments are capable of detecting and measuring magnetic fields in the samples which are several orders of magnitude weaker than the Earth's magnetic field. In order to determine the remnant magnetism, the orientation of the magnetism preserved in the rock at the time of its formation, the effects of any magnetic field resident in the sample due to alteration or weathering of minerals must be removed. This is achieved by putting the sample in a space shielded from the present-day field and either heating it up or subjecting it to the field of an alternating current; the result of this is that the magnetism remaining in the sample is relict from an earlier stage in its history, hopefully the time at which the rock was formed. The demagnetization process is normally carried out in a series of temperature or field strength steps and the remnant magnetism measured after each step until the orientation of the field measured does not change between successive steps; this allows the temperature or field conditions under which the remnant magnetism formed to be determined. The remnant magnetism recorded in separate samples at the same site is compared to ensure statistical significance of the result.

20.2.4 Application of magnetostratigraphy

The long procedure of sampling and analysis required to obtain a reversal stratigraphy for a succession of sedimentary rocks means that this technique is normally only used when other (biostratigraphic) methods cannot be used or a high-resolution stratigraphy is required. Magnetostratigraphy is often employed in continental successions which lack age-diagnostic flora or fauna and which cannot be biostratigraphically dated.

A major drawback of magnetostratigraphy is that only two 'signals', normal and reverse remnant magnetic polarity, are recorded in the rocks. It is therefore essential to have some sort of tie-point to the geological time-scale. This may be provided by absolute dating of a unit, such as a lava, within the successions, or biostratigraphic information which can be used to relate a point in the succession to the time-scale. The reversals identified in the studied section can then be matched to the established reversals on time charts. An important proviso is that any time gaps in the record provided by the succession are recognized and accounted for: a period of normal or reversed polarity may not be represented if there is no sedimentation or if the deposits of that period are removed by erosion. Once a reversal stratigraphy has been established in part of a sedimentary basin, correlation within the basin is possible by matching the reversal patterns at other localities, again taking any evidence for breaks in the sedimentary record into account (Hailwood 1989).

20.3 Other dating methods

There is a continual search for new techniques for dating rocks. Two of the most important for use in pre-Quaternary rocks are strontium isotopes and fission track analysis. Strontium dating involves isotopes but it is not an absolute dating technique and should be distinguished from absolute rubidium–strontium dating *(20.1.2)*.

20.3.1 Strontium isotopes

Strontium is chemically similar to calcium and is found in small quantities in many limestones as strontium carbonate. There are two common isotopes of strontium, ^{86}Sr and ^{87}Sr, and analysis of carbonates through the Phanerozoic has indicated that the ratio between these two isotopes in sea water has changed through time (De Paolo & Ingram 1985; Hess *et al.* 1986). A strontium isotope curve has been constructed using information from carbonate minerals formed from sea water and which have not been subsequently altered or recrystallized. By comparing the $^{87}Sr/^{86}Sr$ ratio in calcite from a sample of unknown age with the established curve, it is possible to determine the age of the sample.

There are two important factors to be considered when dating using strontium isotope ratios. First, a particular $^{87}Sr/^{86}Sr$ ratio is not unique to a particular date as that same ratio may have existed a number of times: some other control on the age of the rock is required in order to constrain the part of the curve to be used for comparison (Fig. 20.5). Second, only carbonate minerals which have been formed from sea water and which have not been subsequently recrystallized can be used. Many organisms make shells of aragonite, but these cannot be used because aragonite recrystallizes to calcite through time. Only organisms which precipitate calcite in their

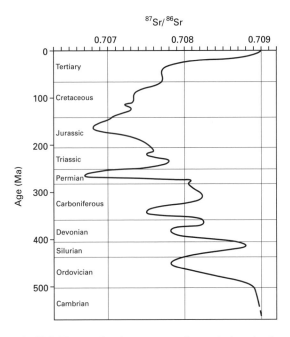

Fig. 20.5 The strontium isotope curve: changes in the ratio of the isotopes ^{86}Sr and ^{87}Sr through geological time. (After Faure 1986; copyright ©1986. Reprinted by permission of John Wiley & Sons, Inc.)

shells or skeletons can be used and these must be free from alteration.

20.3.2 Fission track dating

Radioactive decay of isotopes involves the emission of a form of radiation in the processes of formation of the daughter isotope. The emission of high-energy alpha particles by radioactive decay within a crystal results in a trace of the path of the alpha particle as it disrupts the crystal lattice. For a particular mineral there is a temperature (the *blocking temperature*) above which the crystal lattice anneals, leave no trace of the passage of the alpha particle. At lower temperatures the trace of the alpha particle is left as a fission track which can be detected and measured. By measuring these fission tracks in minerals it is possible to determine the time when the rock passed through the blocking temperature (Hurford & Green 1982, 1983). The two minerals most commonly used for fission track analysis are apatite and zircon: apatite has a blocking temperature of 90°C and zircon around 300°C.

The main use of fission track analysis is to determine the uplift and exhumation history of a body of rock. Temperature increases with depth, so if the geothermal gradient *(17.8.1)* is known, this technique makes it possible to determine the age at which the body of rock was raised above the depth of the blocking temperature.

20.4 Dating in the Quaternary

A number of additional techniques are used in dating the more recent history of the Earth. These methods are restricted to use in the last few tens or hundreds of thousands of years and cannot be used for dating further back in geological time.

20.4.1 Carbon-14 dating

Carbon-14 or *radiocarbon* dating is probably the best known of all radiometric dating methods because it is widely used in archaeology for the dating of bone and charcoal. The principle of this technique is that living organisms continually take up carbon from the environment which includes the isotope ^{14}C. This radioactive isotope is continually produced by cosmic bombardment of ^{14}N in the atmosphere. When the organism dies the ^{14}C in it radioactively decays at a rate determined by its half-life of 5730 years (Table 20.1). By measuring the

proportion of ^{14}C present in a sample of once-living tissue the time elapsed since it died and stopped taking up ^{14}C from the atmosphere can be determined.

From a geological point of view, the main limitation of this technique is that with such a short half-life the levels of ^{14}C present in a sample start to become too low to be detected with accuracy in materials older than 50 000 years, although modern analytical techniques are stretching this back to 100 000 years. There is also an error introduced by the assumption that the levels of ^{14}C in the environment have been constant through this period; it is now known that the ^{14}C levels have varied through time and some corrections have to be made to the dates obtained.

20.4.2 Uranium series dating

The decay series involving isotopes of uranium, thorium and lead is complex and includes short-lived isotopes which are appropriate for dating in the Quaternary. The ^{230}Th–^{234}U part of the decay series is most useful as trace amounts of these elements in carbonate minerals can be used to date material up to about 350 000 years old.

20.4.3 Luminescence techniques

The thermoluminescence and optical stimulating luminescence techniques are based on the ionizing effects of natural radioactivity from minerals containing potassium, thorium and uranium. Electrons released by ionizing radiation become trapped in the crystal lattices of some minerals and accumulate through time. By measuring the 'dose' received by a mineral the age can be calculated, provided that the level of radioactivity in the surrounding rock or sediment is known. Thermoluminescence techniques have been used to date calcite in stalagmites and in archaeology. Optical stimulating luminescence relies on the fact that sunlight 'bleaches' the luminescence signal so it is possible to determine the time elapsed since a grain was last exposed to sunlight, that is, buried in a sediment. The optical stimulating luminescence technique can be used to date sand tens to hundreds of thousands of years old.

20.4.4 Amino-acid racemization

All living organisms contain amino acids and these compounds exist in two geometric forms, 'L' and 'D'. In living organisms the L form is dominant but when the tissue dies the process of *racemization* occurs, converting the L form into the D form until they are in equilibrium. The rate at which this occurs depends on the nature of the organism and is sensitive to temperature. In hot conditions racemization may take a few thousand years, but in colder conditions it may take a hundred to a thousand times longer. This technique can only be used on material where the rate of racemization is known and the temperature can be determined.

20.4.5 Annual cycles in nature

Two techniques fall into this category: tree rings and glacial lake varves. Seasonal variations in the rate of growth of trees leave rings in the wood, the thickness of which depends on the length of the growing season. Climatic variations can be picked out in the pattern of rings, and by matching the ring patterns in very old living trees, a tree ring chronology reflecting climate fluctuations has been extended several thousand years back into the Holocene. These have been useful in calibrating the ^{14}C dating method.

Varves are millimetre-scale laminae in the deposits of glacial lakes which are caused by the seasonal influx of sediment during the summer melt. Counting these laminae back from the present in a lake deposit provides an indication of the age of the deposits.

20.5 Correlation

The concept of correlation has different implications depending on the type of stratigraphic units being considered in the process. If the units are lithostratigraphic then the correlation is between lithological units which may be diachronous. To obtain a correlation between rocks which has some chronostratigraphic significance, biostratigraphic, allostratigraphic and magnetostratigraphic units need to be considered along with radiometric dates where appropriate (Table 20.3).

20.5.1 Lithostratigraphic correlation

In almost any area of outcrop of rocks there are gaps in exposure. Regions devoid of vegetation and soil cover in arid deserts or high latitudes provide continuous exposures which may be kilometres across but even in these situations the complete three-dimensional extent of lithostratigraphic units is not seen. In vegetated regions and areas with extensive superficial cover by Pleistocene glacial deposits bedrock exposures are normally isolated and separated by kilometres of no exposure.

The lithological properties of the units are used to establish similarities and differences between separated exposures and determine whether they can be considered to belong to the same formation or not. Boundaries are established by comparison with the type sections *(18.4.2)* of the formations. The same or similar changes in lithological characteristics are sought and used to

Table 20.3 The types of stratigraphic unit used in correlation.

Lithostratigraphic unit
A body of rock which is distinguished and defined by its lithological characteristics and its stratigraphic position relative to other bodies of rock

Biostratigraphic unit
A body of rock which is defined and characterized by its fossil content

Allostratigraphic unit
A body of rock defined by its position relative to unconformities or other correlatable surfaces ('sequence stratigraphy')

Magnetostratigraphic unit
A body of rock which exhibits magnetic properties which are different from those of adjacent bodies of rock in the stratigraphic succession

Chronostratigraphic unit
A body of rock which has lower and upper boundaries which are each isochronous surfaces

determine where formation or member boundaries should be placed throughout an area. In a complete analysis of a region, the positions of all formation boundaries in all parts of the area are determined as precisely as the exposures allow. This is a fundamental part of the practice of producing a geological map. The spatial distribution of lithostratigraphic units is established on the basis of known and inferred formation boundaries. The attitude (dip and strike) of bedding planes within formations is used to establish the tectonic structure of the rock units and elucidate the pattern of folds and faults. However, because formations are defined in terms of their lithological characteristics and not their ages, the correlation of lithostratigraphic units in the context of a geological map does not provide anything more than general information about the temporal relationships between different parts of the area.

20.5.2 Biostratigraphic correlation

Speciation events and widespread extinctions may be recognizable in the stratigraphic record and the units they define may be considered to approximate to chronostratigraphic units. The confidence with which they can be considered to represent isochronous events increases if the distance over which correlation is attempted is relatively small, say tens or hundreds of kilometres. There are, however, a number of practical considerations to be borne in mind when attempting biostratigraphic correlation.

Biostratigraphy can provide a fine scale of resolution only in circumstances where the rate of sedimentation was relatively slow and the frequency of speciation events in the zone taxa relatively high. If a biozone is estimated to have lasted a million years the complete zone is more likely to be represented in a single exposure of sediments of a deep marine setting where sedimentation rates are slow (a few millimetres per thousand years) than in a shallow marine environment where rates are likely to have been higher (decimetres to metres per thousand years).

It is fairly obvious but important to remember that the rocks must contain the appropriate fossils. This will be dependent upon the environment of deposition because it may not have been suitable for the critical taxa. The appearance or disappearance of a zone fossil may be due to changes in environment rather than be of stratigraphic importance. Another factor is the post-depositional history of the rock because fossil material such as calcium carbonate can be altered or dissolved by diagenetic processes.

20.5.3 Magnetostratigraphic correlation

A fundamental problem with magnetostratigraphy is that the principal magnetic property measured in rocks is the polarity of the Earth's magnetic field at the time of formation of the rock. Although the declination can also be measured, this reflects the latitude of the rock at the time of formation and this will only change slowly through time as the rock moves on a tectonic plate. Polarity changes relatively rapidly but has only two positions, normal and reversed. A pattern of changes from normal to reversed stratigraphy in a succession of rocks at one place can only be compared with a pattern at another if there is an independent means of providing some tie between particular normal or reversed magnetostratigraphic units. Such a tie may be a radiometric date, a marker bed which is assumed to be isochronous, biostratigraphic or allostratigraphic information. Once the general framework is established, magnetostratigraphic correlation provides a detailed means of chronostratigraphic correlation within the area (Talling & Burbank 1993).

20.5.4 Allostratigraphic correlation

Sequence stratigraphic principles (Chapter 21) may be used to establish a chronostratigraphy within an area if key horizons within the successions are assumed to be isochronous. A sequence boundary *(21.2.2)* represents an event in geological time when sea level in the area fell and erosion occurred before sedimentation during rising sea level. This event of sea level fall is not instantaneous and it takes time for erosion to occur, so the sequence boundary is not a truly isochronous surface. However, all the beds above the sequence boundary are younger than the beds below it so the surface does have a chronostratigraphic significance. Horizons which represent the maximum flooding surfaces *(21.2.4)* during the cycle of sea level rise and fall have a similar chronostratigraphic significance.

Sequence boundaries can potentially be assigned a geochronological age if the time of the event which caused the sea level fall is known. Attempts have been made to establish a global eustatic sea level curve which purports to show the timing of rises and falls in worldwide sea level *(21.8)*. The theory is that if a sequence boundary, or series of sequence boundaries, in a succession can be related to the global sea level curve then those surfaces can be dated. This is based on the assumptions that the established eustatic curve is truly global and that there are no other causes of sea level rise or fall. As discussed in the next chapter *(21.8.7)* these assumptions are probably not reasonable in many cases and correlation using the eustatic sea level curve alone is not a viable method.

20.5.5 Radiometric age dating and correlation

General stratigraphic relations and isotopic ages are the principal means of correlating intrusive igneous bodies. Geographically separate units of igneous rock can be shown to be part of the same igneous suite or complex by determining the isotopic ages of the rocks at each locality. Radiometric dating can also be very useful for demonstrating correspondence between extrusive igneous bodies.

The main drawbacks of correlation by this method are the limited range of lithologies which can be dated and problems of precision of the results. For example, if two lava beds were formed only a million years apart and there is a margin of error in the dating methods of 1 million years then correlation of a lava bed of unknown affinity to one or the other is not certain.

20.5.6 Other correlation techniques

In addition to the stratigraphic techniques mentioned above there are additional methods which may be applicable in certain circumstances.

PROVENANCE STRATIGRAPHY

Detrital grains and clasts in clastic sediments are derived from source areas of high ground around the basin of sedimentation which are being eroded. This is the provenance of the detrital material *(5.5)*. Through time different rock units in the source areas are exposed by erosion and the types of clastic material supplied to the basin change. The appearance of new types of grains or gravel-sized material in the succession of sediments will mark the time when that new source area lithology was exposed and eroded. These distinctive clast types will not appear everywhere in the basin at the same time and there may be reasons why they do not occur in some depositional environments. However, this correlation by 'provenance stratigraphy' can be useful in circumstances where other correlation methods are not available — for example, in thick, unfossiliferous continental clastic successions (DeCelles 1988). The clasts used in sandstones are commonly heavy mineral grains *(2.4.2)* which can be characteristic of particular source area lithologies.

CHEMOSTRATIGRAPHY

This is another technique which is only normally attempted when all other methods of correlation fail. It is really an extension of lithostratigraphy using the elemental composition of deposits instead of lithological characteristics. The bulk chemistry of deposits may depend on a variety of factors, including the nature of the source area for clastic detritus and the composition of waters entering a closed basin. Sometimes changes in bulk chemistry appear to have occurred across the basin at the same time and by analysis of the chemical composition of different rock units some correlation is possible (Racey *et al.* 1995).

Further reading

Aïssauoi, D.M., McNeill, D.F. & Hurley, N.F. (eds) (1993) *Applications of Palaeomagnetism to Sedimentary Geology*, 216 pp. Society for Sedimentary Geology Special Publication **49**.

Blatt, H., Berry, W.B. & Brande, S. (1991) *Principles of Stratigraphic Analysis*, 512 pp. Blackwell Scientific Publications, Oxford.

Dalrymple, G.B. (1991) *The Age of the Earth*, 422 pp. Stanford University Press, Palo Alto, CA.

Eicher, D.L. (1976) *Geologic Time*, 2nd edn. Prentice Hall, Englewood Cliffs, NJ.

Faure, G. (1986) *Principles of Isotope Geology*, 2nd edn, 589 pp. Wiley, New York.

Hailwood, E.A. (1989) *Magnetostratigraphy*, 84 pp. Geological Society of London Special Report **19**.

Hawkesworth, C.J. & van Calsteren, P. (1992) Geological time. In: *Understanding the Earth, A New Synthesis* (eds G.C. Brown, C.J. Hawkesworth & R.C.L. Wilson), pp. 132–144. Cambridge University Press, Cambridge.

Prothero, D.R. & Schwab, F. (1996) *Sedimentary Geology: An Introduction to Sedimentary Rocks and Stratigraphy*, 575 pp. W.H. Freeman, New York.

21 Sequence stratigraphy and sea level changes

The sequence stratigraphy approach to the analysis of successions of sedimentary rocks was developed in the 1970s by geologists and geophysicists working for Exxon in the USA. Starting with papers published in 1977, Vail, Mitchum and colleagues set out a scheme for considering stratigraphy in terms of depositional sequences bound by unconformities. This was originally developed for analysis of subsurface seismic reflection data but it was shown how the methods could be applied to all scales of stratigraphic analysis. Strictly speaking, this concept was not new, as large unconformity-bounded packages had been recognized and described in the North American craton over a decade before (Sloss 1963). However, the sequence stratigraphy approach and the global sea level curve developed at the same time by the Exxon group prompted a minor revolution in sedimentology and stratigraphy. In this chapter the principles that underlie sequence stratigraphy are presented and the application of the methodology explained. To some extent the concepts are based on long-established ideas from traditional stratigraphy, but several new terms have been introduced as part of the new approach and these require some explanation. Once the jargon has been pushed aside, the sequence stratigraphy approach will be seen as a common-sense way of looking at successions of sedimentary rocks. The global sea level curves presented by the Exxon group and others are a more contentious issue. The possible causes of global sea level fluctuations and the use of a sea level curve as a correlative tool are considered separately in the last section.

21.1 Introduction

Sediments can only accumulate in the environments described in earlier chapters if there is space for them to be deposited and preserved. In a broad sense, regions where sediment is accumulating are referred to as sedimentary basins (Chapter 23). The entire fill of a basin is known as a *basin fill succession*. It may consist of several kilometres of sedimentary rock deposited over tens of millions of years. At the smaller scale of individual depositional environments, the space locally available within the sedimentary basin is often referred to as the *accommodation space* (Jervey 1988). This refers to the potential at any point for sediment to accumulate. The sediment may form in layers, lenses or other geometries depending on the processes controlling deposition in that environment. These bodies of sediment may ultimately be preserved as sedimentary rocks.

In the absence of accommodation space, an estuary, for example, would act only as a zone of transfer of water and sediment from the continent to the marine realm, with equal amounts of sediment entering and leaving the estuarine environment. If accommodation space is available sands and muds can build up in the estuary to form a deposit. The thickness of sediments formed during a given period of time will depend partly on the accommodation space made available during that period. It will also depend on the amount of sediment supplied by the river to the estuary. As will become apparent in the following sections, the balance between accommodation space and *sediment supply* (the total bedload, suspended load and dissolved material brought to a particular point, plus *in situ* biogenic productivity) determines the character of the stratigraphy developed.

21.1.1 Creation of accommodation space

In the shallow marine realm accommodation space can

be considered to be created by a rise in sea level relative to the sediment–water interface (the sea bed) at that point. A relative sea level rise can be achieved either by the sea level itself rising or by the sea bed subsiding. The causes of sea level rise, such as melting of ice caps to raise the level of the seas world-wide *(21.8.2)*, may be fundamentally different from the causes of subsidence (stretching, flexure or cooling of the lithosphere), but from the frame of reference of a crab sitting on the sea bed the effect is identical (Fig. 21.1). In shallow marine settings the upper limit of the accommodation space is defined by sea level and it is assumed that sediment would build up to sea level provided that there was sufficient sediment supply. At abyssal depths accommodation space is not a factor because the amount of water above the ocean floor does not directly influence sedimentation at that point, although sea level does influence sediment supply to the deep oceans.

The concept of accommodation space also applies to some continental environments. In lakes it is water level, and the responses of lake sedimentary environments to changes in the water level will parallel those in the marine realm. In fluvial environments base level is a more abstract concept. Accumulation of sediment occurs in rivers and on floodplains above sea level but only to a point which appears to be controlled by the local gradient of the river. Accommodation space therefore can be considered to be controlled by an equilibrium profile which exists along the line of the river. If subsidence occurs and the land surface drops, sediment may accumulate along the river and on the floodplain until it reaches this equilibrium profile and further sediment is carried on downstream. Areas of accumulation of wind-blown sand are controlled by the relationship between the geomorphology, wind direction and sediment supply: accommodation space concepts do not apply in any direct sense in aeolian environments, although the level of the water table plays a role in limiting the extent of wind deflation (Kocurek 1996).

21.1.2 Aggradation, progradation and retrogradation

If the rates of creation of accommodation space and supply of sediment at a point are considered relative to each other several possible situations exist (Fig. 21.2).

If the space created and the volume of sediment to fill it are exactly in balance then the deposits will *aggrade*, that is, build up vertically at that point. Along a shoreline where the relative sea level rise is matched by the supply of sediment, the beach will stay in the same position through time (Fig. 21.2a). The shoreline does not shift landwards or seawards and all the facies belts remain fixed in position up through the stratigraphic section. These conditions of *aggradation* rarely exist for long in nature because of changing rates of sea level rise and fall.

Under conditions where the sediment is being supplied at a higher rate than accommodation space is being created, the deposits may not only build up but upon filling the space available also build outwards (*prograde*). This geometry of deposition is a pattern of *progradation* (Fig. 21.2b). Along a coast the beach will build out into foreshore areas and the shoreline will move basinwards. In a vertical section beach sediments will be preserved on top of foreshore deposits. The facies belts offshore will also show an upward change to shallower facies in all depositional settings. This is also referred to as a *regression* of the depositional environment.

In sequence stratigraphic terminology a distinction is drawn between a regression due to sediment supply exceeding the rate of creation of accommodation space and a *forced regression* generated by a relative sea level

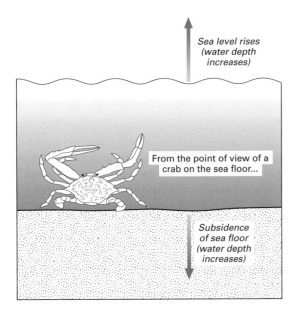

Fig. 21.1 The concepts of accommodation space. From the point of view of a crab sitting on the sea floor, it amounts to the same thing if the overall sea level has risen or the sea floor has subsided: the distance to the water surface has increased, an increase in accommodation space.

Within figure:
Sea level rises (water depth increases)

From the point of view of a crab on the sea floor...

Subsidence of sea floor (water depth increases)

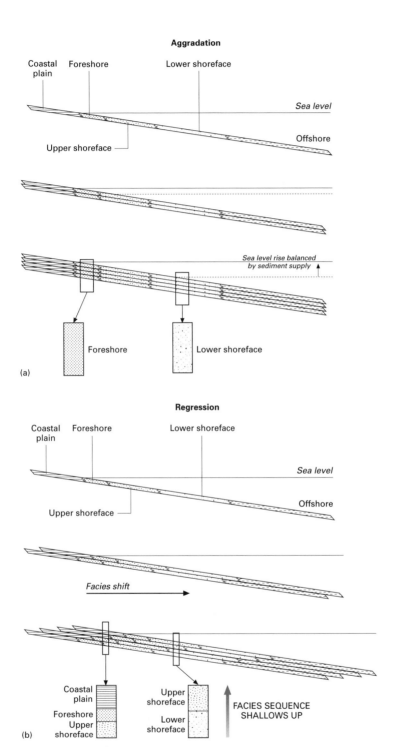

Fig. 21.2 Patterns of strata relative to sediment supply and sea level. (a) Aggradation occurs when there is a balance between the rates of sediment supply and the rise of sea level (creation of accommodation space). (b) A pattern of progradation forms when the rate of supply of sediment exceeds the rate of sea level rise, resulting in a regression of the shoreline. (*Continued*)

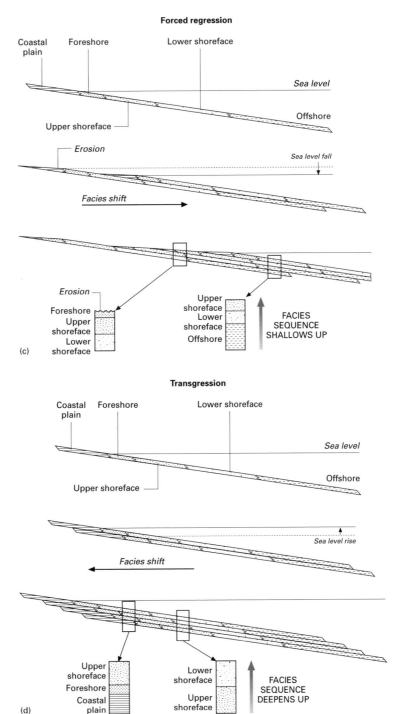

Fig. 21.2 (*Continued*) (c) A relative fall in sea level causes a shift of facies basinwards, a 'forced regression'. (d) When the rate of creation of accommodation space exceeds sediment supply there is a transgression of the sea and a retrogradation pattern results.

fall (Fig. 21.2c). Just as a relative sea level rise can be due to absolute sea level rise or subsidence of the sea bed, a sea level fall can be due to an absolute fall in global sea level or uplift of the sea bed due to tectonism. The crab sitting on the sea bed would only be aware of a reduced amount of water above it. These may be considered to be situations when the rate of creation of accommodation space is negative. In these circumstances, the shift in facies basinwards is accompanied by erosion in the shallower environments in response to the lowering of base level. A 'downstepping' geometry of deposits is characteristically formed during a forced regression (Fig. 21.1c).

Retrogradation (Fig. 21.2d) of the depositional environments due to the rate of creation of accommodation space exceeding the sediment supply is more commonly known as *transgression*. Sedimentation on a beach is unable to keep pace with the relative rise in sea level, the beach is flooded by the sea and the shoreline moves landwards to create a new beach. The same *landward shift* in facies is seen in all the other facies belts offshore, with an upward change to deeper water facies in vertical profile.

21.2 Depositional sequences and systems tracts

The patterns of stratigraphy created by the balance between the supply of sediment and the rate of creation of accommodation space (both positive and negative) form the basis for the division of a basin fill succession. A depositional sequence is a package of sediment deposited during a particular period in the basin history.

21.2.1 Depositional sequences

A *depositional sequence* is defined as a stratigraphic unit bounded at its top and base by unconformities or their correlative conformities (Fig. 21.3) (Mitchum 1977). It represents a period of deposition between two episodes of significant sea level fall. Using the sequence stratigraphy approach depositional sequences are the primary means of dividing up a basin fill succession. If the relative changes in sea level are basin-wide events the sequence boundaries may be recognized in different vertical sections through the basin stratigraphy and used as a means of correlation provided that the same sequence boundary can be confidently recognized in different places.

21.2.2 Sequence boundaries

A relative sea level fall of several tens of metres will have a dramatic effect on a shallow marine environment. Rivers may respond to the fall in base level by cutting down into their floodplain and also into the newly exposed shelf area in the process of adjusting to a new equilibrium profile. The exposed beach, foreshore and shoreface will be subject to erosion on the coastal plain. This erosion will be preserved as an unconformity surface in the stratigraphic record. Further basinwards, the relative sea level fall may not have exposed the outer parts of the shelf to create an erosion surface but shallow water deposits of the beach and shoreface will overlie outer shelf deeper water facies (Fig. 21.3). Facies indicating upward shallowing will be preserved in the stratigraphic record. If the sea level fall is as great as the depth of water at the shelf edge, an unconformity may develop over the whole shelf. In sequence stratigraphic terminology this surface created by the sea level fall is called a *sequence boundary*. A sequence boundary normally consists of an unconformity and its *correlative conformity*, the equivalent surface in the outer shelf where there was no erosion (Van Wagoner *et al.* 1988).

21.2.3 Systems tracts

Between the events of relative sea level fall which create the sequence boundaries there are normally periods of sea level rise. The sediments preserved between the sequence boundaries are deposited during this period and will show patterns of transgression, aggradation and progradation reflecting the sea level changes. Different *systems tracts* can be recognized by the patterns and related to stages in the cycle of rise and fall in sea level (Posamentier *et al.* 1988; Van Wagoner *et al.* 1988). As a first step an idealized model of sea level fluctuations is assumed with a simple sinusoidal curve to describe the rise and fall and a constant sediment supply.

The *transgressive systems tract* formed during the period of most rapid rise in sea level will be characterized by a general *landward shift* in facies belts—that is, shallow water facies are overlain by deeper water

Fig. 21.3 (*Facing page*) The Exxon scheme of depositional sequences, sequence boundaries and systems tracts related to relative fall and rise of sea level on a margin with a distinct shelf edge. (Based on Van Wagoner *et al.* 1990; AAPG ©1990, reprinted by permission of the American Association of Petroleum Geologists.)

MARGINS WITH A SHELF EDGE

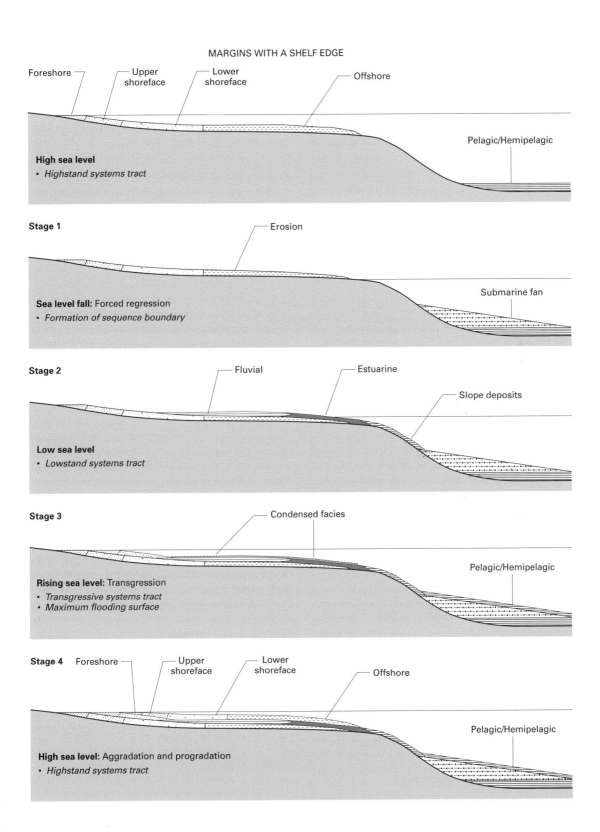

deposits. The rate of creation of accommodation space will be greater than the supply of sediment and outer shelf deep water mudstones will overlie inner shelf shallow marine sandstones, for example. Modification to the surface formed by subaerial erosion may occur as sea level rises during transgression. Wave action on the beach and in shallow water reworks the surface as it is flooded and evidence for subaerial exposure such as the presence of a palaeosol may be destroyed. These surfaces of marine reworking during transgression are called *ravinement surfaces* (Swift 1968; Emery & Myers 1996).

As the rate of sea level rise decreases the rate of creation of accommodation space becomes balanced by the sediment supply and aggradation occurs. When sea level is reaching its peak little accommodation space is being created and the sediment starts to prograde. A vertical profile through this part of the stratigraphy would indicate that the environment of deposition became shallower upwards. This succession of deposits showing aggradation and progradation formed during a period when the sea level is relatively high is called the *highstand systems tract.*

A fall in sea level exposes the inner shelf and initiates erosion to form a sequence boundary. This erosion continues as the sea level falls to its lowest point, generating additional sediment which is deposited in deeper water. Two different systems tracts may form during the period of low sea level. If the sea level falls to or below the level of the edge of the shelf, the whole shelf area is exposed to erosion and most of the sedimentation occurs on the slope and basin beyond the edge of the shelf. These deep water deposits make up the bulk of the sedimentation to form a *lowstand systems tract.* Some sediment also accumulates in the outer parts of the shelf in valleys and canyons during this lowstand period.

If the margin does not show a distinct break but is gently sloping down into deeper water (a *ramp margin*: Fig. 21.4) there will be a *basinward shift* in facies belts during the period of low sea level with shallow water facies overlying deeper water deposits. The deposits of this stage of the cycle between the onset and end of sea level fall have been referred to as the *falling stage systems tract* (Plint 1996) or *forced regressive wedge systems tract* (Hunt & Tucker 1992). The deposits will consist of a progradational succession of shallow marine deposits which fill the available accommodation space.

21.2.4 Maximum flooding surfaces

There is one further key feature of the stratigraphic

geometries formed during a cycle of sea level rise and fall and this is the horizon which represents the furthest landward extent of marine conditions. The *maximum flooding surface* separates the transgressive and highstand systems tracts and may be recognized in a vertical succession by a change from retrogradation to aggradation/progradation (Emery & Myers 1996). In the mid-shelf area the surface itself may be recognized by evidence for very slow rates of sedimentation. During transgression deposits accumulate in the inner shelf area where accommodation space is being created rapidly, leaving the more distal area starved of sediment. Under these conditions firmgrounds, hardgrounds (*11.7.2*) and glauconitic layers (*11.6.1*) form and may be used to recognize a maximum flooding surface. These surfaces may be recognized basin-wide in some circumstances.

Some authors have used the maximum flooding surfaces as the primary bounding horizons in the division of a basin fill succession. Galloway (1989) defined 'genetic stratigraphic sequences' as units bound by maximum flooding surfaces and included the unconformity surface within the sequence. This approach is preferred by some, but whatever the merits of the Galloway scheme the existence of two different definitions of sequences of sediments causes confusion. The Exxon scheme (Van Wagoner *et al.* 1990) is by far the most widely used because the widespread erosion surface at the sequence boundary is often more easily recognized than the maximum flooding surface.

21.3 Facies patterns in depositional sequences

There are two main idealized models for deposition during cycles of sea level rise and fall. The first is where the sea level fall is to the shelf edge and a deep water lowstand systems tract forms. The second involves a ramp margin or a smaller fall in sea level relative to the shelf bathymetry and only part of the shelf is exposed. These are both referred to as 'type 1' sequences by the Exxon group and are defined as such by having evidence of a relative fall of sea level. A third case, originally called 'type 2' sequences in the Exxon terminology (Van Wagoner *et al.* 1990), is a situation where the sea level is continually rising but the rate of sea level rise is

Fig. 21.4 (*Facing page*) Systems tracts on shelves with a ramp geometry. (Based on Van Wagoner *et al.* 1990; AAPG ©1990, reprinted by permission of the American Association of Petroleum Geologists.)

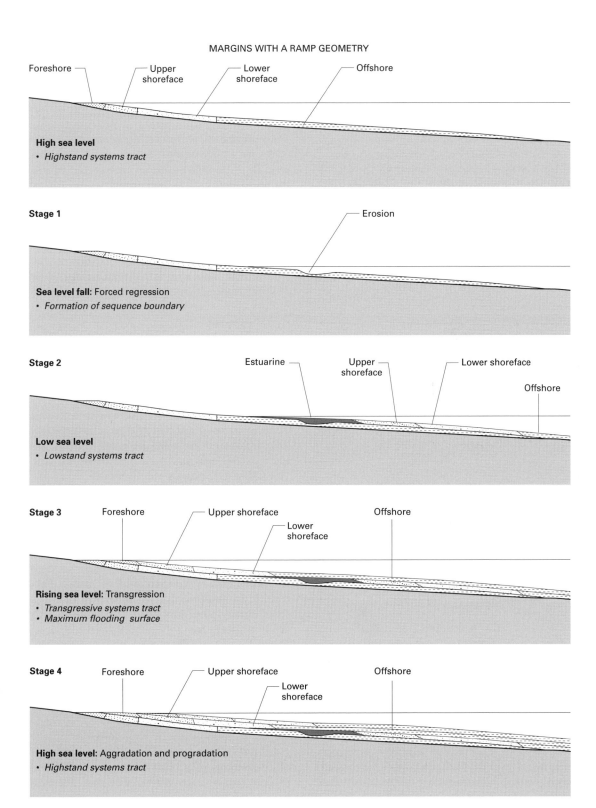

MARGINS WITH A RAMP GEOMETRY

Foreshore — Upper shoreface — Lower shoreface — Offshore

High sea level
• *Highstand systems tract*

Stage 1 — Erosion

Sea level fall: Forced regression
• *Formation of sequence boundary*

Stage 2 Estuarine — Upper shoreface — Lower shoreface — Offshore

Low sea level
• *Lowstand systems tract*

Stage 3 Foreshore — Upper shoreface — Offshore — Lower shoreface

Rising sea level: Transgression
• *Transgressive systems tract*
• *Maximum flooding surface*

Stage 4 Foreshore — Upper shoreface — Offshore — Lower shoreface

High sea level: Aggradation and progradation
• *Highstand systems tract*

changing through time (e.g. in a basin where the rate of tectonic subsidence is high). Patterns of sedimentation allow the succession to be divided into systems tracts but without erosional unconformities bounding the depositional sequences. The lowstand systems tract is replaced by a shelf margin systems tract. These were originally called 'type 2' sequences in the Exxon terminology (Van Wagoner *et al.* 1990), although the difficulties of recognizing a sequence boundary without an unconformity have meant that this term is now rarely used (Van Wagoner 1995).

21.3.1 Margins with a shelf edge (Fig. 21.3)

STAGE I: SEQUENCE BOUNDARY FORMATION AND LOWSTAND SEDIMENTATION

A rapid sea level fall leads to the exposure of and erosion across the shelf. The sediment supplied by rivers bypasses the shelf and is directly deposited on the basin floor. The main process of transport of coarse clastic material in deep water will be turbidity currents which may deposit sediment on submarine fans *(15.3)*. Lowstands are therefore characterized by deposition on submarine fans which show evidence of higher rates of sedimentation during this period than during transgressive and highstand stages.

STAGE 2: LOWSTAND SEDIMENTATION

During the period of sea level lowstand, sediment supplied from the hinterland and by erosion of the shelf will continue to be deposited. It will start to build back up the continental slope as a blanket of slope deposits and accumulate in valleys and canyons cut in the shelf.

STAGE 3: TRANSGRESSIVE SYSTEMS TRACT

As sea level starts to rise fluvial deposits fill incised valleys in the more proximal parts of the shelf and where these valleys are flooded by the sea sediment accumulation occurs in estuaries. The first deposits above a sequence boundary in the shelf succession are therefore commonly fluvial and estuarine sediments.

With continued sea level rise, accommodation space is quickly generated, the shoreline shifts landwards and the sediment supplied is deposited in shallow marine environments close to the shore. Basinwards the shelf area receives little or no sedimentation and condensed facies

develop. These characterize the surface of maximum transgression or maximum flooding surface which marks the top of the transgressive systems tract.

STAGE 4: HIGHSTAND SYSTEMS TRACT

The decreased rate of creation of accommodation space allows the shallow marine facies to aggrade and then prograde as the shoreline moves basinwards. The relative sea level may still be rising during this period but at a slow rate compared to the sediment supply. At the coastline, sediment is building up and out faster than the sea level rise and further transgression does not occur. In the mid- and outer shelf area, the onset of highstand sedimentation is marked by the progradation of shallower marine facies over the condensed facies of the maximum flooding surface.

The cycle then returns to the first stage as relative sea level starts to fall.

21.3.2 Margins with a ramp geometry (Fig. 21.4)

STAGE I: FALLING STAGE SYSTEMS TRACT

The sea level fall only exposes part of the shelf, limiting the area of erosion and formation of an unconformity surface. Sedimentation continues as the sea level falls and the facies belts shift basinwards across the shelf. This falling stage systems tract is characterized by progradational geometries as the accommodation space is filled by sediment supplied from the hinterland and by erosion of the inner shelf area. The sequence boundary is the surface above the falling stage systems tract and is marked by a landward shift in facies.

The stratigraphic relationships developed during transgression and highstand are similar to those described above in stages 3 and 4 for margins where the sea level falls to the shelf edge.

21.4 Sequences in carbonate depositional environments

The effects of sea level change on shelves dominated by carbonate deposition are similar to those on clastic shelves in many ways. It is possible to identify sequence boundaries and maximum flooding surfaces as horizons which separate different patterns of sedimentation and systems tracts can be recognized. The most important factor which makes the response to sea level change in a shelf area distinctly different in carbonate shelves is the mechanism of sediment supply. The geometry of carbonate shelves is also distinct in cases where the rim of a shelf creates a steeper slope at the shelf edge break than is normally seen at the edges of clastic shelves (Schlager 1992).

21.4.1 Effects of sea level rise on carbonate platforms

The shallow waters of the shelf are a 'carbonate factory' *(14.5)*. The total amount of carbonate sediment generated on the shelf is determined by the area of shallow water where benthic and framework-building organisms can thrive. Periods of transgression and highstand are times when the amount of sediment generated is greater because of the larger areas of shallow, flooded shelf. The rate at which carbonate sediment is generated in favourable conditions of warm, clear, shallow seas can exceed 1 m per thousand years (Tucker & Wright 1990). This is as fast as or faster than the rate at which sea level is believed to change under most circumstances. Therefore on rimmed carbonate shelves accommodation space may often be filled as fast as it is created. The position of the shelf edge break may remain fixed during transgression and highstand, and facies will not shift across the shelf in the manner recognized in clastic shelf environments. Thick successions of shallow marine limestone are developed during the transgressive and highstand stages of a depositional sequence.

Exceptions to this general pattern occur when sea level rises rapidly due to abrupt melting of ice caps or local, fault-generated subsidence *(21.8)*. When carbonate productivity cannot keep pace with the rate of creation of accommodation space the shelf is 'drowned' *(14.8.3)* (Schlager 1992). Maximum flooding surfaces which show this change of facies from shallow to deep water facies are relatively easy to recognize in a stratigraphic succession. In addition to the effects of rapid relative sea level rise, carbonate productivity on the shelf can be reduced if the environment cools or the waters become murkier due to increased suspended sediment load. In either case the environment becomes more 'stressed' and organisms which are sensitive to temperature and light levels (many corals) and benthic filter feeders which are 'poisoned' by suspended particles cease to flourish.

21.4.2 Effects of sea level fall on rimmed carbonate shelves

A relative fall in sea level which exposes the whole of a rimmed shelf (creating a type 1 sequence boundary) has a profound effect on the amount of carbonate deposited. The 'carbonate factory' is reduced to a small zone at the top of the steep slope down from the rim of the shelf during the lowstand (Fig. 21.5). The extent to which the carbonates exposed on the shelf are eroded will depend on the climate. Under humid conditions solution may result in a karstic surface but in arid settings an evaporitic coastal sabkha may develop in places. Mechanical erosion to produce clastic detritus is often less significant than on exposed clastic shelves, resulting in less extensive lowstand fans and wedges than seen at the base of the slope in clastic shelf systems. However, both debris cones and turbidite fans can form during the sea level lowstands. The record of the exposure of the shelf as a karstic or evaporitic surface will be the signature of the sequence boundary in a carbonate depositional sequence.

21.4.3 Carbonate shelf and ramp geometries

The rate of sediment aggradation on the shelf is typically much greater than it is in the basin because of the contrast between the high rates of carbonate production in shallow water compared to pelagic deposition and the supply of redeposited detritus during lowstands. Between the shelf and basin lies the slope, a zone made up largely of redeposited material, especially where there is a supply of reef talus from a fringing reef. This slope tends to become more accentuated through time as more sediment is added at the top than at the bottom and develops into a surface dipping 10–15° and with hundreds of metres of relief. These surfaces may be recognized on seismic reflection profiles as distinct clinoforms *(22.2.2)* (Emery & Myers 1996).

The responses to sea level change on a gently sloping carbonate ramp are more analogous to those on a clastic shelf. Slow changes in sea level cause facies belts on the ramp to shift evenly up or down the slope to form diachronous units of depth-defined facies—for example, a shoreward migration of facies during transgression. A rapid sea level rise may result in a ravinement surface developing during transgression and abrupt lateral and vertical facies transitions in the stratigraphic succession.

CARBONATE PLATFORM (RIMMED SHELF)

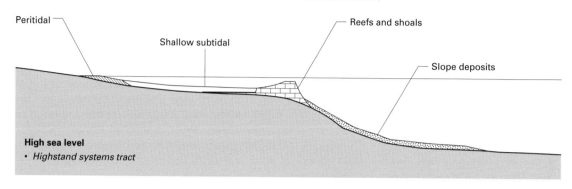

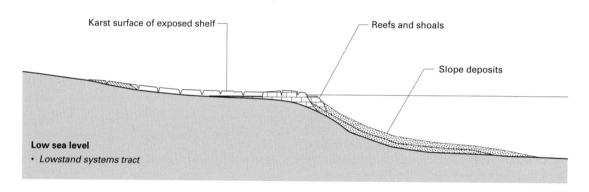

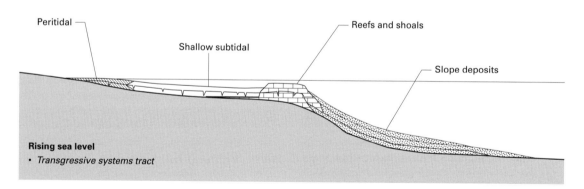

Fig. 21.5 Patterns of sequence development on a carbonate margin. (Based on Sarg 1988; courtesy of the Society for Sedimentary Geology.)

21.5 Subdivision of depositional sequences and systems tracts: parasequences

The trends of progradation, retrogradation and aggrada-

tion described above for the systems tracts of depositional sequences are the gross patterns which can be identified. The thickness of a depositional sequence will depend on the absolute rate at which accommodation space is being created in the basin. This will in turn be primarily controlled by the rates of tectonic and thermo-tectonic subsidence *(21.8.3)*. Depositional sequences may therefore vary from a few tens of metres thick to a thousand metres or more. A 'typical' depositional sequence formed on a passive margin *(23.2.4)* is of the order of a hundred to a few hundred metres thick (Van Wagoner *et al.* 1990). Within this stratigraphic package the details of the patterns are complex and indicate that changes in sea level do not follow the simple curve used to develop the sequence model. A finer division of the stratigraphy is possible if trends over metres and tens of metres are considered.

21.5.1 Parasequences and sedimentary cycles

When, for example, a progradational package in a highstand systems tract is considered in detail it is found to consist of a series of smaller packages of beds, each of which shows an upward shallowing, progradational trend. In the terminology of sequence stratigraphy a group of beds showing a shallowing-upward trend is called a *parasequence*, defined as a relatively conformable succession of genetically related beds bounded by marine flooding surfaces and their correlative surfaces (Van Wagoner *et al.* 1990). A parasequence formed in a shallow marine setting (Fig. 21.6) may consist of mudstone at the base, representing offshore deposition, overlain by lower and then upper shoreface sediments and culminating in sands deposited in the foreshore and beach environment. A return to deeper water conditions is marked by mudstone of a deeper water facies overlying the beach/foreshore sandstone beds. Parasequences are identifiable in carbonate successions on the basis of both lithofacies and biofacies characteristics which indicate upward shallowing. A single parasequence is typically metres to tens of metres thick.

Before the advent of sequence stratigraphic terminology these groups of beds would have been referred to as *cycles* of sedimentation. There is a good argument for continuing to use the term 'sedimentary cycle' in a descriptive sense because by definition a parasequence is formed under conditions of changing relative sea level. A sedimentary cycle may be formed by mechanisms other than sea level change, for example, pulses of sedimentation in clastic deposits or climatic fluctuations in carbonates. A cycle may only be considered to be a parasequence once the control by sea level change has been demonstrated.

The bounding surfaces of parasequences are defined as surfaces of flooding which represent a relative sea level rise. These are called *marine flooding surfaces*, and care must be taken not to confuse them with the maximum flooding surfaces *(21.2.4)* which mark the top

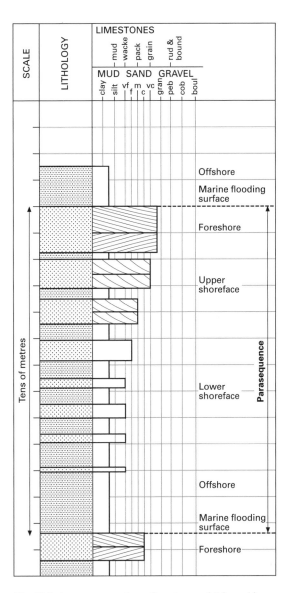

Fig. 21.6 A parasequence (or sedimentary cycle) formed in a shallow marine environment by sediment filling up accommodation space.

of transgressive systems tracts. As a number of parasequences are present within a depositional sequence, the episodes of sea level change which result in parasequences must occur at a higher frequency than the rises and falls which generate the whole depositional sequence. The curve of sea level fluctuation during the deposition of a sequence therefore consists of a high-frequency oscillation superimposed on the longer-period

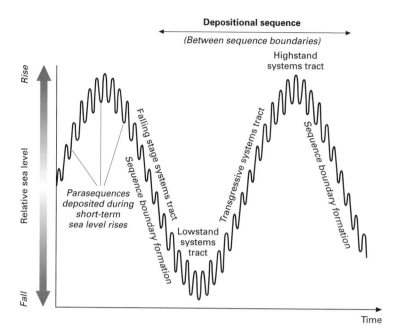

Fig. 21.7 A curve of short-term sea level fluctuation superimposed on a longer-term sea level cycle results in a complex pattern of relative sea level rise and fall: parasequences may form as a consequence of relative sea level rise at different stages in this cycle.

sinusoidal wave (Fig. 21.7). These curves may be generated by the combination of different orders of sea level fluctuation (third- and fourth-order sea level variations: *21.8.5*).

This more complex curve predicts that there can be periods of relative sea level fall or no change in sea level (*stillstand*) during a time of overall sea level rise. The transgressive systems tract therefore may be made up of a number of parasequences, each of which shows upward shallowing, but with each marine flooding surface the sea level rises further. The deposits at the base of the next parasequence represent deeper water facies than are seen at the base of the preceding parasequence. The complete pattern in the systems tract is therefore one of retrogradation (transgression) made up of a series of landward-stepping parasequences. In a regressive succession the superimposed shorter-term sea level curve results in periods when there is either a stillstand or a relative sea level rise allowing a parasequence to build up.

Each systems tract is therefore likely to be composed of a set of parasequences. In most cases each parasequence represents an upward shallowing of depositional environment. The trends through a set of parasequences are recognized by considering the relative shift in the position of the shoreline when going from one parasequence up to another (Fig. 21.8).

21.6 Sequence stratigraphy and depositional environments

So far the effects of rise and fall in sea level on sedimentation have been considered in an idealized, conceptual way. If the sequence approach is to be a practicable tool for the analysis and interpretation of stratigraphy the response of different sedimentary environments to relative sea level change must be considered. In theory it should be possible to recognize the signatures of sea level rise and fall across almost all depositional environments from rivers and coastal and shelf areas to continental slopes and ocean basins. If an event such as a relative sea level fall or rise can be recognized across all these environments it provides a means of correlation which is potentially of higher precision than most other correlation techniques.

21.6.1 The cycle of a depositional sequence

The effects of the cycle of sea level rise and fall across a transect of clastic depositional environments from rivers to the deep sea is here considered in a series of stages (Fig. 21.3).

SEA LEVEL FALL

Rivers may respond to a sea level fall by adjusting to the

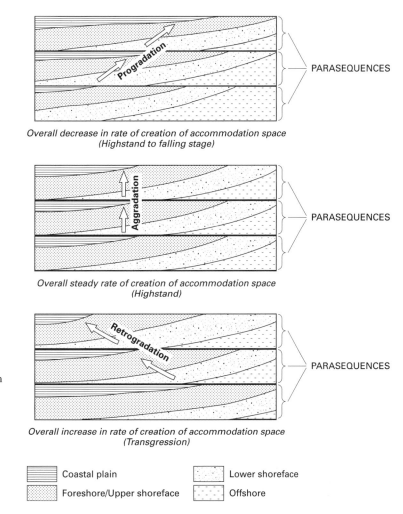

Fig. 21.8 Parasequence sets may form patterns of progradation, aggradation and retrogradation according to the overall rate of creation of accommodation space on the longer-term sea level cycle. (Based on Van Wagoner *et al.* 1990; AAPG ©1990, reprinted by permission of the American Association of Petroleum Geologists.)

Overall decrease in rate of creation of accommodation space
(Highstand to falling stage)

PARASEQUENCES

Overall steady rate of creation of accommodation space
(Highstand)

PARASEQUENCES

Overall increase in rate of creation of accommodation space
(Transgression)

PARASEQUENCES

Coastal plain   Lower shoreface

Foreshore/Upper shoreface   Offshore

new base level which, along with grade and rate of sediment supply, controls the equilibrium profile of the river. There may be a tendency for the rivers to incise into their own floodplain and into the exposed area which had been under water prior to sea level fall. The shallow marine environment may respond by a basin-ward shift in the facies belts or the elimination of a distinct shelf environment if the sea level falls below the shelf edge. Erosion of any exposed shelf areas will occur in clastic systems along with karst formation on exposed carbonate deposits. The reduced accommodation space on the shelf and the increased sediment supply due to erosion will lead to higher rates of deposition in the deeper environments beyond the shelf edge. Submarine fan systems will experience more frequent turbidite

deposition resulting in the progradation of submarine fan lobes. The sequence boundary in deep water is marked by the progradation of coarser turbidite facies.

SEA LEVEL RISE: TRANSGRESSION

As sea level rises from its lowest level, aggradation will occur in the incised river channels, with the fluvial deposits confined to valley fills. Drowned incised river valleys are sites of estuarine sedimentation. The coarser sediments will be deposited in these rivers and estuaries. An incised valley cut into shelf sediments and filled with fluvial or estuarine deposits is a good indicator of a sequence boundary. Some aggradation starts to occur on the outer shelf and the slope beyond the shelf edge as a blanket of relatively fine-grained material. In carbonate-forming rimmed shelf environments areas of primary carbonate sedimentation are restricted to a narrow zone of shallow water.

In the fluvial tract, aggradation in the incised valleys leads to them becoming filled and deposition of fluvial channel and floodplain sediments spreads out over a wider area. A marked landward shift in the depositional environments is seen across the whole shelf area as the coastal plain is inundated by sea water. Parasequence sets show a retrogradational pattern. Coarse sedimentation is confined to the nearshore areas and little fine material reaches the middle and outer parts of the shelf. These areas are starved of sediment and condensed facies form, marking the maximum flooding surface at the top of the transgressive systems tract. In deeper water only pelagic and hemipelagic deposition will occur. The increased area of shallow marine conditions expands the carbonate factory and, if sea level rise is slow under favourable conditions for benthic and reef-forming organisms, carbonate production may keep pace with transgression. An aggradational pattern of facies is therefore seen unless the shelf is drowned and pelagic facies are deposited on the outer shelf during transgression.

HIGHSTAND OF SEA LEVEL

Aggradation in the fluvial tract will continue as long as base level is rising, but the rate of vertical accretion on the floodplain will drop as the sea level reaches its highest point. In the shallow marine environment the rate of creation of accommodation space is matched or exceeded by the sediment supply. The coastline starts to move basinward as a regressive succession is formed across the shelf area. At any point on the shelf there should be a general trend towards shallower facies higher in the section. Supply of detritus to the deeper areas beyond the shelf may increase as the progradation advances towards the shelf edge. Progradation of reefs and shelf-margin carbonate sand bodies occurs during highstand with the build-up of shallow marine facies over the forereef and slope deposits as the slope apron also progrades.

21.6.2 Responses of depositional environments to relative sea level change

Normally only a portion of the complete depositional tract is available for study. It is therefore useful to have some criteria for the recognition of relative sea level changes from deposits where there is information from only a restricted range of settings.

FLUVIAL ENVIRONMENTS

In a succession of entirely continental deposits the effects of sea level changes will be subtle and difficult to recognize (Fig. 21.9a). The most distinctive feature will be the incision of a valley due to base level fall and subsequent fill by river channel sediments during the lowstand. However, it is not easy to demonstrate that the scour at the base of a channel or stack of channels was formed by base level fall as opposed to the autocyclic processes of channel migration and avulsion. The proportion of overbank sedimentation compared to deposition in channels will show trends which can be related to the sea level. During lowstands there will be a high concentration of channel deposits whilst the increased accommodation space available during transgression and highstands will allow more overbank accretion to take place.

Palaeosols *(9.7.2)* can be useful as soil development is partly controlled by the level of the water table. Features characteristic of well-drained soils, such as evidence of oxidizing conditions, will tend to develop during periods of lowstand. During highstands the water table will be higher, resulting in soils which are waterlogged, and organic material will be better preserved under these conditions. However, local climate and floodplain topography are generally more significant controls on soil formation.

CLASTIC SHELF ENVIRONMENTS

Coastal and shallow shelf environments are the most sensitive to relative sea level change because the facies belts are largely controlled by depth (Fig. 21.9b). Water depth increases of a few metres can be readily identified in some coastal deposits by, for example, bioturbated shoreface sands overlying backshore deposits containing rootlets. Sea level falls which cause a forced regression result in a superposition of shallow on deeper facies, such as upper shoreface sands directly on mid-shelf

Fig. 21.9 (*Facing page*) The response of sea level changes in different depositional environments: (a) fluvial environments; (b) clastic shelf environments; (c) carbonate shelf environments; (d) deep marine environments.

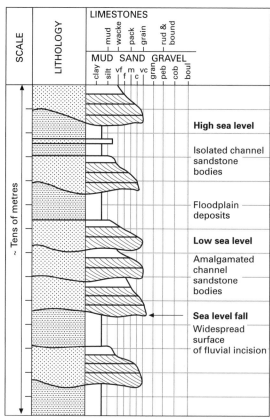

(a)

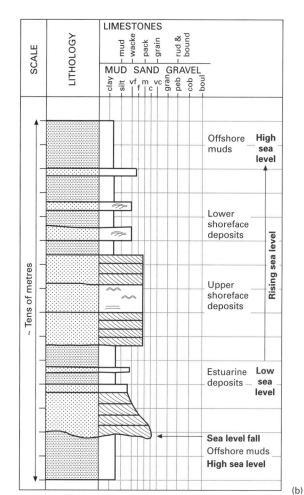

(b)

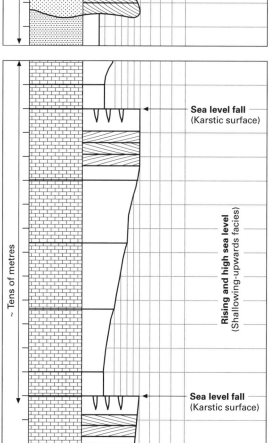

(c)

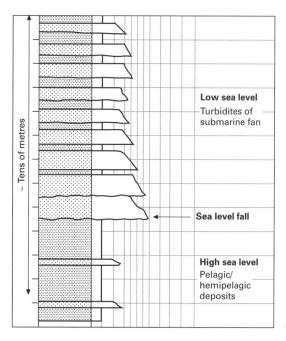

(d)

mudstones. Evidence of small changes in sea level is most easily seen in wave-dominated shorelines which have a microtidal regime. Facies have a limited depth range and hence vertical changes in depositional facies can be interpreted in terms of sea level change.

In macrotidal shallow marine environments the changes in sea level which occur with each tide may obscure small relative sea level changes. Sands are deposited on shallow marine bars and in channels along tidally dominated shorelines. Both bars and channels may be localized by the tidal flow. The facies are therefore not so sensitive to depth because both coarse sands and fine muds can be deposited at the same water depth, depending on the positions of channels and bars. A vertical change from bioturbated muds to cross bedded sands may represent upward shallowing but may also result from the lateral migration of a tidal channel. Filling of a tidal channel results in a fining-upward succession from channel sand to mud deposited on tidal flats. The simple trends seen in wave-dominated deposits of shallowing upward accompanied by coarsening upward are therefore more complicated in tidal settings.

Recognition of relative sea level changes in outer shelf environments depends on the magnitude of the rise or fall. A fall which results in the exposure of the outer shelf or a reduction in water depth to shoreface or shallower water deposition can be recognized by the erosion surface or abrupt shallowing of facies. In water depths greater than 100 m sea level changes of a few tens of metres can be difficult to recognize. At these depths deposition is largely mud out of suspension and storm-driven sands. These facies are not very sensitive to sea level changes. The storm sands tend to become thinner more distally, but are also dependent on the strength of the storm, so there is not a consistent relationship between water depth and thickness of sand beds.

CARBONATE SHELF ENVIRONMENTS

Falls in sea level which result in the exposure of a carbonate deposit are normally readily identifiable from the evidence of such features as solution, karstification, soil development and subaerial evaporite formation (Fig. 21.9c). Such features may be seen along the shorelines of carbonate ramps and may be additionally identifiable in shelf rim reefs, in carbonate sand bodies and in the shelf lagoon if the fall was of sufficient magnitude. Smaller falls in sea level can result in a restriction of circulation to the shelf lagoon leading to hypersaline conditions in

arid regions and reducing salinity under humid climates. Either way, the change in water chemistry will be reflected in the organisms in the lagoon as biodiversity is reduced under these conditions.

The effects of a sea level rise on carbonate deposition are dependent on the geometry of the carbonate platform and the rate of sea level rise. Rapid sea level rises result in abrupt changes from shallow carbonate facies to pelagic limestones on both ramps and rimmed shelves. Sea level rises at a rate which local biogenic carbonate productivity can match may be difficult to recognize at all in reefs. In intermediate- to high-energy environments vertical transitions from wackestone/packstone facies to packstone/grainstone are likely to reflect upward shallowing in response to the increase in accommodation space.

DEEP MARINE ENVIRONMENTS

Changes in relative sea level can only be interpreted from changes in the quantity and nature of sediment supplied to the deep marine environment (Fig. 21.9d). During periods of lowstand there is less accommodation space on the shelf and more sediment is deposited in deep marine systems such as on submarine fans. At highstand the sediment is nearly all deposited on the shelf and mostly pelagic and hemipelagic sediment reaches the deeper water. In a vertical succession through submarine fan deposits packages of sandy turbidites would be interpreted as lowstand deposits and mudstones would be considered the deposits of the transgressive and highstand periods. In practice, turbidite deposition is frequently restricted to lobes which only cover part of the submarine fan whilst other parts of the fan receive only pelagic sedimentation. This leads to a danger of misinterpreting pulses of turbidite sedimentation due to the formation and abandonment of lobes in terms of sea level changes. In addition, these relationships assume a relatively wide shelf: narrower shelves may shed material into the deep marine environment at all stages of sea level.

21.7 Practical sequence stratigraphy

Once facies analysis of a suite of sedimentary rocks has been carried out *(5.2.1)* it is a logical next step to consider the succession in terms of changes in relative base level. This provides an insight into the history of sedimentation and the evolution of depositional environments in the sedimentary basin. Expressing the facies responses to these changes of relative sea level in sequence stratigraphic terms provides a framework for this stratigraphic analysis even if the terminology is somewhat cumbersome.

Considering a single vertical section, it is possible to determine the relative depth of deposition of the facies associations recognized and the environments that they are considered to represent (Fig. 21.10). This can then be translated into a relative sea level curve for that section. Trends in this curve may then be interpreted in terms of transgression and regression and periods of highstand and lowstand of sea level. Particular attention should be given to the recognition of surfaces which may represent sequence boundaries or maximum flooding surfaces as these surfaces are useful marker horizons when correlation between sections is attempted.

Two or more separate stratigraphic sections within a basin may be compared by considering the relative sea level curves for each section. It may be possible to correlate the sections on the basis of the sequence boundaries and maximum flooding surfaces or, with less confidence, the marine flooding surfaces bounding parasequences. Correlation on this basis should be carried out with caution and can only be considered reliable if (a) exposure of the outcrop or spacing of borehole information is close enough to ensure correct correlation of surfaces or (b) other stratigraphic information is available (biostratigraphic or magnetostratigraphic data from each section) to provide a temporal framework for the depositional sequences.

21.7.1 Departures from the idealized models

It is inevitable that nature does not behave in quite the ordered way illustrated in the idealized models for facies trends in depositional sequences. The following are a few of the situations which can result in patterns different from the models presented in the preceding sections.

RAPID SEA LEVEL RISE

If the rate of relative sea level rise is high compared to the carbonate or clastic sediment supply, there may be no deposits representing the transgressive phase in all or part of the area. The sequence boundary may be overlain directly by a set of aggradational and progradational parasequences representing highstand conditions. In these circumstances the maximum flooding surface will be superimposed on the sequence boundary.

SHORT-PERIOD SEA LEVEL FLUCTUATIONS

A systems tract may be poorly developed if there is insufficient time for sediments to accumulate during that phase of the sea level curve. For example, there may be little or no deposition at lowstand if the sea level fall is immediately followed by a sea level rise. Highstand deposits may be absent if the period of stillstand following transgression and before sea level fall is too short for sediment to accumulate.

CHANGES IN SEDIMENT SUPPLY

The models for clastic sequences assume a constant rate of sediment supply. This condition is not met in most sedimentary basins because of changes in sediment supply due to tectonic and climatic controls. A drop in sediment supply will result in a reduced thickness or absence of sediments representing that phase. In carbonate environments sediment supply is directly linked to the area of shelf flooded during transgression and highstand.

HIGH SUBSIDENCE RATES

If the rate of tectonic subsidence is high the succession will not show any evidence for relative falls in sea level due to other mechanisms such as global eustatic variations (21.8). This is because the rate of subsidence exceeds the rate of eustatic sea level fall. The result is a continuous sea level rise which only shows changes in the rate of rise. In these situations lowstands do not occur and the succession will consist only of transgressive and highstand systems tracts. High subsidence rates can result in consistently deepening-upward facies transitions, recognizable in carbonate-forming environments by fauna and lithofacies reflecting increasing water depth; subsidence rates higher than carbonate production can match are required.

21.8 Causes of sea level fluctuations

From the preceding section it is clear that sea level changes have a profound effect on stratigraphy in almost all depositional environments, so what causes the sea level to rise and fall as it clearly has so frequently in the

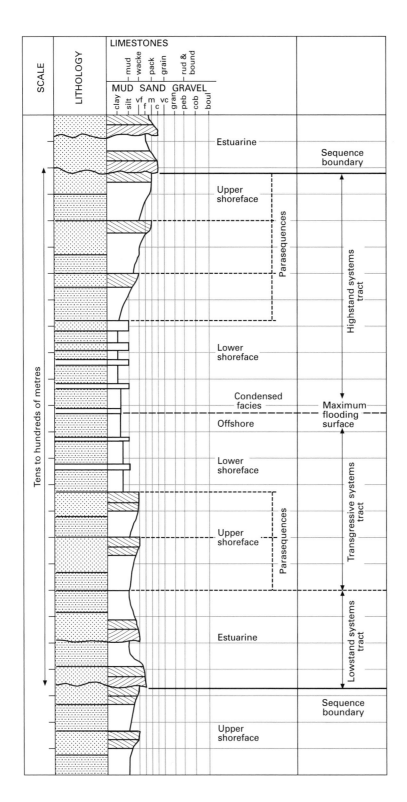

Fig. 21.10 Interpretation of a succession of sedimentary rocks in terms of sequences and systems tracts.

past? (Fig. 21.11). Relative rise or fall of the sea level at a certain place may be due to the land being lowered or raised by local tectonic forces, or a result of absolute changes in global sea level. Sea level fluctuations which are global are known as *eustatic* sea level rises or falls (Jervey 1988; Emery & Myers 1996). Evidence for these comes from a comparison of sea level changes which have affected the coastlines on different continents at the same time.

21.8.1 Local changes in sea level

Tectonic forces and related thermal effects acting on the margins of continents result in the land mass being raised or lowered with respect to sea level. In rifted basins, blocks move down in the rift, but along the flanks blocks may be uplifted: this can give rise to either relative rises or falls in sea level according to which faulted block the coast is situated on. Along passive margins at the edges of ocean basins the continental crust is cooling and contracting, resulting in a relative rise in sea level along these margins. Relative rises and falls in sea level may occur also at active continental margins where ocean crust is being consumed at a subduction zone. In all these cases, the sea level changes are localized to the region affected by the thermal or tectonic event. In the case of rift basins this may be a region only a few kilometres across, but on passive margins the subsidence may affect the whole margin of the continent. The influence of these localized events is not transmitted to other coastlines around the world.

21.8.2 Glacio-eustasy

The most immediate and obvious cause for global changes in sea level is melting and freezing of the polar ice caps. It has been calculated that melting of the ice currently forming the cap on the Antarctic continent would cause the sea level to rise by as much as 150 m (Plint *et al.* 1992). Such melting would be caused by a change in the world climate to warmer conditions. Melting of the northern ice cap would have a smaller influence on sea level because only a small proportion of the ice is on the Greenland continental mass. Floating ice does not change the level of the sea when it melts because the mass above water level is compensated by the reduction in density when the whole body of ice changes to water.

There is abundant evidence that the ice sheets at the poles expanded and contracted in volume during the Quaternary, resulting in world-wide changes in sea level of tens of metres. Glacial deposits from the Pleistocene record periods when much greater areas of northern Eurasia and the North American continent were covered by ice sheets during glacial accretion episodes and in some places morphological features such as raised beaches and river terraces testify to periods of higher sea level in interglacial periods of glacial wastage. The fluctuations between glacial and interglacial conditions are attributed to periodic cooling and warming of the world climate during the Pleistocene. The connection between climate change, glacial accretion/wastage and global sea level changes is well established as a *glacio-eustatic* mechanism in the Quaternary (Chappell & Shackleton 1986; Matthews 1986).

Glacio-eustasy is therefore a mechanism which can explain global sea level changes for periods of Earth history when there were ice caps at the poles to store water during cooler climate periods. The volumes of water involved are sufficient to cause rises and falls in sea level of over 100 m. The time period over which glacio-eustasy operates is also significant because climate changes apparently occur very quickly (Plint *et al.* 1992). The resulting sea level change may take place over a few thousand years and the interval between glacial and interglacial periods in the Pleistocene is in the order of tens to hundreds of thousands of years. Climatically driven glacio-eustasy can therefore provide quick and frequent global sea level fluctuations.

21.8.3 Thermo-tectonic causes of sea level change

A number of factors related to global tectonics can be considered as mechanisms to cause global sea level changes. The configuration of the tectonic plates around the globe is constantly changing, with long-term patterns of amalgamation to form supercontinents such as Gondwana in the late Palaeozoic and subsequent dispersal as the continental mass broke up into smaller units. During continental amalgamation the formation of orogenic belts creates large areas of high ground from which detritus is eroded and deposited in the world's oceans. This sediment may be removed from the oceans by subduction or is accreted on to continental margins at ocean trenches *(23.3)*.

These tectonic processes result in changes in global sea level for a variety of reasons: for example, a higher sediment flux into the oceans will displace water and

CAUSES OF SEA LEVEL CHANGE

Local tectonics
e.g.

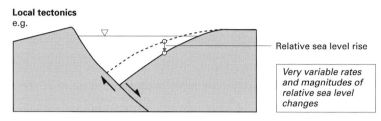

———— Relative sea level rise

Very variable rates and magnitudes of relative sea level changes

Continental ice caps

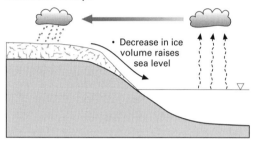

• Increase in ice volume lowers sea level

• Decrease in ice volume raises sea level

Around 100 m sea level change over 100 ka

Global scale thermo-tectonic

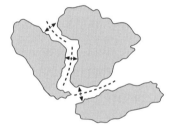

• Formation and breakup of supercontinents

• Changes in rates of formation of ocean crust

10–100 m sea level change over 10–100 Ma

Sea water temperature

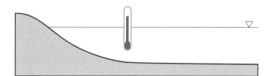

• Changes in sea water temperature cause thermal expansion/contraction

Centimetres to a few metres change over hundreds to thousands of years

Exchange with water on continents

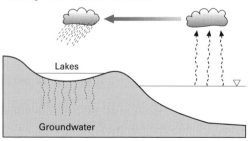

Lakes

Groundwater

Centimetres to metres change over hundreds to thousands of years

Fig. 21.11 Some of the possible causes of sea level change, the approximate magnitudes of change and the rates at which it will occur.

cause sea level rise. The most important processes are the effect of changes in the rate of spreading at mid-ocean ridges and changes in the total lengths of those spreading centres around the world. During supercontinent breakup new oceanic spreading centres develop. Young oceanic crust recently formed at a spreading centre is hot and relatively buoyant and mid-ocean ridges are at about 2000–2500 m water depth. It therefore takes up more space in the ocean basins than older, colder ocean crust which sinks to a lower level allowing 4000–5000 m of water above it. If the total length of spreading ridges in the world's oceans is great, more of the space in the ocean basins will be taken up by the ocean crust and this will cause the ocean water to spill further on to the continental margins (Fig. 21.12). These will therefore be periods of higher sea level. Conversely during periods of supercontinent formation the total length of spreading ridges will be minimized and result in more capacity in the ocean basins for the water and the sea level falls. Changes in the rate of sea floor spreading also cause eustatic sea level changes for much the same reasons. When spreading rates are high there is more hot crust in the ocean basins and sea level rises. Slower spreading rates result in a fall in sea level.

It has been calculated that sea level changes of 100 m or more could be produced by these tectono-eustatic mechanisms (Plint *et al.* 1992). However, these changes in global tectonics take place slowly. The cycle of supercontinent amalgamation and breakup takes hundred of millions of years and the changes in spreading rates probably occur over tens of millions of years. The rates of rise or fall in sea level generated by these mechanisms would therefore be slow.

21.8.4 Other causes of global sea level change

Glacio-eustasy is climatically controlled, and in addition to changing the proportion of ice on polar ice caps there is a second, more direct effect of changes in global temperature on the world's oceans. As water is warmed up the volume increases by thermal expansion. An increase in global air temperature would result in a warming of the oceans, although a deep water circulation is required to affect all the ocean waters. However, it has been calculated that a rise in temperature of several degrees Celsius would only result in a few metres of eustatic sea level rise, so these thermal volume changes in the oceans would have a very limited effect. A similarly small change of sea level is predicted from changing the proportion of the world's water (hydrosphere) which is resident on the continents in rivers, lakes and groundwater (Plint *et al.* 1992).

21.8.5 Cyclicity in changes in sea level

Patterns in changes in sea level through the Phanerozoic had been recognized for many years but the biggest impact on sedimentology and stratigraphy resulted from a publication in 1977 by Vail and co-workers from Exxon oil company (Vail *et al.* 1977). In the 1977 work and a later paper in 1988 (Haq *et al.* 1988), the Exxon group presented a global sea level curve for parts of the Phanerozoic which indicated that there were several orders of cyclicity superimposed upon each other.

FIRST-ORDER CYCLES

Looking at smoothed global sea level over the whole of

Slow mid-ocean ridge spreading

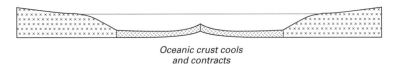

*Oceanic crust cools
and contracts*

Fig. 21.12 The effects of rates of sea floor spreading on the capacity of ocean basins to hold water: at times of relatively rapid spreading at the mid-ocean ridge there is more warm, buoyant crust in the basin which takes up space in the basin and this leads to the water in the basin spreading out on to the continental shelves.

Fast mid-ocean ridge spreading

*Sea water displaced onto
continental shelves*

*More hot, buoyant oceanic crust
occupies more space in the
ocean basin*

the Phanerozoic (Fig. 21.13) the main trends are a rise during the Cambrian to a peak in the early Ordovician, followed by a steady decline through to the end of the Palaeozoic. A slow rise in the Jurassic followed by a steep rise during the Cretaceous culminated in a peak in global sea level in the Late Cretaceous which has been followed by a steady fall to present-day levels. The magnitudes of these fluctuations are disputed, with estimates for the difference between Late Cretaceous levels and today ranging from 300 m to less than 150 m.

There is a strong correlation between this curve and the patterns of continental dispersal and amalgamation through the Phanerozoic (Vail *et al.* 1977; Worsley *et al.* 1984). Supercontinent breakup occurred in the early Palaeozoic and was followed by a long period of dispersal of continents prior to amalgamation into Pangaea in the Permian. Sea level rose in the Cambrian as the new spreading centres formed between the continental masses and then fell again as the continents regrouped during the late Palaeozoic. Breakup of Pangaea and in particular the dispersal of the fragments which made up Gondwana led to the formation of new active ocean ridges between the continents of South America, Africa,

Antarctica, Australia and India. Sea level rose sharply during this period until continents started to amalgamate again in the late Cretaceous with the collision between India and Eurasia and further west the closure of the Tethys Ocean to form the Alpine mountain belt.

SECOND-ORDER CYCLES

Superimposed on the first-order cycles which show a duration of hundreds of millions of years is a pattern of rises and falls with a duration of tens of millions of years. The shape of the curve is apparently strongly asymmetrical, with gradual rises in sea level followed by sharp falls. This asymmetry may partly be a result of the way the curve has been constructed from stratigraphic data: sediment accumulates during periods of sea level rise but a fall may be marked by a time of non-deposition and erosion which may appear to be instantaneous. The increasing frequency of second-order cycles from the early Palaeozoic to the Cenozoic may also be a function of the data: there is a lot more detailed stratigraphic information available for recent periods of Earth history than for the Palaeozoic.

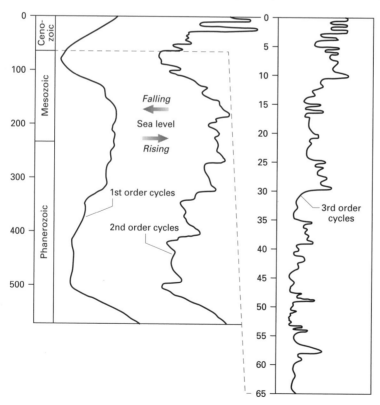

Fig. 21.13 First-, second- and third-order sea level cycles. (After Vail *et al.* 1977; AAPG ©1977, reprinted by permission of the American Association of Petroleum Geologists.)

The causes of second-order sea level changes are a topic of much speculation. The most widely accepted theory is that changes in the rates of spreading at mid-ocean ridges can account for changes in ocean water levels on a scale of tens of millions of years (Hallam 1963; Pitman 1978). The shorter-term global cycles in the Neogene may be accounted for by long-term trends in glaciation and deglaciation. Other explanations for changes in sea level at this scale focus on tectonic processes at continental margins such as changes in subsidence rate at passive margins and uplift where there is compression at the edge of a basin. These tectonic processes may account for regional, relative sea level fluctuations but not global eustatic changes.

THIRD-ORDER CYCLES

Rises and falls of sea level with a magnitude of several tens of metres and a periodicity of 1–10 million years are recognized throughout the Phanerozoic stratigraphic record (Fig. 21.13). Shallow marine sediments are very sensitive to changes in sea level of this magnitude and display shifts in facies which reflect changing water depth. Close inspection of strata deposited in these environments from many parts of the world and different ages reveal that such sea level changes are common features. Using a combination of subsurface and outcrop data the Exxon group constructed a curve which documents fluctuations every few million years which they considered to be synchronous global sea level changes.

Once again, there is no general agreement on the mechanisms which may cause sea level fluctuations at this third order of cyclicity (Plint *et al.* 1992). The tectono-eustatic mechanisms for changing the lengths and volumes of material in mid-ocean spreading centres can be ruled out because they act too slowly. Thermal expansion and contraction the world's sea water cannot generate the magnitude of change indicated by the stratigraphic record, and nor can changes in the volume of water in and on the continents.

Glacio-eustasy is a viable mechanism because it can generate the appropriate magnitude of sea level change in a short period of time (Vail *et al.* 1977; Haq *et al.* 1988). However, glacio-eustatic sea level fluctuations which have been well documented in the Quaternary took place over periods of only a few hundred thousand years, significantly faster than the third-order cycles. Even if glacially driven sea level changes can be considered to have acted more slowly in the past, there is a problem with

advocating this mechanism throughout the Phanerozoic. Evidence from glacial deposits and calculations of palaeotemperatures indicate that major glaciations occurred during the Ordovician–Silurian, the Carboniferous–Permian and the Neogene–Quaternary *(24.7)*. During the warmer periods in between, there may have been ice sheets at one or other of the poles, but there is considerable doubt over whether there have always been ice caps present. The mid-Cretaceous was one of the warmest periods of the Phanerozoic, and the presence of temperate floras near the poles during that period suggests that there were no permanent ice caps at that time.

Other mechanisms which may generate the frequency and magnitude implied by the third-order cycle curve are applicable only to individual basins. Changes in the regional tectonic stresses acting on a basin may result in basin-wide subsidence or uplift (Cloetingh 1988) although the rates at which these changes occur are not well known.

21.8.6 Short-term changes in sea level

In addition to the above three orders of cycle, shorter-term changes in sea level have also been recognized. These are fourth-order cycles of 200 000–500 000 years' duration and fifth-order cycles lasting 10 000–200 000 years. According to the Exxon models, the magnitude of sea level change in these cycles ranges from a few metres to 10–20 m, although short-term sea level changes in the Quaternary have much higher magnitudes. Evidence for frequent relative changes in sea level of about 10 m has been found in successions throughout the stratigraphic record, and in sequence stratigraphy terminology these are referred to as parasequences.

A very detailed record of sea level changes has been established for the Quaternary and related to estimates of palaeotemperature from the oxygen isotope record. These indicate that changes in global climate caused periodic wasting and accretion of ice masses with a cyclicity of tens to hundreds of thousands of years. These global climatic variations have been related to the behaviour of the Earth in its orbit around the Sun and changes in the axis of rotation. Three orbital rhythms were recognized and their periodicity calculated by the mathematician Milankovitch after whom these cycles are now commonly known as *Milankovitch cycles* (Fig. 21.14) (Friedman *et al.* 1992). The longest-period rhythm, approximately 100 000 years, is due to changes of the eccentricity of the

Changes in the eccentricity of the Earth's orbit around the Sun

100 ka cycle

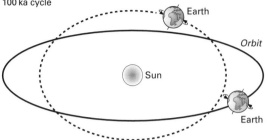

Changes in the obliquity (tilt) of the Earth's axis of rotation

41 ka cycle

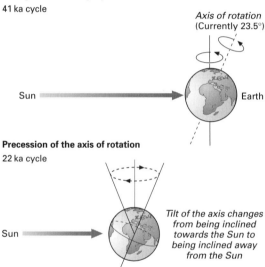

Precession of the axis of rotation

22 ka cycle

Fig. 21.14 Milankovitch cycles: the eccentricity of the Earth's orbit of the Sun, changes in the obliquity of the axis of rotation of the Earth and the precession of the axis of rotation may result in global climatic cycles on the scale of tens of thousands of years.

Earth's orbit around the Sun—the orbit is elliptical and changes its shape with time. The rotation of the Earth shows two patterns of variation. The axis of rotation is oblique with respect to the plane of the Earth's rotation about the Sun and the angle of tilt changes over a period of 40 000 years between 21.5° and 24.5°. The shortest rhythm is 21 000 years, caused by the precession or 'wobble' in axis of rotation analogous to the behaviour of a spinning top. These three cycles, working independently and in combinations, are believed to exert fundamental controls on the global climate, leading to cycles of global warming and cooling.

The application of Milankovitch cyclicity to the pre-Quaternary sea level changes poses two principal problems. First, the climatic cycles have been demonstrated to influence the glacio-eustatically driven sea level changes but if ice caps were not present in parts of the Phanerozoic they would not have affected sea level during those periods. Second, there is some doubt whether the cyclicity of the precession and obliquity of the axis of rotation have been stable and maintained the same period through Earth history. The astronomical behaviour of the planet may have a more chaotic behaviour over geological time.

21.8.7 Global synchroneity of sea level fluctuations

The sea level curves were presented by the Exxon group (Vail *et al.* 1977; Haq *et al.* 1987) as global eustatic signatures. The group claimed that the effects of falls in sea level could be seen as sequence boundaries occurring at the same time in successions in different parts of the world. The sequence boundaries have been dated and considered to mark episodes of global eustatic sea level fall. The dates are presented to a precision as great as a hundred thousand years. As an example, during the Oligocene sea level falls are documented at 36 Ma (the base of the Oligocene), 33 Ma, 30 Ma, 28.4 Ma, 26.5 Ma and 25.5 Ma on the 1987 version of the cycle chart.

Questions have been raised over the precision of the dating of the rocks which record these events. The record of sea level changes is inevitably found in sedimentary rocks which cannot be dated directly by radiometric methods. Age determinations are made on the basis of biostratigraphic data which are then related to the geological time-scale. Miall (1992, 1997) pointed out that the errors in dating by these methods are often in the order of hundreds of thousands of years. These errors increase further back in time because of the resolution of radiometric dating techniques *(20.1)*. Therefore, the errors are of the same order of magnitude as the third-order cycles, which may be as short as 500 000 years and commonly last around 1 million years. Miall (1992) calculated that if dating errors were taken into account even a randomly generated pattern of rises and falls in sea level could be made to match the global cycle chart presented by the Exxon group. These considerations cast doubt on whether the cycle chart can be demonstrated to be truly global because it is difficult to prove that a sequence boundary in the Cretaceous of Alberta, Canada, occurred at the same time as a sequence boundary in the Cretaceous of Sussex, England.

Given these questions over the global nature of the cycle chart and the problems of finding a mechanism which will generate sea level changes of the appropriate magnitude at the right frequency, the concept of high-frequency global eustasy must be used with caution. There is little doubt that the signature of global eustasy is present in parts of the stratigraphic record (e.g. in the latter half of the Tertiary), as the effects are seen all over the world: a major sea level fall in the mid-Oligocene has been recognized on many continents. In other parts of the stratigraphic record the synchroneity of sea level fluctuations remains unproven and in these instances it is not appropriate to use the chart as a correlation or dating tool (Miall 1997).

21.8.8 Sea level changes and sequence stratigraphy

The concepts and applications of sequence stratigraphy as outlined in this chapter can be considered completely independently of the global cycle chart. Any problems with demonstrating the synchroneity of recorded sea level changes do not negate the methodology of considering stratigraphic successions in terms of base level changes. In any given basin, or part of a basin, base level changes will exert a fundamental control on the stratigraphy and the effects of these changes can be used as a way of analysing the succession of sediments. The causes of base level change in that place can be considered as a separate issue, as can any consideration of correlation between successions in different basins on the basis of sea level changes. Sequence stratigraphy and global eustasy are separate issues: the first a technique in stratigraphy, the second a possible cause of sea level changes at any given place.

Further reading

Emery, D. & Myers, K.J. (eds) (1996) *Sequence Stratigraphy*, 297 pp. Blackwell Science, Oxford.

Miall, A.D. (1997) *The Geology of Stratigraphic Sequences*, 433 pp. Springer-Verlag, Berlin.

Nummedal, D., Pilkey, O.H. & Howard, J.D. (eds) (1987) *Sea Level Fluctuations and Coastal Evolution*, 267 pp. Society of Economic Paleontologists and Mineralogists Special Publication **41**.

Plint, A.G., Eyles, N., Eyles, C.H. & Walker, R.G. (1992) Control of sea level change. In: *Facies Models: Response to Sea Level Change* (eds R.G. Walker & N.P. James), pp. 15–26. Geological Association of Canada, St Johns, Newfoundland.

Van Wagoner, J.C., Mitchum, R.M., Campion, K.M. & Rahmanian, V.D. (1990) *Siliciclastic Sequence Stratigraphy in Well Logs, Cores and Outcrop: Concepts for High Resolution Correlation of Time and Facies*, 55 pp. Methods in Exploration Series **7**. American Association of Petroleum Geologists, Tulsa, OK.

Wilgus, C.K., Hastings, B.S., Posamentier, H.W., Ross, C.A. & Kendall, C.G.St.C. (eds) (1988) *Sea Level Changes: An Integrated Approach*, 407 pp. Society of Economic Paleontologists Special Publication **42**.

Wilson, R.C.L. (1992) Sequence stratigraphy, an introduction. In: *Understanding the Earth, A New Synthesis* (eds G.C. Brown, C.J. Hawkesworth & R.C.L. Wilson), pp. 388–414. Cambridge University Press, Cambridge.

Subsurface stratigraphy

Exploration for coal, oil and gas in the last few decades has resulted in the development of a branch of geology concerned with the analysis of stratigraphy, sedimentology and structure in the subsurface. These approaches to looking at rocks principally involve geophysical techniques and the same basic concepts of stratigraphy apply. The wealth of information from the subsurface provided by intensive exploration in some areas by oil companies has led to a better understanding of the stratigraphy of sedimentary basins and also to the concepts of sequence stratigraphy.

22.1 Seismic reflection profiles

The main geophysical technique for the investigation of subsurface geology is the analysis of the reflection of sound waves which penetrate rocks below the ground or under the sea. A *seismic reflection profile* (Fig. 22.1) is generated by providing a series of artificial shock waves which are reflected by surfaces within rocks and recording the returning waves (McQuillin *et al.* 1984).

22.1.1 Generating a seismic reflection profile

The source of the energy at the surface is provided by a number of different mechanisms. On land, explosives may be used but it is now more common to use a 'vibraseis' set-up, a vehicle or group of vehicles which vibrate at the

Fig. 22.1 Example of a seismic reflection profile.

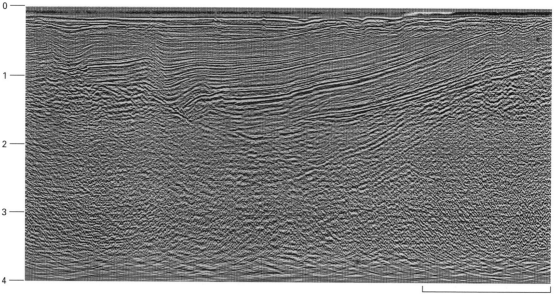

5 km

surface at an appropriate frequency to generate shock waves. At sea the sound energy is provided by an 'airgun', a device which builds up and releases compressed air with explosive force. The returning sound waves are detected by a string of microphones known as *geophones* on land and *hydrophones* at sea. The data for the full seismic profile are produced by moving the source of energy and the recording instruments along the line of the survey, which may be tens or even hundreds of kilometres long.

A sound wave is partially reflected when it encounters a boundary between two materials of different density and sonic velocity (the speed of sound in the material). The product of the density and sonic velocity of a material is the *acoustic impedance* of that material. A strong reflection of sound waves occurs when there is a strong contrast between the acoustic impedance of one material and that of another. In geological terms there is a strong reflection of the sound waves at the contact between two rocks which have different acoustic properties, such as a limestone and a mudstone. In general, crystalline or well-cemented rocks have a higher sonic velocity than clay-rich or porous lithologies.

A seismic profile is presented as a line consisting of a series of closely spaced vertical traces. Each trace is a record of the acoustic impedance contrasts which generated reflections recorded at the surface. In order to visualize them, the peaks on the right-hand side of each trace representing high contrasts are filled in black (Fig. 22.2). When thousands of these traces are put next to each other, lines of strong impedance contrast (*reflectors*) show as black lines on the profile.

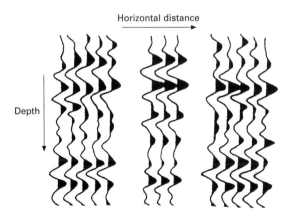

Horizontal distance →

Depth ↓

Fig. 22.2 Detail of an individual seismic trace: the right-hand side of the curve is filled in black on the trace.

Before the results of a seismic reflection survey are presented as a profile, a considerable amount of manipulation of the signals collected is carried out. This *processing* of the digital data by computer is carried out to make the patterns on a seismic reflection profile easier to interpret (Emery & Myers 1996). More sophisticated digital processing has led to greater resolution on seismic profiles and enhancement of the signals has made it possible to image features at depths of over 5 km.

22.1.2 Seismic profiles compared to geological cross-sections

At a first glance there is a lot in common between a seismic reflection profile and a cross-section compiled from surface outcrop data. Layers may be seen on the profile, unconformities, folds and faults may be picked out and contrasts in the detailed pattern of the reflectors suggest that different rocks may be identified on a seismic reflection profile. Whilst all these features can indeed be related to stratigraphic and structural features seen in rocks, comparison and interpretation must be carried out with caution because there are important differences too.

First, there is a question of scale. In dealing with outcrop, a geologist is accustomed to looking at beds centimetres to metres thick and features tens to hundreds of metres across are considered large-scale. These large-scale features are the smallest things that can be picked out on a seismic profile. The wavelength of a sound wave at 100 Hz is about 15 m, so layers of rock less than a few tens of metres thick cannot be resolved seismically (Emery & Myers 1996). The units defined by reflectors are packages of beds, not individual beds.

Second, a contact between two rock units will not show up on a seismic profile if there is no acoustic impedance contrast between them. A thick sandstone and a conglomerate body might be easily distinguished in outcrop, but if they have the same acoustic properties, the contact between the two would not be imaged as a reflector.

The character of some units on a profile may give some indication of the lithology and facies (see below) but interpretation of the layers in terms of a stratigraphy of rock units can only be carried out with any confidence if the succession imaged has been drilled. The reflectors can then be related to the rock units encountered in the borehole.

The horizontal scale on a seismic profile is normally in kilometres and the units on the vertical scale will usually be in *two-way travel time*, the time taken for the sound waves to reach the reflectors and return to the surface. The depth is hence measured in seconds and milli-seconds and the relationship between this and actual depth is non-linear (see depth conversion: *22.5*). The angles of layering and the attitudes of faults on a profile scaled in seconds are not true angles and the relative thicknesses of layers are distorted unless a depth conversion is carried out.

22.2 Interpreting seismic reflections

By tracing reflectors across profiles it is possible to recognize stratal relationships which are on the scale of hundreds to thousands of metres. When traced and marked, these form a framework for the interpretation of the whole succession of rocks imaged on the profile.

22.2.1 Continuous reflectors

A well-defined reflector marks a boundary between two layers of different acoustic impedance. For this reflector to be continuous over kilometres it must mark a change in lithological characteristics of the same extent, and changes in lithology in a sedimentary succession result from changes in depositional environment. A widespread change in depositional environment can result from events such as a rise in sea level: sandy, shallow water deposits may be replaced by muddy, deeper water sediments over a wide area. A similar widespread change may occur when a carbonate shelf environment receives an influx of terrigenous material and the lithology depo-sited changes from limestone to mudstone. In deeper water the progradation of a sandy submarine fan lobe over muddier turbidites may also mark a change in depo-sitional style over a wide area. Continuous reflectors therefore may be seen as markers which indicate a signi-ficant, widespread change in deposition in the basin. For

this reason, prominent reflectors are often considered to represent time-lines, isochronous surfaces, within a basin fill succession, although extreme care should be exercised in making this assumption where there are complex stratigraphic relationships or where reflectors merge. Changes in depositional environment usually occur over a period of time because events such as trans-gressions which result in retrogradation of facies (*21.1.2*) do not occur instantaneously.

22.2.2 Clinoforms

Inclined surfaces bounding stratal packages on seismic reflection profiles are referred to as *clinoforms* (Mitchum *et al.* 1977). Their pattern suggests a progradational geometry of packages of sediment building out into deeper water. However, it may be misleading to try to relate clinoforms directly to the depositional geometries. Away from localized slopes of alluvial fans and coarse-grained deltas, depositional slopes in most clastic envi-ronments are very low. Fluvial deposits mainly occur on slopes of less than 1°, fine-grained delta fronts may have slopes of a couple of degrees, shelves slope at angles between 0.05° and 1.5°, a 'steep', depositional continen-tal slope may be a few degrees and the basin floor on to which submarine fans build out is essentially flat. All these clastic depositional slopes are a lower angle than suggested by the clinoforms on seismic profiles. In car-bonate environments, a forereef slope may be 25° or more and large-scale clinoforms can be developed in these settings, but not all clinoforms recognized on seismic profiles are composed of limestone. One pos-sible explanation for clinoform geometries in some clastic strata is an effect produced by differential compaction: sandstone has a much lower initial porosity than mudstone and is therefore compact to a lesser degree on burial. Units which grade from sandstone to mudstone would tend to taper distally upon compaction, resulting in inclined surfaces on a large scale (Fig. 22.3).

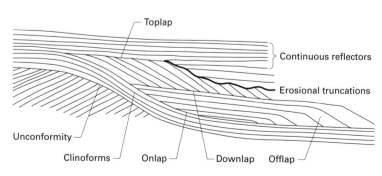

Fig. 22.3 Reflector patterns and reflector relationships on seismic reflection profiles.

22.2.3 Unconformities

The unconformity surface itself will not be represented by a reflector unless there is a consistent change in lithology across it to create an acoustic impedance contrast. Instead, an unconformity may be identified on a seismic reflection profile by the presence of *reflector terminations*, the points at which relatively continuous reflectors end (Mitchum *et al.* 1977). Some terminations are not related to unconformities (see below) but result from the shapes of the stratal packages. The terminology used to describe these features is much the same as that used for stratal relationships in outcrop, but once again the difference in scale must be emphasized.

22.2.4 Erosional truncation

If the surface of truncation is at a high angle to the orientation of the layers it intersects, erosional truncations are relatively easy to recognize (Fig. 22.3). They are assumed to result from the removal of packages of beds by subaerial erosion, or possibly submarine erosion in some cases. They are most distinct where the underlying layers have been uplifted and tilted prior to erosion. A truncation surface caused by the incision of a river valley into shelf strata following a sea level fall may also be recognized, but only if the incision is several tens of metres and therefore enough to be resolved on seismic profiles. Low-angle erosional truncations may be difficult to identify.

22.2.5 Onlap

This relationship forms where there is a clear topography at the edge of or within the basin. Reflectors indicate that stratal packages are banked up against this topography, with the younger layers successively covering more of the underlying unit and sometimes covering it completely. Geometries of this type may form by the drowning of a topography (Fig. 22.3). The breaks in the sedimentary record represented by unconformities are also often considered to be time-lines within the stratigraphy. In detail, an unconformity may represent a series of events over a period of time. To begin with, there may be a long time period between the erosion and subsequent deposition above that erosion surface. Deposition may not occur across the whole unconformity at one time, as is the case with onlap. This allows the possibility of erosion occurring at a more elevated point of the unconformity surface at the same time as

deposition is occurring on that same surface further basinwards.

22.2.6 Other reflector relationships

Downlap. This term is used to describe inclined surfaces which terminate downwards against a horizontal surface (Fig. 22.3). This geometrical relationship is rarely seen in the smaller scale of outcrop because steeply inclined bedding surfaces are uncommon—forereef slopes (*14.7.2*), and Gilbert-type deltas (*12.3*), are the only notable exceptions. The existence of downlap surfaces at seismic scale therefore requires some explanation. One possibility is that the inclined stratal units (clinoforms: *22.2.2*) are packages of sandstone which pass laterally into mudstone: differential compaction between the two lithologies would result in a thinner unit of amalgamated mudstone which could not be separated into packages.

Toplap. Inclined reflectors which have upper surfaces which terminate against a horizontal surface create a pattern which is described as toplap (Fig. 22.3). This relationship occurs where there is a succession of packages of sediment which prograde basinwards, without any aggradation.

Offlap. This relationship refers to a pattern of reflectors, rather than a reflector termination. Offlap is a pattern of stratal packages which build upwards and outwards into the basin (Fig. 22.3).

22.3 Structural features on seismic profiles

Folds and faults can also be recognized on seismic reflection profiles, and much of our knowledge of the relationship between structural features seen at the surface and structures several kilometres deep in the crust has come from the interpretation of seismic reflection profiles.

A fault surface is not often seen on a seismic line as a distinct reflector. Even if there is an acoustic impedance contrast across the fault, steep structures dipping at more than 30° or 45° are poorly imaged by conventional seismic surveys. This is because the reflected sound waves return to the surface at a high angle and are not picked up by the recording array. Faults are normally recognized by the displacement of continuous reflectors. If distinctive individual reflectors can be recognized on both sides of the fault, the direction and amount of displacement can be determined. Large-scale folds can be identified on seismic profiles although steep limbs are poorly imaged for the same reasons as discussed for steep fault surfaces.

If the vertical scale on a seismic profile is in two-way time the shapes of the faults and folds seen must be treated with some caution as the vertical scale is non-linear with respect to true depth. Depth conversion of the stratigraphy must be carried out to provide the correct geometries.

22.4 Seismic facies

It is possible to make a preliminary interpretation of the type of rock and sometimes the depositional facies from the patterns of reflectors on a seismic profile (Mitchum *et al.* 1977; Friedman *et al.* 1992). Such interpretations should be corroborated by lithological data from boreholes and by combining borehole and seismic data.

The form of the reflectors can give some clues to depositional setting. Continuous reflectors suggest an environment which is relatively stable with periodic changes, such as a shelf affected by sea level changes or a deep basin with periodic progradation of submarine fan lobes. In continental environments lateral facies patterns tend to be complex as rivers change course and wide-spread surfaces are less common, so a discontinuous reflector pattern results. Some lithologies are characterized by an absence of parallel reflectors. For example, salt and other evaporites tend to have a 'chaotic' pattern (random reflectors) or 'transparent' pattern (lacking internal reflectors). A basement of metamorphic or igneous rocks generally lacks regular reflectors. The geometry of units bounded by reflectors can also give an indication of the depositional setting. Estuarine or fluvial deposits may be underlain by an erosional truncation and confined to a valley fill. Large reefs may be picked out by their morphology and chaotic to transparent internal reflectors.

22.5 Relating seismic profiles to geological cross-sections

Before the structural and stratigraphic relationships seen on a seismic profile can be related directly to a geological cross-section the vertical scale must be transformed into a linear depth in metres. This *depth conversion* requires a knowledge of the acoustic characteristics of all the stratigraphic units from the surface down to the chosen limits of interpretation of the profile. These acoustic characteristics, the sonic velocity of the layers, will vary with lithology and depth, increasing as more compacted lithologies are encountered at greater depth. The sonic velocity of the stratigraphic units can be obtained from measurements made in boreholes through the section *(22.6.3)*. Using this information, the two-way time through an interval can be converted into a true thickness for that interval. If carried out in a series of steps for each unit a cross-section can be constructed with true thicknesses.

22.5.1 Extending to three dimensions

A full three-dimensional picture of the subsurface geology can be created by matching and correlating grids of seismic lines. A seismic survey for exploration purposes may consist of a set of cross-cutting lines spaced kilometres or tens of kilometres apart to build up a general picture of the area. Once an area has been targeted for more detailed study the spacing between seismic lines may be less than a kilometre and the interpolation between lines is improved.

New techniques have been developed which allow a full three-dimensional data set to be recorded in an area. The results of a *3-D seismic survey* can be processed to provide an image of any vertical or horizontal line across the area. Images of horizontal lines, or *time-slices*, approximate to a map of the area. Using depth-converted data from grids of conventional seismic profiles or 3-D surveys, maps can be drawn of horizons or packages of strata within the area covered. For example, the depth of the basement can be shown as a map if the contact between the basement and the basin fill has been identified on all the seismic profiles in the area. Variations in the thickness of a particular unit may be shown as a map from information about the position of the top and bottom of that unit on the various profiles.

These techniques make it possible to consider stratal units in three dimensions in a way which is rarely, if ever, possible from outcrop data alone. This has greatly improved the understanding of the large-scale stratigraphy and structure of sedimentary basins.

22.6 Borehole stratigraphy and sedimentology

The stratigraphic and structural relationships interpreted

from seismic reflection profiles can only be transformed into a full geological picture by the addition of information on lithology and lithofacies. This can be provided by drilling through the succession. Information about the subsurface geology comes from a number of sources.

22.6.1 Core

A cylinder of rock can be cut by a drill bit and brought back up to the surface. These bottom-hole cores provide the greatest detail of the lithologies present, the small-scale sedimentary structures, body and trace fossils. Recovery is often incomplete, with preservation of only part of the succession drilled, and the core may be broken up during drilling. In commercial drilling for hydrocarbons, bottom-hole cores are normally only cut through horizons of particular interest and it is rare for a core to be taken all the way down the borehole as coring is relatively expensive. Cores are cut lengthwise into slabs to reveal the details of the sedimentary and biogenic structures. Sidewall cores may also be taken: these are less expensive, but their small size (3–5 cm across) makes them less useful.

22.6.2 Cuttings

When a borehole is being cut, a fluid (drilling mud) is pumped down to the drill bit to keep the bit cool and lubricated, to maintain pressure in the borehole and to flush out the chips of rock cut by the bit. These cuttings are typically 1–5 mm in diameter and are sieved out of the drilling mud at the surface. Recording the lithology of these drill chips (*mud-logging*) provides information about the rock types, but information about bed thickness is very approximate and details such as sedimentary structures are not preserved. Microfossils *(19.3.2)* and palynomorphs *(19.3.3)* can be recovered from cuttings and used in biostratigraphic analysis of the succession in the borehole.

22.6.3 Wireline logs

There are now a wide range of instruments, *geophysical logging tools*, which are lowered down a borehole to record the physical and chemical properties of the rocks. One or more of these instruments will be mounted on a device called a *sonde* which is lowered down the drill hole (on a *wireline*) once the drill string has been removed. Data from these instruments are recorded at the

surface as the sonde passes up through the formations (Fig. 22.4). An alternative technique is to fix a sonde mounted with logging instruments behind the drill bit and record data as drilling proceeds. The data recorded by the logging instruments provide information about the lithologies, the porosity and permeability, water and hydrocarbon content of the formations that the borehole has passed through. The principal tools and their applications are briefly listed below. Further details are provided in a number of specialist texts on the subject (e.g. Rider 1986; Hurst *et al.* 1990; Doveton 1994).

Gamma-ray log. This records the natural gamma radioactivity in the rocks. The principal sources of this radioactivity are potassium, uranium and thorium. The main use of this tool is to distinguish between mudrocks, which generally have a high potassium content and hence high natural radioactivity, and sandstone and limestone, both of which normally have a lower natural radioactivity. Organic-rich rocks can also be detected by this method as uranium is often naturally associated with organic material.

Resistivity logs. A range of instruments are used to measure the electrical conductivity of the rocks close to the borehole and deeper into the formation. These provide information about the composition of the pore fluids because hydrocarbons are poor electrical conductors and saline groundwater is a relatively good conductor. Resistivity changes away from the borehole can give an indication of how far the drilling mud has penetrated into the formation, and this gives a measure of the formation permeability.

Sonic log. The velocity of sound waves in the formation is determined by this tool, which comprises a pulsing sound source and receiver microphone. The information provided is used for depth conversion of seismic reflection profiles *(22.5)*. The sonic velocity is dependent upon the lithology and the porosity as more compact materials tend to have higher sonic velocities. Using this tool, if the lithology is known, the porosity can be calculated, or vice versa.

Density logs. These tools work by emitting gamma radiation and detecting the proportion of the radiation which returns to detectors on the tool. The amount of radiation returned is proportional to the electron density of the medium bombarded, and this is in turn

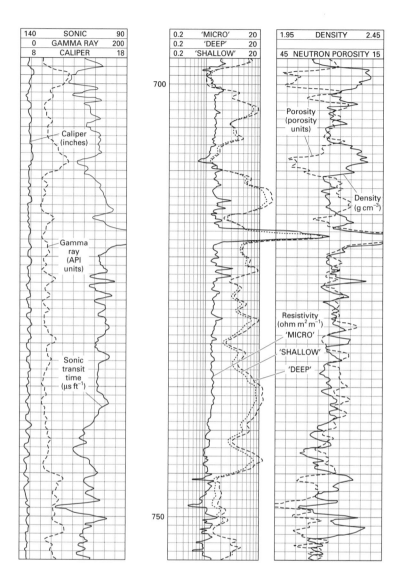

Fig. 22.4 Wireline logging traces produced by geophysical logging tools.

proportional to the overall density of the formation. If the lithology is known, the porosity can be calculated as density decreases with increased porosity.

Neutron log. In this instance the tool has a source which emits neutrons and a detector which measures the energy of returning neutrons. Neutrons lose energy by colliding with a particle of similar mass, a hydrogen nucleus, so this logging tool effectively measures the hydrogen concentration of the formation. Hydrogen is mostly present in the pore spaces in the rock filled by formation fluids, oil or water (which have approximately the same hydro-

gen ion concentration) so the neutron log provides a measure of the porosity of the formation.

Other logging tools in common use are the *dipmeter log*, a multiple resistivity tool which can measure the orientation of dipping surfaces within the rocks, and the *formation microscanner*, which images the cut surface inside the borehole.

22.6.4 Borehole lithology determinations

The natural gamma-ray logging tool provides the most

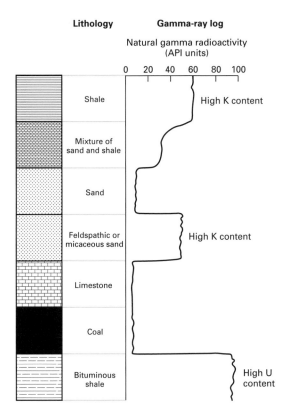

Fig. 22.5 Determination of lithology using information provided by a gamma-ray logging tool. (After Rider 1986, Fig. 7.1; with kind permission from Kluwer Academic Publishers.)

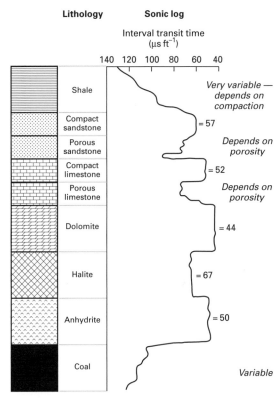

Fig. 22.6 Determination of lithology and porosity using information provided by a sonic logging tool. (After Rider 1986, Fig. 9.1; with kind permission from Kluwer Academic Publishers.)

information by distinguishing between mudrock and either sandstone or limestone. This tool provides a guide to the 'shaliness' of the formation, the gross proportion of clay present. It also picks out organic-rich mudrocks by their anomalously high gamma-ray count (Fig. 22.5). Sandstone beds with high proportions of mica, feldspar, glauconite or heavy minerals have a high natural gamma radiation and can be confused with mudrocks. Under ideal conditions sandstone and limestone can be distinguished by their different density: the porosity must first be established using the neutron log, then the density of the framework grains can be calculated from the bulk density determined from the density or sonic logs (Fig. 22.6). Quartz sand has a density of $2.65\,\mathrm{g\,cm^{-3}}$, calcite $2.71\,\mathrm{g\,cm^{-3}}$ and dolomite $2.87\,\mathrm{g\,cm^{-3}}$. Other lithologies which can be picked out on the basis of their density are halite ($2.03\,\mathrm{g\,cm^{-3}}$), coal ($1.2–1.5\,\mathrm{g\,cm^{-3}}$) and igneous rocks ($2.95\,\mathrm{g\,cm^{-3}}$).

22.7 Subsurface facies analysis

In the absence of information from core, the information from wireline logs can be used to provide a guide to the depositional facies. The main technique is to use the gamma-ray log to pick out trends (Fig. 22.7) (Cant 1992). An increase in gamma count upwards suggests that the formation is becoming more clay-rich upwards, and this may be interpreted as a fining-upward trend. Fining-upward trends are seen in channel fills in fluvial, tidal and submarine fan environments. A coarsening-upward pattern, as seen in prograding clastic shorelines, shoaling carbonate successions and submarine fan lobes, may be recorded as a decrease in natural gamma radiation upwards.

A drawback of relying on wireline logs for facies interpretation is apparent from these examples of trends. It may not be possible to distinguish between, for

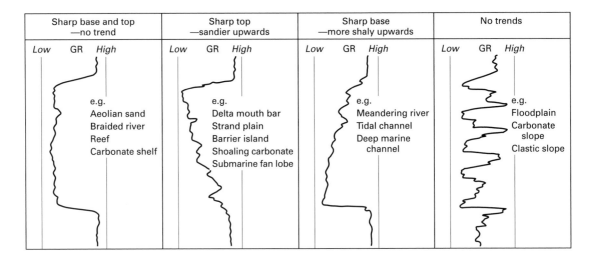

Sharp base and top —no trend	Sharp top —sandier upwards	Sharp base —more shaly upwards	No trends
Low GR *High*	*Low* GR *High*	*Low* GR *High*	*Low* GR *High*
e.g. Aeolian sand Braided river Reef Carbonate shelf	e.g. Delta mouth bar Strand plain Barrier island Shoaling carbonate Submarine fan lobe	e.g. Meandering river Tidal channel Deep marine channel	e.g. Floodplain Carbonate slope Clastic slope

Fig. 22.7 Trends in gamma-ray traces can be interpreted in terms of depositional environment provided that there is sufficient corroborative evidence from cuttings and cores. (After Cant 1992.)

example, submarine fan and shoreline environments on the basis of geophysical data alone. Subsurface facies analysis is therefore most successful when trends identified on the logs can be compared with information from core.

22.8 Use of borehole data

If the information from core, cuttings and wireline logs is carefully integrated a fairly detailed interpretation of the sedimentology and stratigraphy of the drilled succession can be made. One of the advantages of this type of information is that a continuous record through a succession is provided. Such a continuity of information is rarely available in outcrop. Fine-grained units tend to be weathered or covered in vegetation in outcrop and are better represented from subsurface data. Boreholes also provide information from parts of a basin which would not normally be exposed, as natural outcrop tends to be biased towards the edges of basins. The disadvantages of borehole data are that in each case the

information is one-dimensional. Even with closely spaced holes it is not easy to correlate at the detailed scale necessary to generate a three-dimensional facies model. Another problem is that directional data such as palaeocurrent information are very rarely available from the subsurface. This limits the scope for the development of palaeogeographic reconstructions using subsurface data alone as a basis.

Further reading

Cant, D.J. (1992) Subsurface facies analysis. In: *Facies Models: Response to Sea Level Change* (eds R.G. Walker & N.P. James), pp. 27–46. Geological Association of Canada, St Johns, Newfoundland.

Doveton, J.H. (1994) *Geologic Log Interpretation*, 169 pp. Society of Economic Paleontologists and Mineralogists Short Course **29**.

Emery, D. & Myers, K.J. (eds) (1996) *Sequence Stratigraphy*, 297 pp. Blackwell Science, Oxford.

Hurst, A., Lovell, M.A. & Morton, A.C. (eds) (1990) *Geological Applications of Wireline Logs*, 357 pp. Geological Society of London Special Publication **48**.

McQuillin, R., Bacon, M. & Barclay, W. (1984) *An Introduction to Seismic Interpretation*, 287 pp. Graham and Trotman, London.

Rider, M.H. (1986) *The Geological Interpretation of Well Logs*, 175 pp. Blackie, Glasgow.

23 Sedimentary basins

The surface of the Earth can broadly be divided into places where sediment accumulates and areas of erosion or non-accumulation. Sedimentary basins are regions where sediment accumulates into successions hundreds to thousands of metres in thickness over areas of thousands to millions of square kilometres. The North Sea in north-west Europe and the Gulf of California on the North American continent are modern-day examples of basins. The underlying control on the formation of sedimentary basins is plate tectonics and hence basins are normally classified in terms of their position in relation to divergent, convergent and strike-slip tectonic settings. The characteristics of sedimentation and the stratigraphic succession which develops in a rift valley can be seen to be distinctly different from those of an ocean trench, a basin formed adjacent to a volcanic arc or in the periphery of a mountain belt. A stratigraphic succession therefore can be interpreted in terms of plate tectonics providing a larger scale for the consideration of sedimentary rocks. Similarly the sedimentary rocks in a basin provide a record of the tectonic history of the area.

23.1 Tectonics of sedimentary basins

The places around the globe where sediments accumulate into stratigraphic successions are collectively referred to as *sedimentary basins*. This is a rather loose usage of the word 'basin' as there may be build-ups of sediment around or within a continent as well as the fills of topographic and bathymetric lows. Sedimentary basins are conventionally classified and described according to their tectonic setting (Ingersoll 1988). At a simple level three main settings of basin formation can be recognized:
1 basins associated with regional extension within and between plates;
2 basins related to convergent plate boundaries;
3 basins associated with strike-slip plate boundaries.

In the context of these three settings nine main basin types can be recognized (Fig. 23.1). This is necessarily a simplification of more detailed schemes in which over 20 types of basin can be distinguished on the basis of a more detailed consideration of the tectonic setting of both modern and ancient basins (Ingersoll 1988; Busby & Ingersoll 1995). In addition, hybrid forms exist because of the complexities of plate tectonic processes.

In the following discussion the main basin types and the transitions between them are considered in terms of the plate tectonic setting. An elementary knowledge of plate tectonic processes and the nature of continental and oceanic lithosphere is assumed, as is an understanding of the basic terminology of structural geology. A conceptual model for each basin is presented along with a case study.

23.1.1 Sedimentary basins and geosynclines

In the 19th century it was recognized that great thicknesses of sedimentary rocks existed in many parts of the world. To explain these accumulations and the subsequent deformation of the strata in mountain belts some mechanism of subsidence to produce basins and then deform strata was required. Large folds in the crust were envisaged, hundreds of kilometres across. The large downfolds were termed *geosynclines* and the upfolds *geanticlines*. A geosyncline was considered to be a site for sediment accumulation, which gradually sank, at least partly under the weight of the added material. Eventually, according to the theory, the lower parts of the

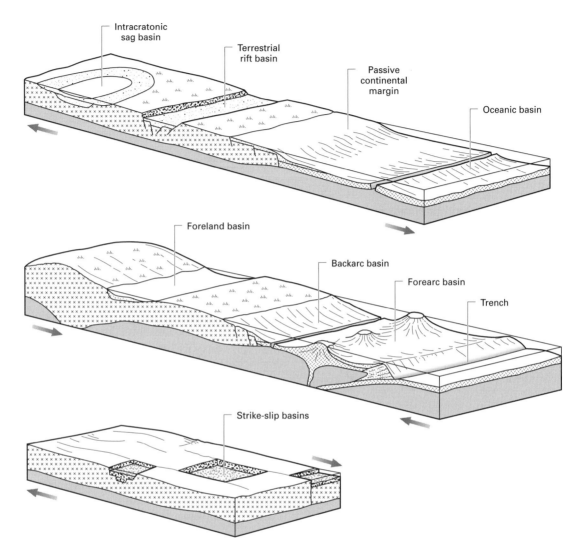

Fig. 23.1 Classification of sedimentary basins in terms of tectonic setting. (Reproduced with the permission of Stanley Thornes Publishers Ltd from Duff (1993) *Holmes' Principles of Physical Geology*, 4th edn.)

geosyncline would become deformed, and igneous intrusions and extrusions would occur.

The problem with the geosynclinal theory is that the forces which cause the subsidence and later deformation of the layers of strata are not adequately explained. The theory of plate tectonics is a more comprehensive way of accounting for the formation of sedimentary basins and the deformation of strata by the movement of crustal plates. The concept of geosynclines and the terminology associated with them (eugeosynclines, miogeosynclines,

and so on: Mitchell & Reading 1986) have now effectively been supplanted by plate tectonic theory and terminology.

23.2 Basins related to crustal extension

Horizontal stresses associated with plate movements create extension along lines of weakness in the lithosphere. These lines may lie within continental crust and may extend into oceanic crust where they are zones of formation of new crust at spreading centres. They terminate at triple junctions where three zones of extension meet. The triple junction is normally associated with a 'hot-spot', an area of increased heat flow in the crust generated by thermal plumes in the mantle. This association suggests that there may be a relationship between the formation of extensional plate boundaries and the location of hot mantle plumes.

Rift basins start to form within continental crust (Fig. 23.1) and the earliest deposits are typically terrestrial, although the basin may be flooded by an adjacent sea. Following the initiation of rifting in continental crust, there are two possible courses for the further development of the area. Rifting may continue and split the continental crust completely to form a proto-oceanic trough. Further extension leads to the formation of an ocean basin flanked by passive margins. Alternatively, the extension may cease and be followed by subsidence over a broad area within the craton to form an intracratonic basin.

23.2.1 Rift basins

In regions of extension continental crust fractures to produce *rifts*, which are structural valleys bound by extensional (normal) faults (Leeder 1995). In extensional regimes the downfaulted blocks are referred to as *graben* and the upfaulted areas as *horsts*. The bounding faults may be planar or listric, and they form asymmetric valleys referred to as *half-graben* which are basins deeper on one side than on the other. The axis of the rift lies more or less perpendicular to the direction of the stress. The structural weakness in the crust may allow magmas to rise from deeper levels, and volcanic activity is commonly associated with rifting of continental crust. Thinning of the continental crust raises the geothermal gradient in the area by bringing hot mantle closer to the surface. This heats up the crust and the flanks of the rift which have not been thinned by extension will be uplifted.

Sediment shed from the adjacent uplifted areas is deposited in *terrestrial rift valleys* by aeolian, fluvial, alluvial and lacustrine processes. Alluvial fans form along the rift margins and river systems may develop at points where large amounts of sediment and water are supplied to the basin. The locations of the sediment bodies are commonly controlled by the structure of the rift, particularly the patterns of the extensional faults at the edges and the tilting of the surface over large areas. The centres of rift basins may be regions of fluvial sedimentation and/or regions of lake formation. The style of lake sedimentation will be climatically controlled (Leeder & Gawthorpe 1987). In *maritime rifts*, which are those flooded by the sea, coarse sediment is deposited as fan deltas at the basin margins and distributed in the

basin by wave, tide, storm and gravity currents. In the absence of terrigenous clastic material, carbonate sedimentation may be dominant, consisting of reefs on the basin flanks and deep water carbonates in the centre (Leeder & Gawthorpe 1987).

EXAMPLE OF A RIFT BASIN

The Gulf of Suez in Egypt is the north-western arm of the Red Sea (Fig. 23.2). It is a rift basin which started to form in mid-Tertiary times due to extension between the Arabian plate and north Africa (Freund 1970; Coleman 1993). The oldest rift deposits are red beds of fluvial sandstone and mudstone which lie unconformably on pre-rift Eocene limestone strata (McClay *et al.* 1998). Basaltic dykes, sills and lavas amongst these red beds indicate that a small amount of igneous activity occurred during the initial rifting phase. Flooding of the rift basin in early Miocene times is indicated by shallow marine facies which directly overlie the red beds. Clastic detritus

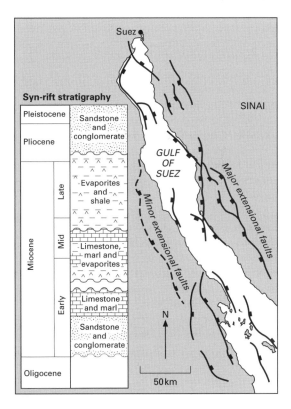

Fig. 23.2 Example of a basin formed by crustal extension: Gulf of Suez, Egypt, structure and stratigraphy. (After Coleman 1993; McClay *et al.* 1998.)

eroded from the flanks of the rift supplied coarse debris to fan deltas along the margins of the basin and sand to tide-influenced shoreface environments. In shallow marine areas protected from clastic input carbonate reefs formed, and in the deeper central parts of the basin muds and turbidite sands accumulated. Widespread late Miocene evaporite beds indicate that restriction of the connection to the marine waters of the Mediterranean occurred at this time. A marine connection was subsequently established with the Red Sea and Gulf of Aden. The Gulf of Suez is no longer actively extending because the movement between the African and Arabian plates is now taken up along the Gulf of Aqaba on the eastern side of the Sinai Peninsula.

23.2.2 Intracratonic basins

After the cessation of rifting within continental crust there is a change in the thermal regime of the area. When continental crust is extended it is thinned, and this brings hotter mantle material closer to the surface. Rifts are therefore areas of high heat flow, a high *geothermal gradient* (rate of change of temperature with depth). When rifting stops the geothermal gradient is reduced and the crust in the region of the rift starts to cool and thicken. Cold rock is denser than hot rock, so as the continental crust cools it contracts and sinks, resulting in *thermal subsidence*. The area around the rift which had formerly been heated develops into a broad area of subsidence within the continental block (*craton*) and becomes an *intracratonic basin* (Klein 1995). Although the precursor rift can be recognized under most intracratonic basins, others appear to lack evidence for the rift stage and their origin is more enigmatic.

Intracratonic basins are typically broad but not very deep and the rate of subsidence due to the cooling of the lithosphere is slow. Fluvial and lacustrine sediments are commonly encountered in these basins although flooding from an adjacent ocean may result in a broad epicontinental sea. Intracratonic basins in wholly continental settings are very sensitive to climate fluctuations as increased temperature may raise rates of evaporation in lakes and reduce the water level over a wide area.

EXAMPLE OF AN INTRACRATONIC BASIN

The Chad Basin in west central Africa is a basin of internal drainage with a central lake, Lake Chad, fed by rivers around the rim of the basin (Fig. 23.3). The size of

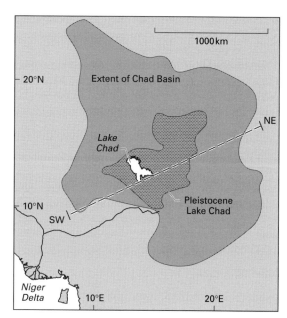

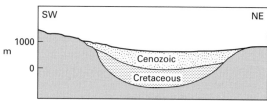

Fig. 23.3 An intracratonic basin: Chad Basin, central Africa, map and cross-section. (After Reading 1982; Mitchell & Reading 1986.)

the lake has fluctuated between its present size and an area almost 10 times as great during periods of cooler, wetter climate in the Pleistocene (Reading 1982). Sediments in the Chad Basin are mostly fine-grained clastic sediments formed on the river floodplains and within the lake. The basin lies adjacent to and partly over the Benue Trough, a rift which formed in the Mesozoic and Tertiary. The Benue Trough formed as the 'third arm' of a triple rift system as part of the spreading of the South Atlantic; this arm stopped widening whilst the other two 'arms' of the rift went on to become the margins of the South Atlantic (Burke 1976).

23.2.3 Proto-oceanic troughs: the transition from rift to ocean

Continued extension within continental crust leads to

thinning and eventual complete rupture. Basaltic magmas rise to the surface in the axis of the rift and start to form new oceanic crust. Where there is a thin strip of basaltic crust in between two halves of a rift system the basin is called a *proto-oceanic trough* (Leeder 1995). The basin will be wholly or partly flooded by sea water by the time this amount of extension has occurred and the trough has the form of a narrow seaway between continental blocks.

Sediment supply to this seaway comes from the flanks of the trough which will still be relatively uplifted. Rivers will feed sediment to shelf areas and out into deeper water in the axis of the trough as turbidity currents. Connection to the open ocean may be intermittent during the early stage of basin formation and in areas of high evaporation the basin may periodically desiccate. Evaporites may form part of the succession in these circumstances.

EXAMPLE OF A PROTO-OCEANIC TROUGH

The Red Sea is the only modern example of a basin in the stage of transition from a rift basin to an ocean basin (Fig. 23.4). Extension in the Red Sea started in the mid-

Tertiary. The southern end of the sea lies at a triple junction with the Gulf of Aden, a small ocean basin, and the East African rift system, an intra-continental rift. The floor of the Red Sea is thinned continental crust in the north and newly formed oceanic crust in the south. This reflects the amount of extension across the basin which is greater in the south than in the north. Rivers feeding the Red Sea from the African and Arabian margins supply clastic detritus, and carbonate sedimentation occurs in the warm waters. In the early stages of basin formation when connection with the open ocean was intermittent, evaporite deposition was important (Coleman 1993).

23.2.4 Passive margins

Passive margins are the regions of continental crust along the edges of ocean basins. The term 'passive' is used in this sense as the opposite to the 'active' margins between oceans and continents where oceanic crust is being subducted. The continental crust is commonly thinned in this region and there may be a zone of transitional crust before fully oceanic crust of the ocean basin is encountered. *Transitional crust* forms by basaltic

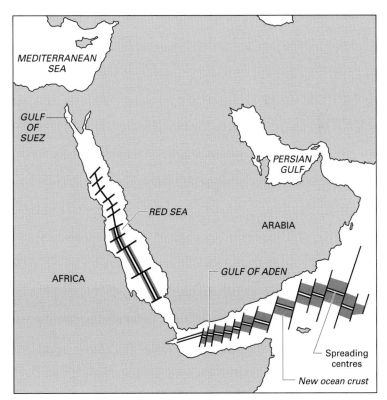

Fig. 23.4 The Red Sea is a proto-oceanic trough, a basin which is transitional between a continental rift basin and an ocean basin.

magmas injecting into continental crust in a diffuse zone as a proto-oceanic trough develops. Subsidence of the passive margin is due mainly to continued cooling of the lithosphere as the heat source of the spreading centre becomes further away, augmented by the load on the crust due to the pile of sediment which accumulates (Einsele 1992).

Morphologically the passive margin is the continental shelf and slope. The source of clastic sediment is from the adjacent continental land mass. Rates of erosion and transport in the land area affect the thickness and distribution of clastic deposits on the passive margin. Adjacent to large rivers very thick piles of sediment build out on to the shelf. On the other hand, shelves adjoining desert areas may supply little material and the margin will be 'starved'. In favourable climatic conditions shelves

Fig. 23.5 The eastern seaboard of North America is a passive margin of the Atlantic Ocean. (Reproduced with the permission of Stanley Thornes Publishers Ltd from Duff (1993) *Holmes' Principles of Physical Geology*, 4th edn.)

are areas of high biogenic productivity and in the absence of terrigenous detritus thick build-ups of carbonate may occur. The large areas of shallow seas on passive margins are very sensitive to the effects of global eustatic sea level changes *(21.8)*.

EXAMPLE OF A PASSIVE MARGIN

The eastern seaboard of North America is the passive margin of the continental crust of the North American continent and the western half of the Atlantic Ocean (Fig. 23.5). Rifting started to separate the continent from Europe and Africa in the Triassic but sea floor spreading did not commence until the Jurassic. The North Atlantic has continued to open since that time. Along the northern part of the margin the Mesozoic and Tertiary succession is dominated by shallow marine clastic sedimentary rocks, but further south the terrigenous supply is less and the warm shallow waters off Florida have been a site of carbonate sedimentation through most of the history of the margin (Sheridan 1974; Steckler *et al.* 1988).

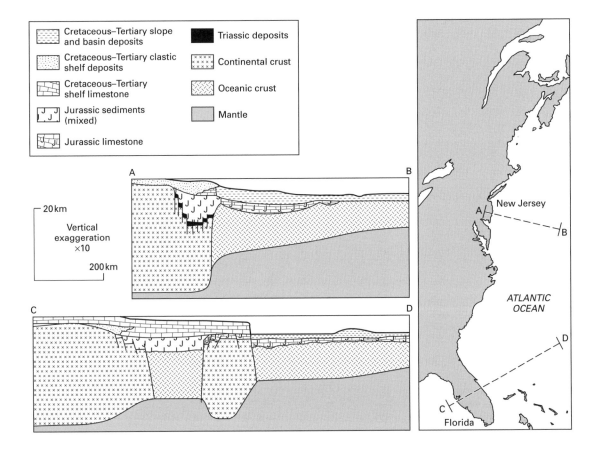

23.2.5 Ocean basins

Basaltic crust formed at mid-oceanic ridges is hot and relatively buoyant. As the basin grows in size by new magmas created along the spreading ridges, older crust moves away from the hot mid-ocean ridge. Cooling of the crust increases its density and decreases relative buoyancy, so as crust moves away from the ridges, it sinks. Mid-ocean ridges are typically at depths of around 2500 m. The depth of the ocean basin increases away from the ridges to 4000–5000 m where the basaltic crust is old and cool.

The ocean floor is not a flat surface. Spreading ridges tend to be irregular, offset by transform faults which create some areas of local topography. Isolated volcanoes and linear chains of volcanic activity such as the Hawaiian Islands form submerged *seamounts* or exposed islands. In addition to the formation of volcanic rocks in these areas, the shallow water environment may be a site of carbonate production and the formation of reefs. In the deeper parts of the ocean basins sedimentation is mainly pelagic, consisting of fine-grained biogenic detritus and clays. Nearer to the edges of the basins terrigenous clastic material may be deposited as turbidites.

EXAMPLE OF AN OCEAN BASIN

The stratigraphy of an ocean basin will not be preserved as a succession of rocks on land as an intact unit. Oceanic crust is denser than continental crust and at most convergent plate boundaries the ocean basin sediments are either subducted or are deformed as they are incorporated into an accretionary prism *(23.3.2)*. It is only in

situations where obduction occurs (see below) that relatively undeformed parts of the ocean basin succession are found exposed on the continents. The stratigraphy of ocean basins is mainly known from drilling the sedimentary succession in present-day ocean basins. The Pacific Ocean is the largest modern ocean basin. Away from the margins pelagic sedimentation dominates in the basin. In equatorial areas biogenic productivity is high and calcareous deposits occur on the basin floor. In more temperate regions the sea floor is below the calcite compensation depth (CCD) *(15.5.1)* and deposits are siliceous muds. Reefs form around volcanic seamounts.

23.2.6 Obducted slabs

Most oceanic crust is subducted at destructive plate margins but there are circumstances under which slabs of ocean crust are *obducted* up on to the over-riding plate to lie on top of continental or other oceanic crust. Outcrops of oceanic crust preserved in these situations are known as *ophiolites* (Gass 1982). Until drilling in the deep oceans became possible in the last few decades, initially with the Deep Sea Drilling Project and later as part of the Ocean Drilling Program, ophiolite complexes provided the only tangible evidence of oceanic crust and deep sea sediments. An ophiolite suite consists of the ultrabasic and basic intrusive rocks of the lower oceanic crust (peridotites and gabbros), a dolerite dyke swarm which represents the feeders to the basaltic pillow lavas which formed on the ocean floor (Fig. 23.6). The lavas are overlain by deep ocean sediments deposited at or close to the spreading centre. If sea floor spreading occurred above the CCD these sediments would have been calcareous

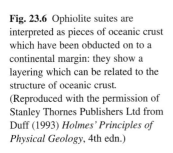

Fig. 23.6 Ophiolite suites are interpreted as pieces of oceanic crust which have been obducted on to a continental margin: they show a layering which can be related to the structure of oceanic crust. (Reproduced with the permission of Stanley Thornes Publishers Ltd from Duff (1993) *Holmes' Principles of Physical Geology*, 4th edn.)

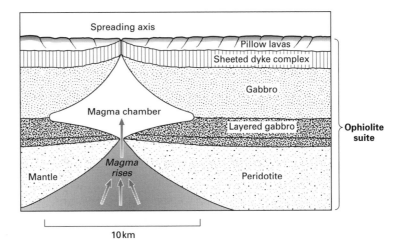

oozes, preserved as fine-grained pelagic limestones. In the absence of carbonates, red clays and siliceous oozes are lithified to form red mudstones and cherts. Concentrations of metalliferous ores are common, formed as hydrothermal deposits close to the volcanic vents.

Three of the best known ophiolites are in Newfoundland, Cyprus and Oman. The Bay of Islands ophiolite in Newfoundland formed on the margins of the Iapetus Ocean and is Palaeozoic in age. The Cyprus Ophiolite and the Semail Ophiolite in Oman are both related to the closure of the Tethys Ocean in the Mesozoic. Ophiolites may represent the stratigraphic succession formed in an ocean basin or the fill of a backarc basin *(23.3.4)*.

23.3 Basins related to subduction

At convergent plate margins involving oceanic lithosphere subduction occurs. The downgoing ocean plate descends into the mantle beneath the over-riding plate, which may be either another piece of oceanic lithosphere or a continental margin. As the downgoing plate bends to enter the subduction zone a trough is created at the contact between the two plates: this is the *ocean trench*. The descending slab is heated as it goes down and partially melts. The magmas generated rise to the surface through the over-riding plate to create a line of volcanoes, a *volcanic arc*. The magmas start to form when

Fig. 23.7 Arc–trench systems at destructive plate margins are sites of sediment accumulation in the trench, forearc and backarc areas. (After Dickinson & Seely 1979; AAPG ©1979, reprinted by permission of the American Association of Petroleum Geologists, modified from Dickinson (1975).)

the downgoing slab reaches 90–150 km depth. The *arc–trench gap* (distance between the axis of the ocean trench and the line of the volcanic arc) will depend on the angle of subduction: at steep angles the distance will be as little as 50 km, and where subduction is at a shallow angle it may be over 200 km (Fig. 23.7).

Arc–trench systems are regions of plate convergence, yet there may be local extension as well as compression in the upper plate. The relative rates of plate convergence and subduction are the governing factor. If convergence is faster than subduction some of the shortening is taken up in the over-riding plate to form a *retroarc foreland basin*. If convergence is slower than subduction at the trench the upper plate is in net extension and an *extensional backarc basin* forms. A balance between the two results in neither compression nor extension and a 'neutral' arc–trench system (Dickinson 1980).

23.3.1 Trenches

Ocean trenches are elongate, gently curving troughs which form where an oceanic plate bends as it enters a subduction zone. The inner margin of the trench is formed by the leading edge of the over-riding plate of the arc–trench system. The bottoms of modern trenches are up to 10 000 m below sea level, twice as deep as the average bathymetry of the ocean floors. They are also narrow, sometimes as little as 5 km across, although they may be thousands of kilometres long. Trenches formed along margins flanked by continental crust tend to be filled with sediment derived from the adjacent land areas. Intra-oceanic trenches are often starved of sediment because the only source of material apart from

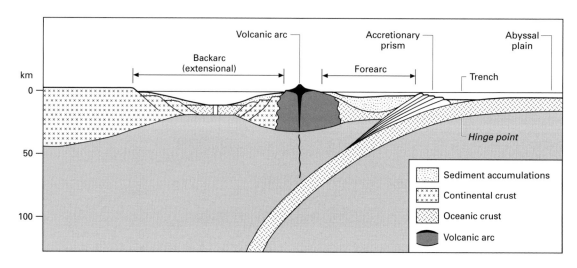

pelagic deposits are the islands of the volcanic arc. Transport of coarse material into trenches is by mass flows, especially turbidity currents which may flow for long distances along the axis of the trench (Underwood & Moore 1995).

EXAMPLE OF A TRENCH BASIN

The Chile Trench off the western coast of South America is the southern part of the Peru–Chile trench system, where Pacific Oceanic lithosphere is being subducted beneath the continental crust of South America. The Chile Trench is over 2500 km long and around 30 km wide. It varies in depth from 7000–8000 m in the northern part to around 5000 m in the south. The amount of sediment in the trench is extremely variable. Where the adjacent land is a warm, arid region in northern Chile the supply to the trench is very sparse and there are only a few hundred metres of sediment. Further south, where the climate is cool and humid sediment supply from the mountains of southern Chile is abundant and there are 2–3 km of sediment in the trench. Transport is principally by turbidity currents. A main axial turbidite channel in the southern Chile Trench stretches for many hundreds of kilometres along the trench with smaller lobes of turbidite deposits fed by canyons (Thornburg & Kulm 1981).

23.3.2 Accretionary complexes

The sedimentary pile accumulated on the ocean crust and in the trench is not necessarily subducted along with the crust at a destructive plate boundary. The pile of sediments may be wholly or partly scraped off the downgoing plate and accrete on the leading edge of the over-riding plate to form an *accretionary complex* or *accretionary prism*. These prisms or wedges of oceanic

and trench sediments are best developed where there are thick successions of sediment in the trench. A subducting plate can be thought of as a conveyor belt bringing ocean basin deposits, mainly pelagic sediments and turbidites, to the edge of the over-riding plate. In some places this sediment is carried down the subduction zone, but in others it is sliced off as a package of strata which is then accreted on to the over-riding plate (Fig. 23.8). Along the Sunda Trench, where the Indian Ocean plate is subducting beneath the continental island of Sumatra, the accretionary prism has grown up to sea level to form a chain of islands between the trench and the coast of Sumatra (Karig *et al.* 1980) (Fig. 23.9). If ocean closure is complete, the accretionary prism can become incorporated into an orogenic belt. This is one of the interpretations of the rocks which make up the Southern Uplands of Scotland: Ordovician and Silurian strata in this area are all interpreted as deep water sediments, arranged in a structural pattern which is consistent with the formation of an accretionary prism on the north side of the Iapetus Ocean (Leggett *et al.* 1982; Leggett 1987).

23.3.3 Forearc basins

The width of a forearc basin will depend on the arc–trench gap, which is in turn determined by the angle of subduction. Its inner margin is the edge of the volcanic arc and the outer limit the accretionary complex formed on the leading edge of the upper plate. The basin may be underlain by either oceanic crust or a continental margin (Dickinson 1995). The thickness of sediments which may accumulate will be partly controlled by the height of the accretionary complex: if this is close to sea level the forearc basin may also fill to that level. Subsidence in the forearc region is due only to the load from the sediment pile.

The main source of sediment is from the region of the

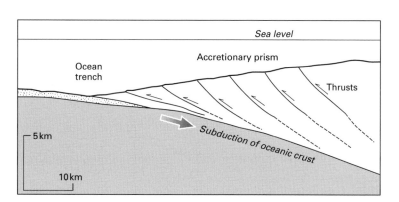

Fig. 23.8 Accretionary prisms form on the inner sides of ocean trenches by the accretion of material from the downgoing plate on to the leading edge of the over-riding plate.

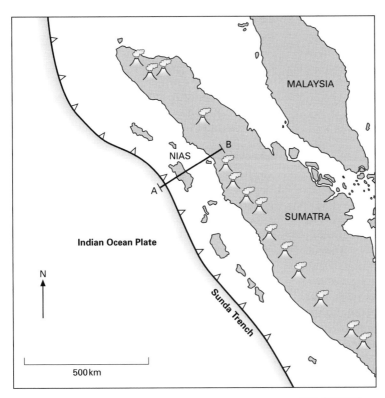

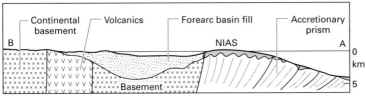

Fig. 23.9 Subduction of Indian Ocean crust along the Sunda Trench has led to the formation of a trench, accretionary prism and forearc basin on the south-western side of the island of Sumatra. (After Karig *et al.* 1980.)

volcanic arc and, if the arc lies in continental crust, the hinterland of continental rocks. Intra-oceanic arcs are commonly starved of sediment because the island arc volcanic chain is the only source of detritus apart from pelagic sediment. Given sufficient supply of detritus a forearc basin succession will consist of deep water deposits at the base shallowing upward to shallow marine, deltaic and fluvial sediments at the top. The upper part of this succession will be absent in forearc basins which lack an outer rim formed by the accretionary complex or are under-supplied with sediment. Volcaniclastic debris is likely to be present in almost all cases.

EXAMPLE OF A FOREARC BASIN

The Indian Ocean subducts along a trench which lies

offshore of the island of Sumatra. This subduction zone is part of a more extensive system which stretches from the Bay of Bengal, past Sumatra and Java to the Timor Sea. The accretionary complex is a large feature and the top is emergent as a string of islands between the trench and Sumatra (Fig. 23.9). A chain of volcanoes along the mountainous axis of Sumatra marks the volcanic arc related to the subduction (Karig *et al.* 1980). The Sumatra forearc basin contains several thousand metres of strata which date back to the early Miocene (Beaudry & Moore 1985). Rivers from the Sumatran highlands are still actively supplying sediment to the basin, building up a coastal plain which passes into deltas which prograde seawards. The basin has filled to sea level in places, supplied by the arc and the continental highlands of Sumatra.

23.3.4 Backarc basins

When the upper plate in an arc–trench system is under extension it rifts in the region of the volcanic arc where the crust is hotter and weaker. Initially the arc itself rifts and splits into two parts, an active arc with continued volcanism closer to the subduction zone and a remnant arc. As extension between the remnant and active arcs continues, a new spreading centre is formed to generate basaltic crust between the two. This region of extension and crustal formation is the backarc basin. The basin continues to grow by spreading until renewed rifting in the active arc leads to the formation of a new line of extension closer to the trench. Once a new backarc basin is formed the older one is abandoned. Extensional backarc basins can form in either oceanic or continental plates (Marsaglia 1995).

The principal source of sediment in a backarc basin formed in an oceanic plate will be the active volcanic arc. Once the remnant arc is eroded down to sea level it contributes little further detritus. More abundant supplies are available if there is continental crust either or both sides of the basin. Backarc basins are typically under-filled, containing mainly deep water sediment of volcaniclastic and pelagic origin.

EXAMPLE OF A BACKARC BASIN

Most of the active intra-oceanic backarc basins lie in the western Pacific. The Sumisu rift is a backarc basin which is part of the Izu–Bonin arc–trench system and lies south of Japan. Oceanic crust of the west Pacific plate is being subducted in a westerly direction in this area and the Sumisu rift lies west of the Bonin Arc (Fig. 23.10). The backarc basin is over 100 km long and 40 km wide. It is around 2000 m deep and has formed during the past 2 million years. A thousand metres of sediment provide a record of the evolution of the basin: the lower third consists of volcaniclastic breccia and flow deposits, whilst the remainder is composed of calcareous mudrock. These two sediment types reflect the two main sources, the active volcanic arc and pelagic sedimentation, the latter having become more important as the basin has evolved (Einsele 1992; Marsaglia 1995).

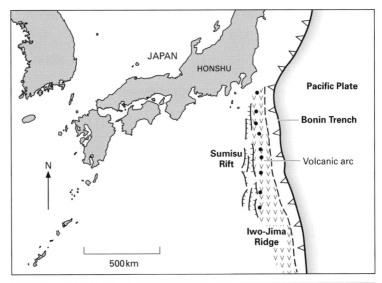

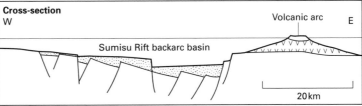

Fig. 23.10 The Sumisu rift is an example of a backarc basin south of Japan. (After Marsaglia 1995.)

23.3.5 Retroarc foreland basins

In compressional convergence regimes the over-riding continental plate shortens by the development of a mountain belt. Magmas from the subduction zone also add material to the upper plate in an arc along the mountain belt. Thickening of the crust results in the upward and outward movement of masses of rock along thrusts and as nappes (Dickinson 1974; Scholl *et al.* 1980). As these thrust slices move on to the continent on the opposite side of the arc to the trench, they add a load on to it by increasing the mass of material on the lithosphere. Loading of the lithosphere causes it to bend and it is as a result of this flexure that a basin forms. These basins are called 'retroarc' because of their position behind the arc and 'foreland' because the mechanism of formation is by flexure of the leading edge of the continent in a similar way to peripheral foreland basins *(23.4.1)*.

The continental crust will be close to sea level at the time the loading commences so most of the sedimentation occurs in fluvial, coastal and shallow marine environments. Continued subsidence occurs due to further loading of the basin margin by thrusted masses from the mountain belt, augmented by the sedimentary load. The main source of detritus is the mountain belt and volcanic arc.

EXAMPLE OF A RETROARC FORELAND BASIN

The eastern Andean basins are present-day examples of basins formed by the convergence between the eastern Pacific oceanic plates and the South American continent (Fig. 23.11). The retroarc foreland basin is up to 200 km wide and contains up to 8000 m of sediment. Subsidence since the early Miocene has been the result of the eastward movement of thrusts from the Andean mountain belt loading the South American foreland. The rocks of the Andes provide the source area for the sediments. Throughout the basin evolution sedimentation has been in a continental environment but the facies show lateral variations due to differences in climate. In arid regions aeolian and playa environments are important, whilst in the wetter regions fluvial facies dominate (Jordan 1995).

23.4 Basins related to continental collision

When an ocean basin completely closes with the total elimination of oceanic crust by subduction, the two continental margins eventually converge. Where two continental plates converge subduction does not occur because the thick, low-density continental lithosphere is too buoyant to be subducted. Collision of plates involves a thickening of the lithosphere and the creation of an *orogenic belt*, a mountain belt formed by collision of plates. The two continental margins which collide are likely to be thinned, passive margins. Shortening initially increases the lithosphere thickness up to 'normal' values before it over-thickens. As the crust thickens it undergoes deformation with metamorphism occurring in the

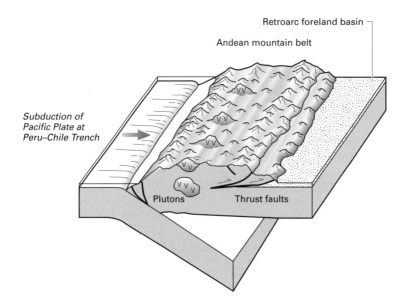

Fig. 23.11 In the eastern part of the Andes in South America lie retroarc basins formed by the loading of the crust by the Andean mountain belt. (After Jordan 1995.)

lower parts of the crust and faulting and folding at shallower levels in the mountain belt. Thrust faults form a *thrust belt* along the edges of the mountain chain within which material is moved outwards, away from the centre of the orogenic belt.

23.4.1 Peripheral foreland basins

The thrust belt moves material out on to the foreland crust either side of the orogenic belt. Under this load the crust flexes to form a *peripheral foreland basin*. The width of the basin will depend on the amount of load and the flexural rigidity of the foreland lithosphere — the ease

with which it bends when a load is added to one end (Beaumont 1981).

In the initial stages of foreland basin formation the collision will have only proceeded to the extent of thickening the crust (which was formerly thinned at a passive margin) up to 'normal' crustal thickness. Although this results in a load on the foreland and lithospheric flexure the orogenic belt itself will not be high above sea level at this stage and little detritus will be supplied by erosion of the orogenic belt. Early foreland basin sediments will therefore occur in a deep water basin with the rate of subsidence exceeding the rate of supply. Turbidites are typical of this stage.

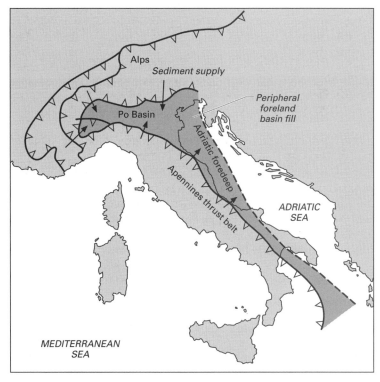

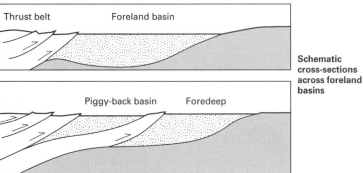

Fig. 23.12 The Po Basin in northern Italy is an example of a peripheral foreland basin. (After Ricci-Lucchi 1986.)

When the orogenic belt is more mature and has built up a mountain chain there is an increase in the rate of sediment supply to the foreland basin. Although the load on the foreland will have increased, the sediment supply normally exceeds the rate of flexural subsidence. Foreland basin stratigraphy typically shallows up from deep water to shallow marine and then continental sedimentation which dominates the later stages of foreland basin sedimentation (Miall 1995). In foreland basins formed during the Tertiary in Europe as part of the Alpine orogeny, the early, turbidite stage of sedimentation is referred to as the 'flysch' stage and the later, shallow marine to continental deposits are referred to as the 'molasse' stage.

Foreland basin stratigraphy is often complicated by the deformation of the earlier basin deposits by later thrusting. These thrusts may subdivide the basin into *piggy-back basins* (Ori & Friend 1984) which lie on top of the thrust sheets and which are separate from the *foredeep*, the basin in front of all the thrusts.

EXAMPLE OF A PERIPHERAL FORELAND BASIN

The Apennines are a mountain chain along the axis of Italy which are part of the Alpine orogenic belt (Fig. 23.12). To the north-east lies the Adriatic Sea which is the foreland area to the north-east-directed thrust movement of the Apennines belt. The northern extension of the Adriatic foreland basin is the Po Basin in northern Italy, which is also marginal to the main Alpine chain. The Adriatic foredeep has been an area of turbidite deposition since the Oligocene, with material derived from

the orogenic belt as well as the foreland of the Balkans on the north-eastern side of the foreland basin. At the northern end of the Adriatic Sea the Po Basin has received sediment from the Alps to the north; fluvial and shallow marine deposition is currently occurring in this area (Ricci-Lucchi 1986).

23.5 Basins related to strike-slip plate boundaries

If a plate boundary is a straight line and the relative plate motion purely parallel to that line there would be neither uplift nor basin formation along strike-slip plate boundaries. However, such plate boundaries are not straight, the motion is not purely parallel and they consist not of a single fault strand but of a network of branching and overlapping individual faults. Zones of localized subsidence and uplift create topographic depressions for sediment to accumulate and the source areas to supply them (Christie-Blick & Biddle 1985).

23.5.1 Strike-slip basins

Most basins in strike-slip belts are generally termed *transtensional* basins and are formed by a number of mechanisms (Fig. 23.13). The overlap of faults can create regions of extension between them known as *pull-apart basins*. Such basins are typically rectangular or rhombic in plan with widths and lengths of only a few kilometres or tens of kilometres. They are unusually deep, especially compared to rift basins. Where there is a branching of faults a zone of extension exists between

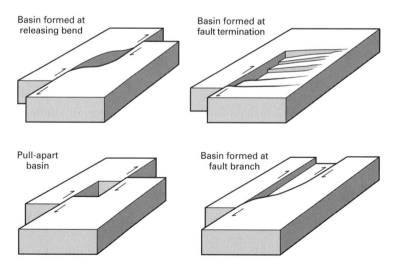

Fig. 23.13 Basins formed at strike-slip plate boundaries and in regions of intra-continental strike-slip tectonics.

the two branches forming a basin. The curvature of a single fault results in bends which are either restraining (locally compressive) or releasing (locally extensional): releasing bends form elliptical zones of subsidence and small basins. Basins may also form adjacent to fault terminations.

The exact mechanism of formation of basins in strike-slip belts is variable but there are a number of common characteristics (Reading 1980; Nilsen & Sylvester 1995). They are relatively small, usually in the range of a hundred to a thousand square kilometres, and often contain thicker successions than basins of similar size formed by other mechanisms. Subsidence is usually rapid and several kilometres of strata can accumulate in a few million years. Typically the margins are sites of deposition of coarse facies (alluvial fans and fan deltas) and these pass laterally over very short distances to lacustrine sediments in continental settings or marine deposits. In the stratigraphic record, facies are very varied and show lateral facies changes over short distances.

EXAMPLE OF A STRIKE-SLIP BASIN

A left-lateral strike-slip fault system runs approximately north–south up the valley of the River Jordan from the head of the Gulf of Aqaba (Fig. 23.14). This is a plate margin which separates the small Palestine block on the west side from the Arabian plate to the east. Within this plate boundary zone lies the Dead Sea, a basin bound on both sides by steep strike-slip faults. Sediment is supplied to the Dead Sea Basin by the Jordan River, by aeolian input and by alluvial fans which have formed on both flanks of the transtensional basin from material shed from the high ground either side (Zak & Freund 1981). The region is one of low rainfall and high rates of evaporation, and the lake at the centre of the basin, the Dead Sea, is highly saline and 400 m below the level of the Mediterranean Sea.

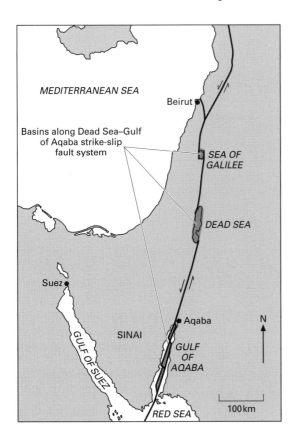

Fig. 23.14 The region extending from the Dead Sea to the Gulf of Aqaba is a zone of strike-slip tectonics and the formation of sedimentary basins. (After Zak & Freund 1981.)

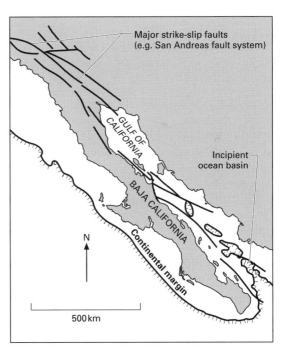

Fig. 23.15 The Gulf of California may be considered to be a hybrid basin, controlled by a combination of extensional and strike-slip tectonics. (Reproduced with the permission of Stanley Thornes Publishers Ltd from Duff (1993) *Holmes' Principles of Physical Geology*, 4th edn. After Kelts 1981.)

Rift basin

Ocean basin

Arc and trench formation

Ocean closure

Mountain belt

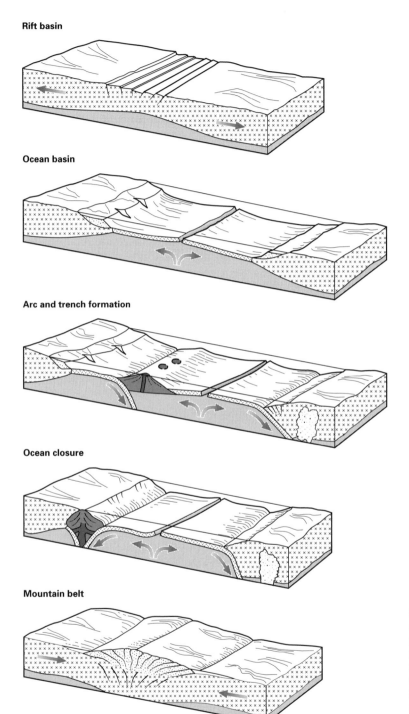

Fig. 23.16 The Wilson cycle of extension to form a rift basin and ocean basin followed by basin closure and formation of an orogenic belt. (Reproduced with the permission of Stanley Thornes Publishers Ltd from Duff (1993) *Holmes' Principles of Physical Geology*, 4th edn. After Wilson 1966.)

23.6 Complex and hybrid basins

Not all basins fall into the simple categories outlined above because they are the product of the interaction of more than one tectonic regime. This most commonly occurs where there is a strike-slip component to the motion at a convergent or divergent plate boundary. A basin may therefore partly show the characteristics, say, of a peripheral foreland but also contain strong indicators of strike-slip movement. Such situations exist because plate motions are commonly not simply orthogonal or parallel and there is oblique convergence or extension between plates in many parts of the world.

EXAMPLE OF A HYBRID BASIN

A modern example of a basin which is controlled by both extensional and strike tectonics is the Gulf of California (Fig. 23.15). This basin lies at the southern end of the San Andreas strike-slip fault zone. To the south and west lies a spreading ridge within the eastern Pacific Ocean. It is therefore in a zone of transition from a strike-slip plate boundary to an extensional, oceanic basin. The overall basin characteristics are a mixture of those of a pull-apart basin and those of a proto-oceanic trough. In the northern part of the Gulf of California rhombic sub-basins have formed by strike-slip pull-apart but further south these sub-basins are larger and deeper, floored by newly formed oceanic crust (Kelts 1981).

23.7 Changes in tectonic setting with time

Tectonic forces act slowly on a human time-scale but in the context of geological time the surface of the planet is in a continuous state of flux. Rift basins form and evolve into proto-oceanic troughs and eventually into ocean basins bordered by passive margins. After a period of tens to hundreds of millions of years the ocean basin starts to close, with subduction zones around the margins consuming oceanic crust. Final closure of the ocean results in continental collision and the formation of an orogenic belt. These patterns of plate movement through time are known as the *Wilson cycle* (Fig. 23.16). The whole cycle starts again as the continent breaks up by renewed rifting. Adding the complications of oblique and strike-slip plate motions leads to modifications of this simple cycle (Mitchell & Reading 1986).

The Wilson cycle predicts that at any place tectonic setting changes through time. A piece of continental crust may experience rift extension and then become the passive margin to an ocean; during closure of the ocean that passive margin may become the forearc region related to a subduction zone and finally be incorporated into an orogenic belt upon continental collision. The record of these plate tectonic events is provided by stratigraphy.

The stratigraphy of an area must be considered at different scales. At the scale of mapping formations in an area a succession may be considered in terms of the plate tectonic controls on its formation and interpreted as the product of a certain sedimentary basin type. However, at a larger scale of regional mapping of groups, a stratigraphic succession should be considered not as the deposits of a single type of sedimentary basin but as being formed in an evolving tectonic setting. The stratigraphy of an area may be divided into units formed under different tectonic regimes using the characteristics of the various basin types to interpret the setting at different times.

It is only in the centres of stable continental areas that basins are unchanging over long periods of geological time. The central part of the Australian continent has not experienced the tectonic forces of plate margins for 400 million years and in the latter part of that time a broad intracratonic basin, the Lake Eyre Basin, has formed by very slow subsidence. In regions closer to plate margins basins typically have a life-span of a few tens of millions of years. The backarc basins in the west Pacific appear to be active for 20 million years or so. In contrast, the passive margins of the Atlantic have been sites of sedimentation at the edges of the continents for over 200 million years.

Further reading

Allen, P.A. & Allen, J.R. (1990) *Basin Analysis: Principles and Applications*, 451 pp. Blackwell Scientific Publications, Oxford.

Busby, C. & Ingersoll, R.V. (eds) (1995) *Tectonics of Sedimentary Basins*, 580 pp. Blackwell Science, Oxford.

Einsele, G. (1992) *Sedimentary Basins, Evolution, Facies and Sediment Budget,* 628 pp. Springer-Verlag, Berlin.

Ingersoll, R.V. (1988) Tectonics of sedimentary basins, *Geological Association of America Bulletin* **100**, 1704–1719.

Mitchell, A.H.G. & Reading, H.G. (1986) Sedimentation and tectonics. In: *Sedimentary Environments and Facies* (ed. H.G. Reading), pp. 471–519. Blackwell Scientific Publications, Oxford.

24 The Earth through geological time

Modern processes and environments are our keys to understanding the formation of sedimentary rocks and their arrangement in stratigraphic patterns, but there are limitations to this approach. First, the magnitude of events and rates of processes which occur today may have been different in the past. Second, the complex interactions of chemical, physical and biological processes which control depositional environments have produced situations in the past which are quite different from anything we see today. Through geological time the Earth has undergone a number of changes in the fundamental controls on sedimentation. The arrangement and behaviour of crustal plates has created tectonic settings for which there are no modern analogues. The global climate has gone through long- and short-term cycles of warmer and cooler conditions, affecting sea level and the distribution of sedimentary environments. Extraordinary catastrophic events such as meteorite impacts have occurred, producing unique scenarios locally and world-wide. Most importantly, the evolution of life has resulted in a number of fundamental changes in the biosphere and atmosphere. The evolution of land plants radically altered the physical and chemical processes of the land surface, and the development of an oxygenated atmosphere changed both the chemical and biological environment on the surface of the Earth. Sedimentary rocks and stratigraphy must therefore be considered in terms of this dynamic, evolving environment of the Earth's surface.

24.1 Uniformitarianism: 'the present is the key to the past'

The principle of *uniformitarianism*, 'the present is the key to the past', is an underlying thread in the interpretation of the stratigraphic record in terms of the world around us. At certain levels it is reasonable to expect things to have behaved exactly the same way in the past as they do today. The basic laws of physics and chemistry must be assumed to be constant and inviolate: gravitational attraction must always have occurred between two masses and oxygen must always have been able to react with metals to form oxides. We can therefore reasonably apply these laws to interpret purely physical and chemical phenomena such as the movement of boulders downhill and the formation of minerals.

However, geological processes take place in complex settings where there are a large number of physical, chemical and biological factors, all of which are variable and which do not necessarily produce uniquely predictable results. In the same sense that meteorologists have found that there are many complexly interacting variables controlling weather systems, geologists must also consider Earth processes through time to show 'chaotic' behaviour at certain scales. Rivers change their course by avulsion, but it is not possible to predict exactly where and when the avulsion will occur in a modern river, so the stratigraphy of fluvial sediments will be chaotic at the level of predicting where channel-fill sandstone bodies are likely to be. The same could be said of all sedimentary environments and the strata which result from deposition in those settings. There is therefore a limit to the scale at which we can produce a mathematical model for the processes which govern sedimentology and stratigraphy.

Using observations from modern environments, we can develop general rules and simple mathematical models. For example, we can measure the rates at which pebbles or boulders are moved along a beach by waves and note how far a boulder might move in a storm. These data provide us with some idea of how far or fast these clast sizes of material are moved in modern beach environments and we can apply this to beach sediments in the stratigraphic record. In doing so, we are assuming that the processes in the past were similar in behaviour and magnitude to those we see today. But is this necessarily the case?

24.1.1 Catastrophism

The corollary of uniformitarianism is often considered to be *catastrophism*, the view that extraordinary events have been instrumental in shaping the surface of the Earth. In its most extreme form, catastrophism has been related to creationism and biblical stories such as the 'Flood'. In the 19th century there was much discussion between those who held uniformitarian views, arguing that the Earth had gradually evolved to its present form through processes which we can observe around us today, and the catastrophists, who thought that one or more major events shaped the planet and everything living on it (Hallam 1990). It now seems more probable that in some senses both are correct, with the uniformitarian view of a constancy of physical laws being logical, but with elements of catastrophism to explain some of the events of the past. The key to reconciling catastrophism and uniformitarianism is to consider the magnitude and frequency of events on a geological time-scale.

24.1.2 Periodicity of events

In terms of a human life-span high-magnitude phenomena such as very violent storms, high-energy earthquakes or large volcanic eruptions which occur once every hundred years or more are considered to be rare and extraordinary (Fig. 24.1). On a geological time-scale, events which occur every hundred years will have happened many tens of millions of times through Earth history so are common incidents. In considering the likelihood of a certain event, the human time-frame is too short for us to be good judges of the frequency with which it may occur on a geological time-scale.

Sometimes these high-magnitude events may last only a very short period of time: an earthquake may last only seconds or minutes, a storm will last for hours or days and a volcanic eruption may last weeks or months. However, short-lived, high-magnitude events leave as much of a mark in the geological record as much slower processes. The Toba tuff is a layer of late Pleistocene volcanic ash which can be traced from its volcanic vent in Sumatra in western Indonesia across large areas of the Bay of Bengal and India (Ninkovitch *et al.* 1978). The eruption which produced the Toba tuff may have lasted only a few years, but its mark on the stratigraphy of a wide area is considerable.

24.1.3 Magnitude of processes and events in the past

Every now and then there is a storm or flood or similar

Fig. 24.1 A volcano which has erupted on numerous occasions in recent history on Ternate, eastern Indonesia, causing widespread damage in the towns and villages on the coast at the foot of the volcano.

natural phenomenon which is described as 'the biggest for a hundred years' or some other period of time. We know from these events that there is a great variability in their magnitude and we can conclude that there is every chance that there were even bigger events in the past. Furthermore, the longer the period of time, the chances of a storm, flood, volcanic eruption or earthquake being bigger than anything we could imagine occurring today become higher. Volcanic eruptions are a good example: the Mount St Helens eruption in North America in the early 1980s was a devastating event, but compared to eruptions which occurred in the past million years in western North America, it was relatively insignificant: the Bishop Tuff eruption 740 000 years ago resulted in an ash cloud which deposited material over 1000 km away (Izett & Naeser 1976).

24.2 Catastrophic events: meteorite and comet impacts

Collisions of extra-terrestrial bodies with the Earth are frequent events, but most meteorites are very small, and have little effect other than creating a small crater. However, an impact on the Earth of a large meteorite or comet several kilometres across would be a catastrophe in every sense of the word. Hypothetical considerations of the effect of an impact range from knocking the Earth out of its orbit to global wildfires, tsunami affecting all coastlines and the expulsion of huge quantities of particulate matter into the atmosphere, precipitating a darkening of the skies in a 'nuclear winter'. It is unlikely that all life on Earth would be extinguished but the devastation would be immense.

There is a growing body of evidence to suggest that large meteorites have struck the Earth on numerous occasions in the last billion years. The most celebrated of these impacts is one which is implicated in the mass extinctions at the Cretaceous–Tertiary boundary, the *K–T boundary event*. Many different types of organism became extinct at the end of the Cretaceous, including ammonites and dinosaurs. Evidence that a meteorite impact played a role in the demise of these animals comes from anomalously high levels of the heavy metal element iridium in sediments deposited at that time (iridium is rare on the Earth's surface but is relatively abundant in meteorites) and the presence of quartz grains which show signs of a massive impact ('shocked quartz' grains). A similar meteorite or comet has been suggested for the Permian–Triassic boundary when there was an

even more widespread extinction event. There are now several dozen documented meteorite impact sites around the world at different stratigraphic levels with associated evidence of the event from iridium anomalies and shocked quartz, and some geologists consider that these impact events have played a major role in the evolution of life on Earth (Rampino & Caldeira 1992; Van Andel 1994).

24.3 'Catastrophic' events in sedimentation

Events which we consider to be catastrophic in terms of human experience such as earthquakes, volcanic eruptions, storms and fires are important in sedimentological terms. Deposits in almost all environments may result from these events and be preserved in the stratigraphic record. These events occur at a frequency of hundreds to thousands of years so in terms of a geological time-scale they can simply be considered to be *episodic* processes of sedimentation.

24.3.1 Earthquakes

The direct evidence of earthquakes in the stratigraphic record is the presence of faults in rocks. The shock waves of an earthquake also affect unconsolidated sediment, producing convolute bedding, slumps and dewatering structures *(17.1)*. Some depositional events may also be triggered by earthquakes. In continental environments these may include landslides and rock falls adjacent to steep slopes and debris flows on alluvial fans. Reworking of material as a result of an earthquake shock may occur in many submarine environments, including delta fronts and continental slopes. Redeposition occurs by gravity-driven events such as debris flows and turbidity currents. An earthquake off Newfoundland in 1929 triggered what is now thought to have been a series of turbidity currents *(15.2.2)* (Heezen & Drake 1964). Other events associated with earthquakes are tsunami *(11.3.3)* which may travel across oceans before hitting coasts. Tsunami have low amplitudes in deep water but may increase to tens of metres high in shallow water. They cause extensive coastal flooding and redeposit shoreline sediments far inland (Tarbuck & Lutgens 1997).

24.3.2 Volcanoes

At any time there is likely to be a volcano erupting

somewhere around the Earth. However, most of the eruptions are relatively small, producing a steady stream of lava and/or the ejection of small quantities of ash. Periodically there are more violent eruptions which eject many cubic kilometres of ash and volcanic gases. These larger events are recognizable in the stratigraphic record as thicker and more widespread deposits of volcanic ash sometimes occurring in depositional environments many hundreds of kilometres from the site of the eruption (Izett & Naeser 1976). These ash bands are very useful marker horizons as distinctive beds and have the additional benefit of being potentially datable if not too altered. The effects of a single large volcanic eruption may be felt all over the world. Airborne ash and aerosols (droplets of water containing dissolved sulphates and nitrates) from an eruption can be carried high into the atmosphere where they affect the penetration of radiation from the Sun and can result in temporary global cooling. In comparison to volcanic eruption in geological history the recent large eruptions of Mount St Helens in 1980 and Mount Pinatubo in 1991 were relatively small events, involving between 1 and 10 km^3 of eruptive material: eruptions in Yellowstone in the Quaternary are thought to have produced up to 2500 km^3 of ash (Smith & Braile 1994).

24.3.3 Exceptional storms

Tropical storms, typhoons and hurricanes occur every year in certain parts of the world. Every few years there is one of exceptional violence which causes devastation in the areas affected: storm surges cause coastal flooding, strong winds destroy vegetation and high rainfall raises water levels in rivers and lakes, leading to inland flooding and erosion. Evidence of such storms in the stratigraphic record will be layers of sand within overbank mudstone beds in fluvial deposits, and on alluvial fans each bed deposited from a sheetflood or debris flow may have been the result of heavy rainfall triggering the flow. Storm deposits are widespread in shallow sea successions. In lower shoreface deposits storms are the most important depositional process contributing sand beds to otherwise muddy successions. The magnitude of storm events in the past can only be guessed at although it may be assumed that storms of greater violence than any recorded by man have occurred.

24.3.4 Jökulhlaups and dam bursts

The sudden release of water from within a glacier as a jökulhlaup *(7.3.3)* or from behind a temporarily dammed lake can result in catastrophic flooding downstream. A wall of water tens of metres high may sweep away vegetation and sediment in its path, causing widespread erosion. It may also carry large amounts of sediment, depositing a thick layer of material much coarser than the normal sedimentation in the area (Fig. 24.2). In areas peripheral to the Pleistocene ice sheets there are signs of flood events which carried coarse material hundreds of kilometres down river valleys and across floodplains. These deposits are attributed to the release of vast quantities of glacial meltwater following the burst of a

Fig. 24.2 An alluvial fan formed by a sudden dam burst in a stream feeding into the Roaring River in Colorado, USA: a road was swept away by the flow.

moraine dam which may have been many tens of metres high (Baker 1973). Quaternary deposits in north-west Montana have been interpreted as the products of a natural dam burst causing a catastrophic discharge a hundred times greater than the discharge of the Amazon (Baker & Bunker 1985). Such events may therefore be recognized within continental strata by their distinctly coarser character.

24.3.5 Submarine slumps

Bathymetric mapping and sonar images of the ocean floor around volcanic islands such as Hawaii in the Pacific and the Canary Islands in the Atlantic have revealed the existence of very large-scale slumps. Surfaces interpreted as slump scars *(4.8.3)* can be seen on a scale of kilometres, indicating that slumping events have occurred involving several cubic kilometres of rock. Mass movements on this scale would generate huge tsunami *(11.3.3)* around the edges of the ocean, possibly inundating coastal areas with waves tens of metres high travelling far inland. Evidence for these tsunami may come from deposits of marine debris, such as coral and other distinctively marine material, occurring kilometres inland from the shoreline. The scale of coastal devastation caused by submarine slumps would rival the effects of a meteorite landing in an ocean.

24.3.6 Fires

There is very little documented evidence of the effects of fires in the pre-Quaternary geological record. Nichols and Jones (1992) record evidence for a fire in the early Carboniferous of western Ireland. The volume of fusain (fossil charcoal) in estuarine deposits led them to suggest that the fire would have burnt an area of around $95\,000\,km^2$, an area equivalent to the surface of Ireland but reasonable in the context of some modern fires in Siberia and south-east Asia which have covered similar areas. Enhanced surface erosion occurs in the aftermath of a major fire resulting in more sediment entering drainage systems, so there is evidence of a link between wildfire events and sedimentation, as suggested in this Carboniferous case study.

24.4 Life (biosphere) through time

Plants and animals influence most sedimentary environments and in some cases they both control and provide the source for the deposits. It follows that some types of deposit can only have formed during the time that the relevant organisms existed. The two examples below, from marine and continental environments respectively, are two of the more obvious cases of the effects of the evolution of particular groups of organisms on depositional environments.

24.4.1 Carbonate depositional environments through time

The earliest biogenic carbonate deposits on Earth are stromatolites *(3.1.2)* which exhibit microbial structures considered to have been formed by cyanobacteria. These stromatolites have been found in rocks up to 3500 million years old. The number and diversity of stromatolites as the dominant life forms in carbonate-forming environments increased up to about 1000 Ma and then started to decline. This decline is attributed to the development of *metazoans* (multicellular organisms) in the late Precambrian and the 'explosion' of these more advanced forms of life at about 600 Ma. The microbial mats of cyanobacteria were the foodstuff of creatures such as molluscs, arthropods and many others, leading to a suppression of the abundance of cyanobacteria due to 'grazing' by these animals.

The evolution of complex organisms in the Palaeozoic included reef-building organisms which made it possible for carbonate reef structures and the whole depositional setting of a rimmed carbonate shelf to develop for the first time. The structure of reefs on carbonate platforms has varied through the Phanerozoic according to the dominance of different reef-forming organisms through time. Moreover, reef formation appears to have been more common at certain times (Tucker 1992) (Fig. 24.3).

The nature of pelagic deposits in the oceans has also changed considerably as different planktonic organisms have evolved. Coccoliths *(3.1.2)*, the algal structures which make up nannofossil oozes in modern environments above the calcite compensation depth *(15.5)*, did not become common until the middle Jurassic when these algae evolved. Foraminifera, the other main component of modern calcareous oozes, did not develop a planktonic lifestyle until the Cretaceous, although they had been in existence as benthic forms for a long time. Pelagic limestones are consequently much less common before the middle part of the Mesozoic and Palaeozoic pelagic rocks are dominated by siliceous radiolaria which form chert.

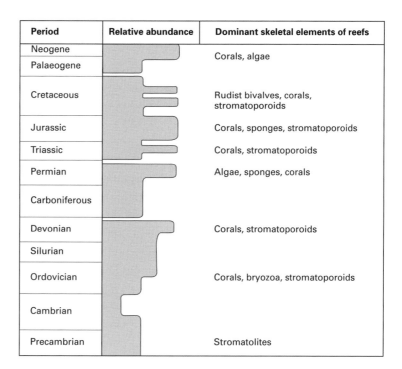

Period	Relative abundance	Dominant skeletal elements of reefs
Neogene		Corals, algae
Palaeogene		
Cretaceous		Rudist bivalves, corals, stromatoporoids
Jurassic		Corals, sponges, stromatoporoids
Triassic		Corals, stromatoporoids
Permian		Algae, sponges, corals
Carboniferous		
Devonian		Corals, stromatoporoids
Silurian		
Ordovician		Corals, bryozoa, stromatoporoids
Cambrian		
Precambrian		Stromatolites

Fig. 24.3 The type and abundance of carbonate reefs has varied through the Phanerozoic. (After Tucker 1992.)

24.4.2 Vegetation through time

Plants are an essential part of most modern continental and marginal marine environments. Plants contribute to soil-forming processes, physical and chemical weathering of bedrock, and influence air flow and overland flow of water. Vegetation cover increases the surface roughness for flows of water and air over the land and reduces erosion. Plant root systems also have a binding effect on soils. However, the nature of the vegetation colonizing the land surface has changed considerably through geological time. Four stages in the development of land plants are significant in terms of sedimentological processes (Fig. 24.4) (Schumm 1968).

1 *Pre-Silurian*: there was no land vegetation at this time and surface run-off, erosion and river bank stability would have been controlled primarily by sediment cohesion; it would be expected that surface erosion was more effective and lateral migration of rivers more rapid in this period.

2 *Silurian to mid-Cretaceous*: the main plant groups during this time were ferns, conifers and lycopods with relatively simple root patterns.

3 *Mid-Cretaceous to mid-Tertiary*: angiosperms

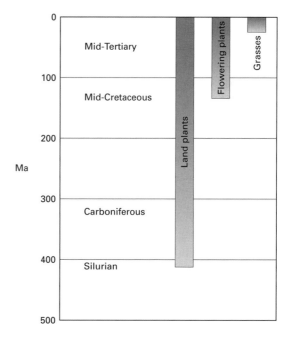

Fig. 24.4 Changes in the types of vegetation through the Phanerozoic. (Data from Schumm 1968.)

(flowering plants) became important in many climatic belts; these shrubs and trees had more complex root systems which were more effective at binding the soil and providing a dense vegetation cover.

4 *Mid-Tertiary to present*: the evolution of grasses meant that there was now a widespread plant type which covered large areas of land surface with a dense fibrous root system which is very effective at binding soil.

These changes in vegetation character through time mean that the uniformitarian approach to the study of fluvial and aeolian sediments needs to be modified. The type and density of present-day vegetation are largely controlled by climatic factors and the impact of cultivation by humans. The natural behaviour of rivers is therefore related to climate and vegetation belts across the continents. Climatic regimes which have relatively stable river channels today may have had much less stable ones in the past because of the absence of vegetation to bind the soil.

A second consequence of vegetation changes though time has been the effect on rates and patterns of weathering and erosion. With reduced plant cover in earlier parts of the Phanerozoic loose detritus on the surface released by weathering processes would have been removed more easily by overland flow of water. Removal of the carapace of detritus exposes more bedrock and enhances the weathering process. It must therefore be assumed that the denudation rates of continental areas of a given relief were generally higher in the past than they are today.

24.5 The atmosphere and air circulation

The composition and behaviour of the Earth's atmosphere is fundamental to most life on the planet and important to both physical and chemical surface processes. The atmosphere has not always been as we know it today: the proportions of the main gases present have evolved and the patterns of air circulation have altered. These changes have influenced chemical processes such as the oxidation of iron compounds, the capacity of the wind to transport sediment and the colonization by plants and animals of both continental and marine environments. A further consequence of changing atmospheric conditions has been the effects on global climate through the history of the Earth *(24.7)*.

24.5.1 Evolution of atmospheric composition

The early atmosphere of the Earth was dominated by

gases which were the products of volcanism, principally nitrogen and carbon dioxide with little if any oxygen. Early life forms were probably akin to the organisms which presently exist around submarine volcanic vents and did not need free oxygen to metabolize. The oxygen content of the atmosphere probably did not exceed a mere trace amount until around 2000 Ma when photosynthetic organisms started to influence the atmospheric composition by taking in carbon dioxide and releasing oxygen. Evidence for a low-oxygen atmosphere prior to 2000 Ma comes from iron-rich strata. Archaean and Early Proterozoic ironstones are typically thin-bedded strata known as *banded iron formations*, consisting of alternations of iron-rich and iron-poor chert beds (Tucker 1991; Einsele 1992). The formation of these bands is attributed to periodic precipitation of iron due to seasonal changes in the oxygen content of the water body (e.g. a lake): ferrous iron in solution would have precipitated if the water came into contact with oxidizing conditions due to blooms of algae or periodic changes in water circulation. This process requires iron to remain in solution for periods of time, a condition which cannot exist with oxygenated waters in an oxygen-rich atmosphere. There are no banded iron formations less than 2000 million years old, and primary ironstones deposited since then are mostly red beds formed in oxidizing continental conditions (Conway-Morris 1992).

By the beginning of the Phanerozoic the atmosphere had reached levels which allowed a diverse animal and plant community to exist in the oceans and shallow seas. Colonization of the land followed, with plants abundant by Carboniferous times. The huge amounts of vegetation represented by Carboniferous coals around the world has led to suggestions that the level of oxygen in the atmosphere may have been higher than the present-day value of 21%. There is, however, a theoretical upper limit of just over 30% atmospheric oxygen as such a rich atmosphere would make vegetation so susceptible to combustion that fires would burn all over the land surface. Carbon dioxide levels have also fluctuated considerably through geological time. Although it makes up only 0.03% of today's atmosphere, changes in concentration are critical because it is a greenhouse gas which strongly influences global climate. It has been calculated that levels of carbon dioxide in the atmosphere may have been between four and 10 times current levels during the mid-Cretaceous (Barron 1992), resulting in a significantly warmer climate during that time (Fig. 24.5).

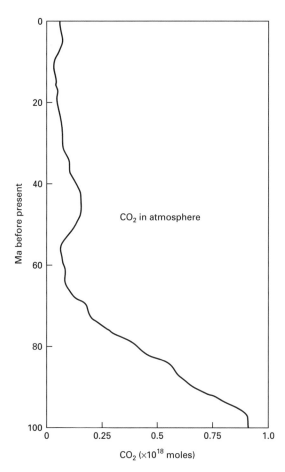

Fig. 24.5 Changes in atmospheric composition through time. (Data from Barron 1992.)

24.5.2 Global air circulation patterns

Air circulation patterns around the globe are driven by the differences in temperature between the cold polar regions and the hot equatorial belt. This basic pattern is modified by the arrangement of large land masses and high mountain ranges. Coastal areas where the wind is predominantly onshore receive high rainfall as the winds have gathered moisture whilst passing over an area of sea. Rainfall is particularly high in mountain areas close to these shores. Winds which have passed over large areas of land lose their moisture content. These wind patterns exert a strong control on the distribution of desert regions. In modern deserts sand blown by the wind is rarely greater than 1 mm in diameter because that is the largest size that the wind can effectively transport. We

might therefore conclude that aeolian sands can never be coarser than 1 mm, but this would not be true if the velocity of winds in the past was higher than it is today because larger grains could be carried in higher-velocity winds *(8.2.1)*.

It has been suggested that the giant draa structures *(8.2.3)* of the Sahara are a relict from the Pleistocene because they appear to be inactive in the present day. These draas may need stronger winds than currently occur in the sub-tropical belt in which the Sahara lies, leading to suggestions that the winds were stronger in the Pleistocene (Glennie 1990). Global wind patterns are controlled by belts of high and low pressure between the hotter equatorial regions and the colder poles. When there is a large ice cap on one or both of the poles, a large area of high pressure builds up. This expanded area of high pressure compresses the pressure belts between the poles and the equator, making areas of high and low pressure closer together. Large pressure gradients result, giving rise to strong winds. In contrast, a weak pressure maximum over the pole, as occurs during warmer, inter-glacial periods, results in lower pressure gradients and weaker winds. It is noteworthy that there is evidence for large draa structures from the Permian of north-west Europe during the time of glaciation in the southern continent of Gondwana (Glennie 1990).

24.6 Plate tectonics through time

The early history of the Earth is poorly represented in the stratigraphic record, with only fragmentary evidence from rocks of the first billion or so years of the planet's history. It is unlikely that the surface of the early Earth bore much resemblance to its present form. Evidence from the Moon suggests that intense meteorite bombardment continued until about 3.8 Ga, and these impacts must have dominated the appearance of the early Earth (Taylor 1992). It is generally thought that the early crust was basaltic in composition and that continental crust formed as a result of differentiation of material at subduction zones. A plate tectonic pattern similar to that seen today, with old cratonic areas forming the cores of continental plates and a relatively small number of major plate boundaries, may have taken hundreds of millions of years to develop.

24.6.1 Global changes in topography and bathymetry

The shape of the continents and oceans has been

controlled by plate tectonics throughout Earth history and every part of the surface of the globe has gone through many changes in that time. Not only have the positions of mountains and lowlands, deep and shallow seas changed through time, but the proportions of different topographic and bathymetric regimes have also changed. There have been periods when the areas of mountains around the world were considerably less than we see today and when large parts of the continents were covered by epicontinental seas.

During the Late Cretaceous the topography on the continents was generally subdued and there were large areas of shallow marine sedimentation with broad shelf areas and extensive epicontinental seas. The amount of terrigenous supply to the seas was much reduced and in the shallow seas there was deposition of carbonate facies. This also coincided with a period of raised global temperatures (Fig. 24.6) enhancing carbonate productivity in the shallow seas. The result was that the Chalk carbonate facies covered vast areas during that period. Ocean circulation, atmospheric circulation and sediment flux into the oceans would have all been quite different from the present-day situation so it is difficult to make direct comparisons between the Late Cretaceous and contemporary environments.

24.6.2 Changes in ocean circulation

The configuration of the tectonic plates around the surface of the Earth affects the patterns of oceanic circulation. These ocean currents distribute cold, nutrient-rich waters from high to low latitudes, affecting biogenic productivity and oceanic and shelf environments. Sea temperature patterns also influence climate belts. Two tectonic events in the mid-Tertiary fundamentally changed the circulation pathways of cold, deep water: the opening of the Drake Passage between South America and the Antarctic Peninsula allowed a current to be set up circling the Antarctic continent; and the convergence between Europe/Asia and the continental masses of Africa, Arabia and India closed the direct connection between the Indian and Atlantic oceans (Van Andel 1994). The present-day relatively mild climate of north-west Europe is a consequence of the ocean currents which bring warm water from the Caribbean area to the northern Atlantic (see Fig. 11.11). In the absence of this current the warm, wet climate in north-west Europe would be replaced by a much colder climate, similar to that experienced in other places of the same latitude such

as Canada. There is speculation that global warming may cause these North Atlantic currents to become weaker, resulting in a colder climate in the coastal parts of north-west Europe.

24.7 Climate through time

The factors which influence global climate, or *climate forcing factors*, are principally the amount of solar radiation which reaches the Earth, the proportion of greenhouse gases in the atmosphere and the physiography of the Earth's surface. The combined effect of these climate forcing factors has been a series of long- and short-term fluctuations in global climate which can be recognized in the stratigraphic record (Frakes 1979).

24.7.1 Astronomical controls on climate

The amount and distribution of solar radiation which reaches the surface is controlled by changes in the orbit of the Earth around the Sun and the axis of rotation of the Earth. These variations are known as Milankovitch cycles *(21.8.6)*, and they result in small changes in global temperature on a relatively short time-scale of tens of thousands of years. In addition, solar luminosity, which is the amount of radiation emitted by the Sun, is variable.

24.7.2 Plate tectonics and palaeoclimates

Tectonics governs global climate in a number of ways. First, there is the relationship between the relief of the continental masses and weathering processes. Chemical weathering processes *(6.5)* involve reactions between silicate minerals in rocks and carbon dioxide, and if there are large surface areas of bedrock exposed to weathering the effect is to remove carbon dioxide from the atmosphere. Orogenic events generate mountain belts where weathering takes place but if large areas of continents have not been uplifted for a long period of time the topography is subdued, and the lower-lying parts of the continents may be marine shelves and epicontinental seas. It is estimated that 17% of the present-day land area was covered by shallow seas in the Cretaceous (Barron 1992) and the general continental topography was very subdued with the consequence that more carbon dioxide was free in the atmosphere instead of being taken up by weathering reactions.

There is also evidence that the world climate has been

strongly influenced by the collision of India with Asia, and more specifically the formation of the Tibetan Plateau (Ruddiman & Kutzbach 1990). Uplift of the Tibetan Plateau to its present position of 4700 m above sea level occurred during the mid- to late Miocene. One consequence of the formation of a large area at a high altitude is an increase in the total amount of weathering. As weathering processes involve carbon dioxide from the atmosphere an increase in these processes could result in a reduction in carbon dioxide levels in the atmosphere. This may have caused a global cooling as the greenhouse effect of carbon dioxide was reduced. Increasing the area of high-altitude land also increases the surface covered by ice and snow. A white land surface has a higher *albedo* (reflectivity), so more of the Sun's energy is reflected back instead of being absorbed.

A further consequence of the formation of the Tibetan Plateau would have been a disruption of the global atmospheric circulation by interrupting the west to east air flow across the continent. The present pattern of the monsoon circulation over the Indian Ocean has probably developed since the uplift in the Himalayas and Tibet. Monsoon winds and rains strongly influence the present-day climate in all the lands which border the Indian Ocean.

24.7.3 Global climate and atmospheric composition

Much of the radiation which reaches the land and sea surface is reflected back out into space unless there are molecules of gases in the atmosphere which act to limit the transmission of this reflected radiation. These are the *greenhouse gases*, principally carbon dioxide and methane. The main inorganic natural source of carbon dioxide is from volcanic eruptions so the amount of volcanicity occurring around the world at any time is a primary control on the global climate. The volcanism is itself determined by plate tectonics as the total length and rate of spreading at mid-ocean ridges where volcanism occurs has changed through geological time *(21.8.3)*. Since the evolution of plant and animal life the carbon dioxide content of the atmosphere has also been strongly influenced by photosynthesis removing the gas from the air and respiration adding to the carbon dioxide concentration. The burning of vegetation and hydrocarbons, either naturally or anthropogenically, also increases the proportion of this greenhouse gas in the atmosphere.

24.7.4 Global climate cycles

The combination of these factors which influence global climate has resulted in periods of Earth history when warm (*greenhouse*) conditions generally prevailed alternating with times when cold (*icehouse*) conditions prevailed (Fig. 24.6). *Palaeotemperatures* can be determined from the oxygen isotopes in the carbonate shells of organisms such as foraminifera because the ratio of oxygen isotopes is dependent on the temperature of the water from which the carbonate was precipitated (Chappell & Shackleton 1986; Matthews 1986). Other indicators of palaeoclimate include sedimentary deposits, such as the extent of evaporite beds suggesting warm conditions and glacial tills during cold periods. The distribution of flora and fauna in relation to the palaeolatitude

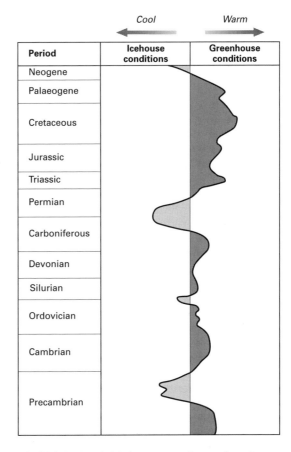

Fig. 24.6 Cycles of global temperature changes alternating between periods of 'greenhouse' and 'icehouse' conditions. (After Doyle *et al.* 1994. Copyright John Wiley & Sons Limited. Reproduced with permission.)

can also be a useful indicator of climate: for example, fossils of alligators in Palaeocene rocks deposited inside the Arctic Circle suggest that conditions in the Arctic then were not as they are today.

Working back from the present day, the Quaternary is part of an icehouse stage which has existed since the mid-Tertiary when extensive ice caps started to occur at the poles. Short-term (Milankovitch) variations in temperature resulted in glacial and interglacial periods during the Pleistocene. The global temperature has generally been falling since the mid-Cretaceous when it may have been at least 5°C and possibly 10°C warmer than it is today. The high temperatures in the mid-Cretaceous are also suggested by the preservation of the remains of relatively delicate plants at very high palaeo-latitudes in both the Arctic and Antarctic regions, making it unlikely that there were permanent ice caps at this time. Dropstones have been reported from marine deposits of this age, but it may be that these came from isolated, seasonal glaciers or were rafted out to sea on tree trunks (Barron 1992).

Icehouse conditions existed towards the end of the Palaeozoic when the Gondwana supercontinent, centred over the south pole, experienced extensive glaciation. Late Carboniferous and Permian strata from many parts of Gondwanaland include glacial tills, particularly in the Karoo basins of southern Africa. The Gondwana glaciation came at the end of a long period of globally warm conditions apart from a brief period in late Ordovician to early Silurian times when there was apparently quite widespread glaciation and the deposition of glacial facies. Before that episode the previous icehouse period had been in the late Proterozoic from 950 to 615 Ma, during which time there were glacial conditions in many parts of the world. The record of palaeoclimate any further back in Earth history is patchy because the evidence from fossils is largely absent and the preservation of sedimentary strata is very fragmentary.

24.8 Planetary behaviour through time

We live on a planet which has an equatorial radius of 6378 km, orbits a star, the Sun, at a distance of 150 million kilometres, and has a single moon orbiting at a distance of 384 000 km. The orbit of the Earth around the Sun takes about 31.5 million seconds, the Moon completes its orbit of the Earth every 2.4 million seconds and the globe rotates on its axis every 86 400 s. The time periods are quoted in seconds because 'day', 'month' and 'year' all have astronomical connotations and the con-

stancy of all these phenomena through geological time can be questioned.

The most convincing evidence for changes in the astronomical behaviour of the Earth–Moon–Sun system comes from considerations of the rate of rotation of the Earth about its axis. Careful examination of well-preserved Devonian corals has indicated that they have daily and monthly growth rings which show patterns consistent with a year of around 400 days (Van Andel 1994). If this evidence is correct then there were more days in the year in the Devonian. If the time taken for the Earth to orbit the Sun is assumed to be constant, the rate of rotation of the Earth must have been faster then and a day would have lasted for 78 750 s instead of 86 400 s. The cause of this decreasing rate of rotation is tidal friction, movements which cause frictional loss of energy within the body of the Earth (Earth tides) due to the gravitational attraction of the Earth–Moon system.

24.9 Gaps in the record

The stratigraphic record provides us with only a few clues with which to build up a history of the Earth. The record is incomplete on all scales, from the years in between the events when layers of sediment are deposited, to hiatuses within successions representing hundreds or thousands of years, to the millions of years represented by major unconformities.

24.9.1 Time represented by bedding planes

There are few situations where sedimentation is a continuous process. Pelagic deposition in the deep oceans is relatively constant but this slow accumulation contributes less to the total volume of material than turbidites which occur every few hundred or thousand years. A bed deposited by a single turbidity current may be decimetres or even metres thick but takes only a matter of hours to form (see Figs 4.26 & 4.27). This episodic character of sedimentation is also apparent in storm deposits on the shelf and in floodplain deposits on land. Even deposition within the channel of a river, delta or estuary is not continuous because sand accumulates by the migration of ripples and subaqueous dunes, which only occurs when the current is at a suitable velocity.

This episodic character of sedimentation means that in almost all clastic sedimentary environments more time is represented by the surfaces between beds, the bedding planes, than the beds themselves (Tipper 1987). In carbonate environments formed by biochemical build-up by organisms such as corals and algae, deposition can be considered to be more continuous and subaqueous evaporite deposition in lakes and lagoons continues provided that the conditions remain the same. However, fluctuations in the environmental conditions due to changes in climate and sea level mean that even chemical and biogenic sedimentation may occur at very variable rates, including periods of non-deposition.

24.9.2 Longer time gaps in a stratigraphic succession

Some bedding surfaces show evidence of having been sites of non-deposition. In the subaerial environment well-developed palaeosols form given sufficient time *(9.7.2)* indicating that the top of the soil profile was constantly exposed. If the rate of sedimentation on the floodplain is relatively high the soil profile does not reach maturity before a new soil forms over the top of it. In shallow seas a number of features may indicate non-deposition. Glauconite is an authigenic mineral which forms at the sea bed *(11.6.1)* and is concentrated on surfaces where clastic and carbonate sediment accumulation is very slow. Phosphates *(11.6.2)* may also form under the same conditions. Cementation of the sea bed starts to occur if sedimentation rates are slow. A moderately cemented sea bed, a 'firmground', will preserve the effects of burrowing organisms as an intensely bioturbated bed; the ichnofacies found in firmgrounds is typically *Glossifungites (11.7.2)*. Where complete lithification occurs a 'hardground' is formed and only boring organisms of the ichnofacies *Trypanites (11.7.2)* occur. The intensity and type of bioturbation are therefore important indicators of breaks in shallow marine sedimentation.

Detailed biostratigraphic analysis can sometimes reveal the presence of significant time gaps in a stratigraphic succession. The absence of a zone fossil may indicate that there are no deposits which represent the time period during which that organism existed. A time gap in the record is a *hiatus* in sedimentation, also sometimes referred to as a *lacuna* (plural *lacunae*). A lacuna may be a result of a period of non-deposition or may be due to erosion of sediment. Erosion of the sea floor can occur as a consequence of a relative sea level fall (a sequence boundary) or during transgression (a ravinement surface). In either case there may be evidence of erosion within the strata and a sharp facies change across the surface.

24.9.3 Major gaps in the record

Most of the surface of the Earth has undergone repeated tectonic changes during the Phanerozoic. Ocean floors have been repeatedly created and destroyed (the oldest ocean floor is Mesozoic), mountain belts have formed and then worn down, and the continental margins have undergone uplift and subsidence. Only the stable cratonic areas in the middle of large continental plates have remained largely unaffected by these tectonic forces. Uplift of strata in an orogenic belt exposes the bedrock to erosion and the removal of material to a thickness of hundreds or thousands of metres. Subsequent subsidence and a cover of younger strata preserve an angular unconformity which may have a time gap of tens to hundreds of millions of years across it (Fig. 24.7). No record of the

Fig. 24.7 An unconformity in western Madagascar between Precambrian metamorphic rocks and Permian sandstone beds representing a time gap of at least 300 million years in the stratigraphic record.

strata which were removed will remain except for clasts in younger sediment which provide some indirect clues. There will also be no direct record of the events in that area whilst erosion occurred. Nowhere is there a complete record of sedimentation since the Palaeozoic without major time gaps due to unconformities or long periods of non-deposition in the cratonic centres.

24.10 The sedimentary record

Sedimentology equips us with the tools to interpret rocks in terms of processes and environments. Whilst an understanding of present-day sedimentary environments provides the key to interpreting sedimentary rocks, the uniformitarian approach must be employed with care, bearing in mind the 'catastrophic' events which have undoubtedly been important. These depositional environments are considered through geological time in the context of stratigraphic principles and analytical techniques. Sedimentological and stratigraphic analysis of the rock record using the material exposed at the surface,

drilled and surveyed in the subsurface, provides us with the means to reconstruct the history of the surface of the Earth as far as the data allow. However, the information we have is very fragmentary, with more gaps than hard data, and there will always be room to improve our knowledge and interpretation of the sedimentary record.

Further reading

Brown, G.C., Hawkesworth, C.J. & Wilson, R.C.L. (eds) (1992) *Understanding the Earth: A New Synthesis*, 551 pp. Cambridge University Press, Cambridge.
Goudie, A. (1992) *Environmental Change*, 3rd edn, 329 pp. Clarendon Press, Oxford.
Hallam, A. (1990) *Great Geological Controversies*, 322 pp. Oxford University Press, Oxford.
Nisbet, E.G. (1991) *Living Earth: A Short History of Life and Its Home*, 286 pp. Harper Collins, New York.
Pickering, K.T. & Owen, L.A. (1994) *An Introduction to Global Environmental Issues*, 390 pp. Routledge, London.
Van Andel, T.H. (1994) *New Views on an Old Planet: A History of Global Change*, 439 pp. Cambridge University Press, Cambridge.

References

Adams, A.E., MacKenzie, W.S. & Guilford, C. (1984) *Atlas of Sedimentary Rocks under the Microscope*, 104 pp. Longman, Harlow.

Aigner, T. (1985) *Storm Depositional Systems*, 174 pp. Lecture Notes in Earth Sciences **3**. Springer-Verlag, Berlin.

Allen, J.R.L. (1964a) Primary current lineation in the Lower Old Red Sandstone (Devonian), Anglo-Welsh Basin. *Sedimentology* **3**, 89–108.

Allen, J.R.L. (1964b) Studies in fluviatile sedimentation. Six cyclothems from the Lower Old Red Sandstone, Anglo-Welsh Basin. *Sedimentology* **3**, 163–198.

Allen, J.R.L. (1965) The sedimentation and palaeogeography of the Old Red Sandstone of Anglesey, North Wales. *Proceedings of the Yorkshire Geological Society* **35**, 139–185.

Allen, J.R.L. (1966) On bedforms and palaeocurrents. *Sedimentology* **6**, 153–190.

Allen, J.R.L. (1968) *Current Ripples*, 433 pp. North-Holland, Amsterdam.

Allen, J.R.L. (1970a) *Physical Processes of Sedimentation*, 268 pp. Allen & Unwin, London.

Allen, J.R.L. (1970b) A quantitative model of grain size and sedimentary structures in lateral deposits. *Geological Journal* **7**, 129–146.

Allen, J.R.L. (1972) A theoretical and experimental study of climbing ripple cross lamination with a field application to the Uppsala esker. *Geografiska Annaler* **53A**, 157–187.

Allen, J.R.L. (1974) Studies in fluviatile sedimentation: implications of pedogenic carbonate units, Lower Old Red Sandstone, Anglo-Welsh outcrop. *Geological Journal* **9**, 181–208.

Allen, J.R.L. (1982) *Sedimentary Structures: Their Character and Physical Basis.* Vol. 1. *Developments in Sedimentology*, 593 pp. Elsevier, Amsterdam.

Allen, J.R.L. (1983) Studies in fluviatile sedimentation: bars, bar complexes and sandstone sheets (low sinuosity braided streams) in the Brownstones (L. Devonian), Welsh border. *Sedimentary Geology* **33**, 237–293.

Allen, J.R.L. (1985) *Principles of Physical Sedimentology*, 272 pp. Unwin Hyman, London.

Allen, J.R.L. (1994) Fundamental properties of fluids and their relation to sediment transport processes. In: *Sediment Transport and Depositional Processes* (ed. K. Pye), pp.

25–60. Blackwell Scientific Publications, Oxford.

Allen, P.A. (1997) *Earth Surface Processes*, 404 pp. Blackwell Science, Oxford.

Anderton, R. (1985) Clastic facies models and facies analysis. In: *Sedimentology, Recent Developments and Applied Aspects* (eds P.J. Brenchley & B.P.J. Williams), pp. 31–47. Blackwell Scientific Publications, Oxford.

Anderton, R. (1995) Sequences, cycles and other nonsense: are submarine fan models any use in reservoir geology? In: *Characterization of Deep Marine Clastic Systems* (eds A.J. Hartley & D.J. Prosser), pp. 5–11. Geological Society of London Special Publication **94**.

Anderton, R., Bridges, P.H., Leeder, M.R. & Sellwood, B.W. (1982) *A Dynamic Stratigraphy of the British Isles: A Study in Crustal Evolution,* 301 pp. Unwin Hyman, London.

Astin, T.R. (1991) Subaqueous shrinkage or syneresis cracks in the Devonian of Scotland reinterpreted. *Journal of Sedimentary Petrology* **61**, 850–859.

Augustinus, P.G.E.F. (1989) Cheniers and chenier plains: a general introduction. *Marine Geology* **90**, 219–229.

Bagnold, R.A. (1946) Motion of waves in shallow water: interaction between waves and shallow bottoms. *Proceedings of the Royal Society, London* **187A**, 1–18.

Bagnold, R.A. (1954) *The Physics of Blown Sand and Desert Dunes*, 2nd edn, 265 pp. Chapman & Hall, London.

Baker, V.R. (1973) *Palaeohydrology and Sedimentology of Lake Missoula Flooding of Eastern Washington*, 79 pp. Geological Society of America Special Paper **144**.

Baker, V.R. & Bunker, R.C. (1985) Cataclysmic late Pleistocene flooding from glacial Lake Missoula: a review. *Quaternary Science Reviews* **4**, 1–41.

Barron, E.J. (1992) Paleoclimatology. In: *Understanding the Earth, A New Synthesis* (eds G.C. Brown, C.J. Hawkesworth & R.C.L. Wilson), pp. 485–505. Cambridge University Press, Cambridge.

Bathurst, R.G.C. (1987) Diagenetically enhanced bedding in argillaceous platform limestones: stratified cementation and selective compaction. *Sedimentology* **34**, 749–778.

Beaty, C.B. (1970) Age and estimated rate of accumulation of an alluvial fan, White Mountains, California, USA. *American Journal of Science* **268**, 50–77.

Beaudry, D. & Moore, G.F. (1985) Seismic stratigraphy and Cenozoic evolution of West Sumatra forearc basin.

American Association of Petroleum Geologists Bulletin **69**, 742–759.

Beaumont, C. (1981) Foreland basins. *Geophysical Journal of the Royal Astronomical Society* **65**, 291–329.

Berner, R.A. (1971) *Principles of Chemical Sedimentology*, 256 pp. McGraw-Hill, New York.

Berner, R.A. (1981) New geochemical classification of sedimentary environments. *Journal of Sedimentary Petrology* **51**, 359–365.

Bernoulli, D. & Jenkyns, H.C. (1974) Alpine, Mediterranean and Central Atlantic Mesozoic facies in relation to the early evolution of the Tethys. In: *Modern and Ancient Geosynclinal Sedimentation* (eds R.H. Dott & R.H. Shaver), pp. 129–160. Society of Economic Paleontologists and Mineralogists Special Publication **22**.

Bhattacharya, J.P. & Walker, R.G. (1992) Deltas. In: *Facies Models: Response to Sea Level Change* (eds R.G. Walker & N.P. James), pp. 157–178. Geological Association of Canada, St Johns, Newfoundland.

Blair, T.C. & McPherson, J.G. (1994) Alluvial fans and their natural distinction from rivers based on morphology, hydraulic processes, sedimentary processes and facies assemblages. *Journal of Sedimentary Research* **A64**, 450–589.

Blatt, H. (1985) Provenance studies and mudrocks. *Journal of Sedimentary Petrology* **55**, 69–75.

Blatt, H., Middleton, G.V. & Murray, R.C. (1980) *Origin of Sedimentary Rocks*, 2nd edn, 782 pp. Prentice Hall, Englewood Cliffs, NJ.

Boothroyd, J.C. (1985) Tidal inlets and tidal deltas. In: *Coastal Sedimentary Environments* (ed. R.A. Davis), pp. 445–533. Springer-Verlag, Berlin.

Boothroyd, J.C. & Ashley, G.M. (1975) Process, bar morphology and sedimentary structures on braided outwash fans, Northeastern Gulf of Alaska. In: *Glaciofluvial and Glaciomarine Sedimentation* (eds A.V. Jopling & B.C. McDonald), pp. 193–222. Society of Economic Paleontologists and Mineralogists Special Publication **23**.

Boothroyd, J.C. & Nummedal, D. (1978) Proglacial braided outwash: a model for humid alluvial fan deposits. In: *Fluvial Sedimentology* (ed. A.D. Miall), pp. 641–668. Canadian Society of Petroleum Geologists Memoir **5**.

Bornhold, B.D. & Giresse, P. (1985) Glauconitic sediments on the continental shelf off Vancouver Island, British Columbia, Canada. *Journal of Sedimentary Petrology* **55**, 653–664.

Bosscher, H. & Schlager, W. (1992) Computer simulation of reef growth. *Sedimentology* **39**, 503–512.

Boulton, G.S. (1972) Modern Arctic glaciers as depositional models for former ice sheets. *Journal of the Geological Society of London* **127**, 361–393.

Bouma, A.H. (1962) *Sedimentology of Some Flysch Deposits. A Graphic Approach to Facies Interpretation*, 168 pp. Elsevier, Amsterdam.

Bouma, A.H. (1972) Fossil contourites in the lower Niesen-flysch, Switzerland. *Journal of Sedimentary Petrology* **42**, 917–921.

Bown, T.M. & Kraus, M.J. (1987) Integration of channel and overbank suites, I. Depositional sequences and lateral relations of alluvial paleosols. *Journal of Sedimentary Petrology* **57**, 587–601.

Boyd, R., Dalrymple, R.W. & Zaitlin, B.A. (1992) Classification of coastal depositional environments. *Sedimentary Geology* **80**, 139–150.

Bridge, J.S. (1993) The interaction between channel geometry, water flow, sediment transport and deposition in braided rivers. In: *Braided Rivers* (eds J.L. Best & C.S. Bristow), pp. 13–72. Geological Society Special Publication **75**.

Bridge, J.S. & Leeder, M.R. (1979) A simulation model of alluvial stratigraphy. *Sedimentology* **26**, 617–644.

Bristow, C.S. (1987) Brahmaputra river: channel migration and deposition. In: *Recent Developments in Fluvial Sedimentology* (eds F.G. Ethridge, R.M. Florez & M.D. Harvey), pp. 63–74. Society of Economic Paleontologists and Mineralogists Special Publication **39**.

Brookfield, M.E. (1992) Eolian systems. In: *Facies Models: Response to Sea Level Change* (eds R.G. Walker & N.P. James), pp. 143–156. Geological Association of Canada, St Johns, Newfoundland.

Bull, W.B. (1972) Recognition of alluvial fan deposits in the stratigraphic record. In: *Recognition of Ancient Sedimentary Environments* (eds J.K. Rigby & W.K. Hamblin), pp. 63–83. Society of Economic Paleontologists and Mineralogists Special Publication **16**.

Burke, K. (1976) Development of graben associated with the initial ruptures of the Atlantic Ocean. In: *Sedimentary Basins of Continental Margins and Cratons* (ed. M.H.P. Bott). *Tectonophysics* **36**, 93–112.

Burley, S.D., Kantorowicz, J.D. & Waugh, B. (1985) Clastic diagenesis. In: *Sedimentology, Recent Developments and Applied Aspects* (eds P.J. Brenchley & B.P.J. Williams), pp. 189–228. Blackwell Scientific Publications, Oxford.

Busby, C. & Ingersoll, R.V. (eds) (1995) *Tectonics of Sedimentary Basins*, 580 pp. Blackwell Science, Oxford.

Calvert, S.E. (1974) Deposition and diagenesis of silica in marine sediments. In: *Pelagic Sediments: On Land and Under the Sea* (eds K.J. Hsü & H.C. Jenkyns), pp. 273–299. International Association of Sedimentologists Special Publication **1**.

Cant, D.J. (1982) Fluvial facies models. In: *Sandstone Depositional Environments* (eds P.A. Scholle & D. Spearing), pp. 115–138. American Association of Petroleum Geologists Memoir **31**.

Cant, D.J. (1992) Subsurface facies analysis. In: *Facies Models: Response to Sea Level Change* (eds R.G. Walker & N.P. James), pp. 27–46. Geological Association of Canada, St Johns, Newfoundland.

Cas, R.A.F. & Wright, J.V. (1987) *Volcanic Successions: Modern and Ancient*, 528 pp. Unwin Hyman, London.

Chappell, J. & Shackleton, N.J. (1986) Oxygen isotopes and sea level. *Nature* **324**, 137–140.

Christie-Blick, N. & Biddle, K.T. (1985) Deformation and basin formation along strike-slip faults. In: *Strike-Slip Deformation, Basin Formation and Sedimentation* (eds K.T. Biddle & N. Christie-Blick), pp. 1–34. Society of Economic Paleontologists and Mineralogists Special Publication **37**.

Clarkson, E.N.K. (1993) *Invertebrate Palaeontology and Evolution*, 3rd edn, 323 pp. Allen & Unwin, London.

Cloetingh, S. (1988) Intraplate stresses: a tectonic cause for third-order cycles in apparent sea level? In: *Sea Level Changes: An Integrated Approach* (eds C.K. Wilgus, B.S. Hastings, C.G.St.C. Kendall, H.W. Posamentier, C.A. Ross & J.C. Van Wagoner), pp. 19–29. Society of Economic Paleontologists and Mineralogists Special Publication **42**.

Coleman, J.M. (1969) Brahmaputra River: channel processes and sedimentation. *Sedimentary Geology* **3**, 129–239.

Coleman, J.M., Prior, D.B. & Lindsay, J.F. (1983) Deltaic influences on shelf edge instability processes. In: *The Shelfbreak: Critical Interface on Continental Margins* (eds D.J. Stanley & G.T. Moore), pp. 121–137. Society of Economic Paleontologists and Mineralogists Special Publication **33**.

Coleman, R.G. (1993) *Geologic Evolution of the Red Sea*, 186 pp. Oxford Monographs on Geology and Geophysics **24**. Oxford University Press, Oxford.

Collinson, J.D. (1969) The sedimentology of the Grindslow Shales and the Kinderscout Grit: a deltaic complex in the Namurian of northern England. *Journal of Sedimentary Petrology* **39**, 194–221.

Collinson, J.D. (1986) Alluvial sediments. In: *Sedimentary Environments and Facies* (ed. H.G. Reading), pp. 20–62. Blackwell Scientific Publications, Oxford.

Collinson, J.D. (1996) Alluvial sediments. In: *Sedimentary Environments: Processes, Facies and Stratigraphy* (ed. H.G. Reading), pp. 37–82. Blackwell Science, Oxford.

Collinson, J.D. & Thompson, D.B. (1982) *Sedimentary Structures*, 194 pp. Allen & Unwin, London.

Coniglio, M. & Dix, G.R. (1992) Carbonate slopes. In: *Facies Models: Response to Sea Level Change* (eds R.G. Walker & N.P. James), pp. 349–373. Geological Association of Canada, St Johns, Newfoundland.

Conkin, B.M. & Conkin, J.E. (eds) (1984) *Stratigraphy: Foundations and Concepts*, 365 pp. Van Nostrand Reinhold, New York.

Conway-Morris, S. (1992) The early evolution of life. In: *Understanding the Earth, A New Synthesis* (eds G.C. Brown, C.J. Hawkesworth & R.C.L. Wilson), pp. 436–457. Cambridge University Press, Cambridge.

Cox, K.G., Price, N.B. & Harte, B. (1974) *The Practical Study of Crystals, Minerals and Rocks*, 245 pp. McGraw-Hill, London.

Curtis, C.D. (1977) Sedimentary geochemistry: environments and processes dominated by involvement in an aqueous phase. *Philosophical Transactions of the Royal Society, London* **A286**, 353–371.

Dalrymple, G.B. (1991) *The Age of the Earth*, 401 pp. Stanford University Press, Palo Alto, CA.

Dalrymple, R.W. (1992) Tidal depositional systems. In: *Facies Models: Response to Sea Level Change* (eds R.G. Walker & N.P. James), pp. 195–218. Geological Association of Canada, St Johns, Newfoundland.

Dalrymple, R.W., Zaitlin, B.A. & Boyd, R. (1992) Estuarine facies models: conceptual basis and stratigraphic implications. *Journal of Sedimentary Petrology* **62**, 1130–1146.

DeCelles, P.G. (1988) Lithologic provenance modelling applied to the Late Cretaceous synorogenic Echo Canyon Conglomerate, Utah: a case of multiple source areas. *Geology* **16**, 1039–1043.

Demaison, G.J. & Moore, G.T. (1980) Anoxic environments and oil source bed genesis. *American Association of Petroleum Geologists Bulletin* **64**, 1179–1209.

De Paolo, D.J. & Ingram, B.L. (1985) High resolution stratigraphy with strontium isotopes. *Science* **217**, 938–941.

Dickinson, W.R. (1974) Sedimentation within and beside ancient and modern magmatic arcs. In: *Modern and Ancient Geosynclinal Sedimentation* (eds R.H. Dott & R.H. Shaver), pp. 1–27. Society of Economic Paleontologists and Mineralogists Special Publication **22**.

Dickinson, W.R. (1975) Potash-depth (K-h) relations in continental margin and intra-oceanic magmatic arcs. *Geology* **3**, 53–56.

Dickinson, W.R. (1980) Plate tectonics and key petrologic associations. In: *The Continental Crust and Its Mineral Deposits* (ed. D.W. Strangway), pp. 341–360. Geological Association of Canada Special Paper **20**.

Dickinson, W.R. (1995) Forearc basins. In: *Tectonics of Sedimentary Basins* (eds R.V. Ingersoll & C.J. Busby), pp. 221–262. Blackwell Science, Oxford.

Dickinson, W.R. & Seely, D.R. (1979) Structure and stratigraphy of fore-arc regions. *American Association of Petroleum Geologists Bulletin* **63**, 2–31.

Dickinson, W.R. & Suczek, C.A. (1979) Plate tectonics and sandstone composition, *American Association of Petroleum Geologists Bulletin* **63**, 2164–2182.

Dott, R.H., Jr & Bourgeois, J. (1982) Hummocky cross stratification: significance of its variable bedding sequences. *Geological Society of America Bulletin* **93**, 663–680.

Doveton, J.H. (1994) *Geologic Log Interpretation*, 169 pp. Society of Economic Paleontologists and Mineralogists Short Course Notes **29**.

Doyle, P., Bennett, M.R. & Baxter, A.N. (1994) *The Key to Earth History: An Introduction to Stratigraphy*, 231 pp. Wiley, Chichester.

Drewry, D.D. (1986) *Glacial Geologic Processes*, 276 pp. Arnold, London.

Duff, P.McL.D. (ed.) (1993) *Holmes' Principles of Physical Geology*, 4th edn, 791 pp. Chapman & Hall, London.

Dunham, R.J. (1962) Classification of carbonate rocks according to depositional texture. In: *Classification of Carbonate Rocks* (ed. W.E. Ham), pp. 108–121. American Association of Petroleum Geologists Memoir **1**.

Eicher, D.L. (1976) *Geologic Time*, 2nd edn, 150 pp. Prentice Hall, Englewood Cliffs, NJ.

Einsele, G. (1992) *Sedimentary Basins, Evolution, Facies and Sediment Budget*, 628 pp. Springer-Verlag, Berlin.

Ekdale, A.A., Bromley, R.G. & Pemberton, S.G. (1984) *Ichnology: The Use of Trace Fossils in Sedimentology and Stratigraphy*, 317 pp. Society of Economic Paleontologists and Mineralogists Short Course Notes **15**.

Eldredge, N. & Gould, S.J. (1972) Punctuated equilibria: an alternative to phyletic gradualism. In: *Models in Paleobiology* (ed. T.J.M. Schopf), pp. 82–115. Freeman, Cooper, San Francisco, CA.

Elliott, T. (1976) Abandonment facies of high-constructive lobate deltas, with an example from the Yoredale Series, *Proceedings of the Geological Association* **85**, 359–365.

Elliott, T. (1986a) Deltas. In: *Sedimentary Environments and Facies* (ed. H.G. Reading), pp. 113–154. Blackwell Scientific Publications, Oxford.

Elliott, T. (1986b) Siliciclastic shorelines. In: *Sedimentary Environments and Facies* (ed. H.G. Reading), pp. 155–188. Blackwell Scientific Publications, Oxford.

Elliott, T. (1989) Deltaic systems and their contribution to an understanding of basin-fill successions. In: *Deltas: Sites and Traps for Fossil Fuels* (eds M.K.G. Whateley & K.T. Pickering), pp. 3–10. Geological Society Special Publication **41**.

Embry, A.F. & Klovan, J.E. (1971) A late Devonian reef tract on north-eastern Banks Island, Northwest Territories. *Bulletin of Canadian Petroleum Geology* **19**, 730–781.

Emery, D. & Myers, K.J. (eds) (1996) *Sequence Stratigraphy*, 297 pp. Blackwell Science, Oxford.

Eugster, H.P. (1985) Oil shales, evaporites and ore deposits. *Geochimica et Cosmochimica Acta* **49**, 619–635.

Eugster, H.P. & Hardie, L.A. (1978) Saline lakes. In *Lakes: Chemistry, Geology, Physics* (ed. A. Lerman), pp. 237–293. Springer-Verlag, Berlin.

Eyles, N. & Eyles, C.H. (1992) Glacial depositional systems. In: *Facies Models: Response to Sea Level Change* (eds R.G. Walker & N.P. James), pp. 73–100. Geological Association of Canada, St Johns, Newfoundland.

Faure, G. (1986) *Principles of Isotope Geology*, 2nd edn, 589 pp. Wiley, New York.

Folk, R.L. (1959) Practical petrographic classification of limestones. *American Association of Petroleum Geologists Bulletin* **43**, 1–38.

Folk, R.L. (1962) Spectral subdivision of limestone types. In: *Classification of Carbonate Rocks* (ed. W.E. Ham), pp. 62–84. American Association of Petroleum Geologists Memoir **1**.

Folk, R.L. (1974) *Petrology of Sedimentary Rocks*, 159 pp. Hemphill, Austin, TX.

Frakes, L.A. (1979) *Climate Through Geologic Time*, 310 pp. Elsevier, Amsterdam.

Francis, P.W., Glaze, L.S., Pieri, D., Oppenheimer, C.M.M. & Rothery, D.A. (1990) Eruption terms. *Nature* **346**, 519.

Freund, R. (1970) Plate tectonics of the Red Sea and Africa. *Nature* **228**, 453.

Friedman, G.M., Sanders, J.E. & Kopaska-Merkel, D.C. (1992) *Principles of Sedimentary Deposits: Stratigraphy and Sedimentology*, 717 pp. Macmillan, New York.

Fritz, W.J. & Moore, J.N. (1988) *Basics of Physical Stratigraphy and Sedimentology*, 371 pp. Wiley, New York.

Froelich, P.N., Bender, M.L., Luedtke, N.A., Heath, G.R. & De Vries, T. (1982) The marine phosphorus cycle. *American Journal of Science* **282**, 474–511.

Galloway, W.E. (1975) Process framework for describing the morphologic and stratigraphic evolution of deltaic depositional systems. In: *Deltas* (ed. M.L. Broussard), pp. 87–98. Houston Geological Society.

Galloway, W.E. (1989) Genetic stratigraphic sequences in basin analysis 1: Architecture and genesis of flooding-bounded depositional units. *American Association of Petroleum Geologists Bulletin* **73**, 125–142.

Gass, I.G. (1982) Ophiolites. *Scientific American* **247**, 122–131.

Gilbert, G.K. (1885) The topographic features of lake shores. *Annual Report of the US Geological Survey* **5**, 75–123.

Glennie, K.W. (1990) Early Permian–Rotliegend. In: *Introduction to the Petroleum Geology of the North Sea* (ed. K.W. Glennie), 3rd edn, pp. 120–152. Blackwell Scientific Publications, Oxford.

Hailwood, E.A. (1989) *Magnetostratigraphy*, 84 pp. Geological Society of London Special Report **19**.

Hallam, A. (1963) Major epeirogenic and eustatic changes since the Cretaceous and their possible relationship to crustal structure. *American Journal of Science* **261**, 164–177.

Hallam, A. (1990) *Great Geological Controversies*, 322 pp. Oxford University Press, Oxford.

Hamblin, A.P. & Walker, R.G. (1979) Storm-generated shallow marine deposits: the Fernie–Kootenay (Jurassic) transition, southern Rocky Mountains. *Canadian Journal of Earth Science* **16**, 1673–1690.

Handford, C.R. (1981) A process-sedimentary framework for characterising recent and ancient sabkhas. *Sedimentary Geology* **30**, 255–265.

Haq, B.U., Hardenbol, J. & Vail, P.R. (1987) Chronology of fluctuating sea levels since the Triassic. *Science* **235**, 1156–1167.

Haq, B.U., Hardenbol, J. & Vail, P.R. (1988) Mesozoic and Cenozoic chronostratigraphy and cycles of sea level change. In: *Sea Level Changes: An Integrated Approach* (eds C.K. Wilgus, B.S. Hastings, C.G.St.C. Kendall, H.W. Posamentier, C.A. Ross & J.C. Van Wagoner), pp. 71–108. Society of Economic Paleontologists and Mineralogists Special Publication **42**.

Hardie, L.A., Smoot, J.P. & Eugster, H.P. (1978) Saline lakes and their deposits: a sedimentological approach. In: *Modern and Ancient Lake Sediments* (eds A. Matter & M.E. Tucker), pp. 7–42. International Association of Sedimentologists Special Publication **2**.

Harland, W.B., Cox, A.V., Llewellyn, P.G., Pickton, C.A.G., Smith, A.G. & Walters, R. (1989) *A Geologic Time Scale*. Cambridge University Press, Cambridge.

Harms, J.C., Southard, J., Spearing, D.R. & Walker, R.G. (1975) *Depositional Environments as Interpreted from Primary Sedimentary and Stratification Sequences*, 161 pp. Lecture Notes. Society of Economic Paleontologists and Mineralogists, Short Course Notes **2**.

Harrell, J. (1984) A visual comparator for degree of sorting in thin and plane sections. *Journal of Sedimentary Petrology* **54**, 646–650.

Haughton, P.D.W. (1989) Structure of some Lower Old Red Sandstone conglomerates, Kincardineshire, Scotland: deposition from late-orogenic antecedent streams? *Journal of the Geological Society of London* **146**, 509–525.

Hazeldine, R.S. (1989) Coal reviewed: depositional controls, modern analogues and ancient climates. In: *Deltas: Sites and Traps for Fossil Fuels* (eds M.K. Whateley & K.T. Pickering), pp. 289–308. Geological Society Special Publication **41**.

Heezen, B.C. & Drake, C.L. (1964) Grand Banks slump. *American Association of Petroleum Geologists Bulletin* **48**, 221–225.

Heezen, B.C., Hollister, C.D. & Ruddiman, W.F. (1966) Shaping of the continental rise by deep geostrophic contour currents. *Science* **152**, 502–508.

Hess, J., Bender, M.L. & Schilling, J.-G. (1986) Evolution of the ratio of strontium-87 to strontium-86 in seawater from the Cretaceous to the present. *Science* **231**, 979–984.

Heward, A.P. (1978) Alluvial fan sequence and mega-sequence models: with examples from Westphalian D–Stephanian B coalfields, Northern Spain. In: *Fluvial Sedimentology* (ed. A.D. Miall), pp. 669–702. Canadian Society of Petroleum Geologists Memoir **5**.

Heward, A.P. (1981) A review of wave-dominated clastic shoreline deposits. *Earth Science Reviews* **17**, 223–276.

Hird, K. & Tucker, M.E. (1988) Contrasting diagenesis of two Carboniferous oolites from South Wales: a tale of climatic influence. *Sedimentology* **35**, 587–602.

Hjulström, F. (1939) Transportation of detritus by moving water. In: *Recent Marine Sediments, a Symposium* (ed. P.D. Trask), pp. 5–31. American Association of Petroleum Geologists, Tulsa, OK.

Hooke, R.Le B. (1967) Processes on arid-region alluvial fans. *Journal of Geology* **75**, 438–460.

Hooke, R.Le B. (1968) Steady state relationships on arid region alluvial fans in closed basins. *American Journal of Science* **266**, 609–629.

Hsü, K.J. (1972) The origin of saline giants: a critical review after the discovery of the Mediterranean evaporite. *Earth Science Reviews* **8**, 371–396.

Hughes, D.A. & Lewin, J. (1982) A small-scale flood plain. *Sedimentology* **29**, 891–895.

Hunt, D. & Tucker, M.E. (1992) Stranded parasequences and the forced regressive wedge systems tract: deposition during base-level fall. *Sedimentary Geology* **81**, 1–9.

Hurford, A.J. & Green, P.F. (1982) A user's guide to fission track dating calibration. *Earth and Planetary Science Letters* **59**, 343–354.

Hurford, A.J. & Green, P.F. (1983) The zeta age calibration of fission track dating. *Isotope Geoscience* **1**, 285–317.

Hurst, A., Lovell, M.A. & Morton, A.C. (eds) (1990) *Geological Applications of Wireline Logs*, 357 pp. Geological Society of London Special Publication **48**.

Ibbeken, H. & Schleyer, R. (1991) *Source and Sediment. A Case Study of Provenance and Mass Balance at an Active Plate Margin (Calabria, Southern Italy)*, 286 pp. Springer-Verlag, Berlin.

Ingersoll, R.V. (1988) Tectonics of sedimentary basins. *Geological Association of America Bulletin* **100**, 1704–1719.

Izett, G.A. & Naeser, C.W. (1976) Age of the Bishop Tuff of eastern California as determined by fission-track method. *Geology* **4**, 587–590.

James, N.P. & Bourque, P.-A. (1992) Reefs and mounds. In: *Facies Models: Response to Sea Level Change* (eds R.G. Walker & N.P. James), pp. 323–348. Geological Association of Canada, St Johns, Newfoundland.

Jenkyns, H.C. & Hsü, K.J. (1974) Pelagic sediments: on land and under the sea — an introduction. In: *Pelagic Sediments: On Land and Under the Sea* (eds K.J. Hsü & H.C. Jenkyns), pp. 1–10. International Association of Sedimentologists Special Publication **1**.

Jervey, M.T. (1988) Quantitative geological modelling of siliciclastic rock sequences and their seismic expressions. In: *Sea Level Changes: An Integrated Approach* (eds C.K. Wilgus, B.S. Hastings, C.G.St.C. Kendall, H.W. Posamentier, C.A. Ross & J.C. Van Wagoner), pp. 47–69. Society of Economic Paleontologists and Mineralogists Special Publication **42**.

Johns, D.R., Mutti, E., Rosell, J. & Seguret, M. (1981) Origin of a thick, redeposited carbonate bed in Eocene turbidites of the Hecho Group, south-central Pyrenees, Spain. *Geology* **9**, 161–164.

Johnson, H.D. & Baldwin, C.T. (1996) Shallow clastic seas. In: *Sedimentary Environments: Processes, Facies and Stratigraphy* (ed. H.G. Reading), pp. 232–280. Blackwell Science, Oxford.

Jones, B. & Desrochers, A. (1992) Shallow platform carbonates. In: *Facies Models: Response to Sea Level Change* (eds R.G. Walker & N.P. James), pp. 277–302. Geological Association of Canada, St Johns, Newfoundland.

Jordan, T.E. (1995) Retroarc foreland and related basins. In: *Tectonics of Sedimentary Basins* (eds R.V. Ingersoll & C.J. Busby), pp. 331–362. Blackwell Science, Oxford.

Karig, D.E., Lawrence, M.B., Moore, G.F. & Curray, J.R. (1980) Structural framework of the forearc basin, NW Sumatra. *Journal of the Geological Society* **137**, 77–91.

Kearey, P. & Vine, F.J. (1996) *Global Tectonics*, 2nd edn, 333 pp. Blackwell Science, Oxford.

Kelly, S. & Olsen, H. (1993) Terminal fans, with reference to Devonian examples. In: *Current Research in Fluvial Sedimentology* (ed. C.R. Fielding), pp. 339–374. Sedimentary Geology **85**.

Kelts, K. (1981) A comparison of some aspects of sedimentation and translational tectonics from the Gulf of

California and the Mesozoic Tethys, northern Penninic margin. *Eclogae Geologicae Helveticae* **74**, 371–338.

Kendall, A.C. (1992) Evaporites. In: *Facies Models: Response to Sea Level Change* (eds R.G. Walker & N.P. James), pp. 375–409. Geological Association of Canada, St Johns, Newfoundland.

Kendall, A.C. & Harwood, G.M. (1996) Marine evaporites: arid shorelines and basins. In: *Sedimentary Environments: Processes, Facies and Stratigraphy* (ed. H.G. Reading), pp. 281–324. Blackwell Science, Oxford.

Kenyon, N.H. (1970) Sand ribbons of European tidal seas. *Marine Geology* **9**, 25–39.

Klein, G.D. (1995) Intracratonic basins. In: *Tectonics of Sedimentary Basins* (eds R.V. Ingersoll & C.J. Busby), pp. 459–478. Blackwell Science, Oxford.

Kocurek, G.A. (1996) Desert aeolian systems. In: *Sedimentary Environments: Processes, Facies and Stratigraphy* (ed. H.G. Reading), pp. 125–153 Blackwell Science, Oxford.

Komar, P.D. (1976) *Beach Processes and Sedimentation*, 429 pp. Prentice Hall, Englewood Cliffs, NJ.

Komar, P.D., Neudeck, R.H. & Kulm, L.D. (1972) Observations and significance of deep water oscillatory ripple marks on the Oregon continental shelf. In: *Shelf Sediment Transport: Process and Pattern* (eds P.J.P. Swift, D.B. Duane & O.H. Pilkey), pp. 601–619. Dowden, Hutchinson and Ross, Stroudsberg, PA.

Krauskopf, K.B. (1979) *Introduction to Geochemistry*, 617 pp. McGraw-Hill, New York.

Krumbein, W.C. & Pettijohn, F.J. (1938) *Manual of Sedimentary Petrography*, 549 pp. Appleton-Century-Crofts, New York. [Reprinted by Society of Economic Paleontologists and Mineralogists, Reprint Series **18**.]

Krumbein, W.C. & Sloss, L.L. (1951) *Stratigraphy and Sedimentation*, 497 pp. Freeman, San Francisco, CA.

Laird, M.G. (1995) Coarse-grained lacustrine fan-delta deposits (Pororari Group) of the north-western South Island, New Zealand: evidence for mid-Cretaceous rifting. In: *Sedimentary Facies Analysis: A Tribute to the Teaching and Research of Harold G. Reading* (ed. A.G. Plint), pp. 197–217. International Association of Sedimentologists Special Publication **22**.

Lajoie, J. & Stix, J. (1992) Volcaniclastic rocks. In: *Facies Models: Response to Sea Level Change* (eds R.G. Walker & N.P. James), pp. 101–118. Geological Association of Canada, St Johns, Newfoundland.

Land, L.S. (1985) The origin of massive dolomite. *Journal of Geological Education* **33**, 112–125.

Leeder, M.R. (1975) Pedogenic carbonate and floodplain accretion rates: a quantitative model for alluvial, arid-zone lithofacies. *Geological Magazine* **112**, 257–270.

Leeder, M.R. (1982) *Sedimentology: Process and Product*, 344 pp. Unwin Hyman, London.

Leeder, M.R. (1995) Continental rifts and proto-oceanic troughs. In: *Tectonics of Sedimentary Basins* (eds R.V. Ingersoll & C.J. Busby), pp. 119–148. Blackwell Science, Oxford.

Leeder, M.R. & Gawthorpe, R.L. (1987) Sedimentary models for extensional tilt-block/half graben basins. In: *Continental Extensional Tectonics* (eds M.P. Coward & J.F. Dewey), pp. 271–284. Geological Society of London Special Publication **28**.

Lees, A. & Butler, A.T. (1972) Modern temperate water and warm water shelf carbonate sediments contrasted. *Marine Geology* **13**, 1767–1773.

Leggett, J.K. (1987) The Southern Uplands as an accretionary prism: the importance of analogues in reconstructing palaeogeography. *Journal of the Geological Society of London* **144**, 737–752.

Leggett, J.K., McKerrow, W.S. & Casey, D.M. (1982) The anatomy of a Lower Palaeozoic accretionary forearc: the Southern Uplands of Scotland. In: *Trench–Forearc Geology: Sedimentation and Tectonics on Modern and Ancient Plate Margins* (ed. J.K. Leggett), pp. 495–520. Geological Society of London Special Publication **10**.

Lewis, D.G. & McConchie, D. (1994) *Analytical Sedimentology*, 197 pp. Chapman & Hall, London.

Lowe, D.R. (1982) Sediment gravity flows II: depositional models with special reference to the deposits of high density turbidity currents. *Journal of Sedimentary Petrology* **52**, 279–297.

McCabe, P.J. (1984) Depositional environments of coal and coal-bearing strata. In: *Sedimentology of Coal and Coal-Bearing Sequences* (eds R.A. Rahmani & R.M. Flores), pp. 13–42. International Association of Sedimentologists Special Publication **7**.

McClay, K.R., Nichols, G.J., Khalil, S., Darwish, M. & Bosworth, W. (1998) Extensional tectonics and sedimentation, Eastern Gulf of Suez, Egypt. In: *Sedimentation and Tectonics of Rift Basins: Red Sea—Gulf of Aden* (eds B.H. Purser & D.W.J. Bosence), pp. 223–238. Chapman & Hall, London.

Macdonald, D.I.M. & Butterworth, P.J. (1990) The stratigraphy, setting and hydrocarbon potential of the Mesozoic Sedimentary Basins of the Antarctic Peninsula. In: *Antarctic as an Exploration Frontier* (ed. B. St. John), pp. 101–125. American Association of Petroleum Geologists Studies in Geology **31**.

Mack, G.H., James, W.C. & Monger, H.C. (1993) Classification of paleosols. *Bulletin of the Geological Society of America* **105**, 129–136.

McKee, E.D. (1979) An introduction to the study of global sand seas. In: *Global Sand Seas* (ed. E.D. McKee), pp. 1–19. United States Geological Survey Professional Paper **1052**.

McKee, E.D. & Ward, W.C. (1983) Eolian environment. In: *Carbonate Depositional Environments* (eds P.A. Scholle, D.G. Bebout & C.H. Moore), pp. 132–170. American Association of Petroleum Geologists Memoir **33**.

McLane, M. (1995) *Sedimentology*, 423 pp. Oxford University Press, Oxford.

McQuillin, R., Bacon, M. & Barclay, W. (1984) *An Introduction to Seismic Interpretation*, 287 pp. Graham and Trotman, London.

Main, I.G. (1993) The Earth's crust and mantle: composition and dynamics. In: *Holmes' Principles of Physical Geology* (ed. P.McL.D. Duff), 4th edn, pp. 599–615. Chapman & Hall, London.

Maizels, J.K. (1989) Sedimentology, palaeoflow dynamics and flood history of jökulhlaup deposits: palaeohydrology of Holocene sediment sequences in southern Iceland sandur deposits. *Journal of Sedimentary Petrology* **59**, 204–223.

Maizels, J. (1993) Lithofacies variations within sandur deposits: the role of runoff regime, flow dynamics and sediment supply characteristics. In: *Current Research in Fluvial Sedimentology* (ed. C.R. Fielding), pp. 299–325. Sedimentary Geology **85**.

Mange, M.A. & Maurer, H.F.W. (1992) *Heavy Minerals in Colour*, 147 pp. Chapman & Hall, London.

Marsagalia, K.M. (1995) Interarc and backarc basins. In: *Tectonics of Sedimentary Basins* (eds R.V. Ingersoll & C.J. Busby), pp. 299–330. Blackwell Science, Oxford.

Masson, D.G. (1994) Late Quaternary turbidite current pathways to the Madeira Abyssal Plain and some constraints on turbidity current mechanisms. *Basin Research* **6**, 17–33.

Masson, D.G., Kidd, R.B., Gardner, J.V., Huggett, Q.J. & Weaver, P.P.E. (1992) Saharan continental rise: facies distribution and sediment slides. In: *Geologic Evolution of Atlantic Continental Rises* (eds C.W. Poag & P.C. de Gracianksy), pp. 327–343. Van Nostrand Reinhold, New York.

Matthews, R.K. (1986) Oxygen isotope record of ice-volume history: 100 million years of glacio-eustatic sea level fluctuation. In: *Inter-Regional Unconformities and Hydrocarbon Accumulation* (ed. J.S. Schlee), pp. 97–107. American Association of Petroleum Geologists Memoir **36**.

Miall, A.D. (1974) Palaeocurrent analysis of alluvial sediments—a discussion of directional variance and vector magnitude. *Journal of Sedimentary Petrology* **44**, 1174–1185.

Miall, A.D. (1978) Lithofacies types and vertical profile models of braided river deposits, a summary. In: *Fluvial Sedimentology* (ed. A.D. Miall), pp. 597–604. Canadian Society of Petroleum Geologists Memoir **5**.

Miall, A.D. (1992) Exxon global cycle chart: an event for every occasion. *Geology* **20**, 787–790.

Miall, A.D. (1995) Collision-related foreland basins. In: *Tectonics of Sedimentary Basins* (eds R.V. Ingersoll & C.J. Busby), pp. 393–424. Blackwell Science, Oxford.

Miall, A.D. (1997) *The Geology of Stratigraphic Sequences*, 433 pp. Springer-Verlag, Berlin.

Middleton, G.V. (1966) Experiments on density and turbidity currents. 3: Deposition of sediment. *Canadian Journal of Earth Science* **4**, 475–505.

Middleton, G.V. (1973) Johannes Walther's law of correlation of facies. *Bulletin of the Geological Society of America* **84**, 979–988.

Middleton, G.V. & Hampton, M.A. (1973) Sediment gravity flows: mechanics of flow and deposition. In: *Turbidites and Deep Water Sedimentation* (eds G.V. Middleton & A.H.

Bouma), pp. 1–38. Pacific Section, Society of Economic Paleontologists and Mineralogists Short Course Notes.

Middleton, G.V. & Hampton, M.A. (1976) Subaqueous sediment transport and deposition by sediment gravity flows. In: *Marine Sediment Transport and Environmental Management* (eds D.J. Stanley & D.J.P. Swift), pp. 197–218. Wiley, New York.

Middleton, G.V. & Southard, J.B. (1978) *Mechanics of Sediment Movement*. Society of Economic Paleontologists and Mineralogists Short Course Notes **3**.

Miller, J.M.G. (1996) Glacial sediments. In: *Sedimentary Environments: Processes, Facies and Stratigraphy* (ed. H.G. Reading), pp. 454–484. Blackwell Science, Oxford.

Mitchell, A.H.G. & Reading, H.G. (1986) Sedimentation and tectonics. In: *Sedimentary Environments and Facies* (ed. H.G. Reading), pp. 471–519. Blackwell Scientific Publications, Oxford.

Mitchum, R.M. (1977) Seismic stratigraphy and global changes of sea level. Part 11: glossary of terms used in seismic stratigraphy. In: *Seismic Stratigraphy—Applications to Hydrocarbon Exploration* (ed. C.E. Payton), pp. 205–212. American Association of Petroleum Geologists Memoir **26**.

Mitchum, R.M., Vail, P.R. & Thompson, S. (1977) Seismic stratigraphy and global changes in sea level. Part 2: the depositional sequence as a basic unit for stratigraphic analysis. In: *Seismic Stratigraphy—Applications to Hydrocarbon Exploration* (ed. C.E. Payton), pp. 53–62. American Association of Petroleum Geologists Memoir **26**.

Morgan, J.P., Coleman, J.M. & Gagliano, S.M. (1968) Mudlumps: diapiric structures in Mississippi delta sediments. In: *Diapirism and Diapirs* (eds J. Braunstein & G.D. O'Brien), pp. 145–161. American Association of Petroleum Geologists Memoir **8**.

Mutti, E. (1992) *Turbidite Sandstones*, 275 pp. Agip Instituto di Geologia, Università di Parma, Milan.

Mutti, E. & Normark, W.R. (1987) Comparing examples of modern and ancient turbidite systems: problems and concepts. In: *Marine Clastic Sedimentology: Concepts and Case Studies* (eds J.K. Leggett & G.G. Zuffa), pp. 1–38. Graham and Trotman, London.

Nemec, W. (1990) Deltas—remarks on terminology and classification. In: *Coarse-Grained Deltas* (eds A. Colella & D.B. Prior), pp. 3–12. International Association of Sedimentologists Special Publication **10**.

Nichols, G.J. (1987a) Syntectonic alluvial fan sedimentation, southern Pyrenees. *Geological Magazine* **124**, 121–133.

Nichols, G.J. (1987b) Structural controls on fluvial distributary systems—the Luna System, Northern Spain. In: *Recent Developments in Fluvial Sedimentology* (eds F.G. Ethridge, R.M. Florez & M.D. Harvey), pp. 269–277. Society of Economic Paleontologists and Mineralogists Special Publication **39**.

Nichols, G.J. (1993) Continental margins and basin evolution. In: *Holmes' Principles of Physical Geology* (ed. P.McL.D. Duff), 4th edn, pp. 698–723. Chapman & Hall, London.

Nichols, G.J. & Hall, R. (1991) Basin formation and Neogene

sedimentation in a backarc setting, Halmahera, eastern Indonesia. *Marine and Petroleum Geology* **8**, 50–61.

Nichols, G.J. & Jones, T. (1992) Fusain in Carboniferous shallow marine sediments, Donegal, Ireland: the sedimentological effects of wildfire. *Sedimentology* **39**, 487–502.

Nickling, W.G. (1994) Aeolian sediment transport and deposition. In: *Sediment Transport and Depositional Processes* (ed. K. Pye), pp. 293–350. Blackwell Scientific Publications, Oxford.

Nilsen, T.H. & Sylvester, A.G. (1995) Strike-slip basins. In: *Tectonics of Sedimentary Basins* (eds R.V. Ingersoll & C.J. Busby), pp. 425–458. Blackwell Science, Oxford.

Ninkovitch, D., Sparks, R.S.J. & Ledbetter, M.J. (1978) The exceptional magnitude and intensity of the Toba eruption, Sumatra: an example of the use of deep-sea tephra layers as a geological tool. *Bulletin of Volcanology* **41**, 286–298.

Normark, W.R. (1970) Growth patterns of deep sea fans. *Bulletin of the American Association of Petroleum Geologists* **54**, 2170–2195.

Normark, W.R. (1978) Fan valleys, channels and depositional lobes on modern submarine fans: characters for recognition of sandy turbidite environments. *American Association of Petroleum Geologists Bulletin* **62**, 912–931.

North, F.K. (1985) *Petroleum Geology*, 607 pp. Allen & Unwin, Boston, MA.

North American Commission on Stratigraphic Nomenclature (1983) North American Stratigraphic Code. *American Association of Petroleum Geologists Bulletin* **67**, 841–875.

Nummedal, D., Pilkey, O.H. & Howard, J.D. (eds) (1987) *Sea Level Fluctuations and Coastal Evolution*, 267 pp. Society of Economic Paleontologists and Mineralogists Special Publication **41**.

Nurmi, R.D. & Friedman, G.M. (1977) Sedimentology and depositional environments of basin centre evaporites, Lower Salina Group (Upper Silurian), Michigan Basin. In: *Reefs and Evaporites — Concepts and Depositional Models* (ed. J.H. Fisher), pp. 23–52. American Association of Petroleum Geologists Studies in Geology **5**.

Oberhänsli, R. & Stoffers, P. (eds) (1988) Hydrothermal activity and metalliferous sediments on the ocean floor. *Marine Geology* (Special Issue) **84**, 145–284.

O'Brien, P.E. & Wells, A.T. (1986) A small, alluvial crevasse splay. *Journal of Sedimentary Petrology* **56**, 876–879.

Oertel, G.F. (1985) The barrier island system. *Marine Geology* **63**, 1–18.

Oomkens, E. (1970) Depositional sequences and sand distribution in the postglacial Rhone delta complex. In: *Deltaic Sedimentation Modern and Ancient* (eds J.P. Morgan & R.H. Shaver), pp. 198–212. Society of Economic Paleontologists and Mineralogists Special Publication **15**.

Ori, G.G. & Friend, P.F. (1984) Sedimentary basins formed and carried piggy-back on active thrust sheets. *Geology* **12**, 475–478.

Ori, G.G., Roveri, M. & Nichols, G.J. (1991) Architectural

pattern of large-scale Gilbert-type delta complexes, Pleistocene, Gulf of Corinth, Greece. In: *The Three-Dimensional Facies Architecture of Terrigenous Clastic Sediments and Its Implications for Hydrocarbon Discovery and Recovery* (eds A.D. Miall & N. Tyler), pp. 207–216. SEPM (Society for Sedimentary Geology) Concepts in Sedimentology and Paleontology **3**.

Orton, G.J. (1996) Volcanic environments. In: *Sedimentary Environments: Processes, Facies and Stratigraphy* (ed. H.G. Reading), pp. 485–567. Blackwell Science, Oxford.

Orton, G.J. & Reading, H.G. (1993) Variability of deltaic processes in terms of sediment supply, with particular emphasis on grain size. *Sedimentology* **40**, 475–512.

Pauley, J.C. (1995) Sandstone megabeds from the Tertiary of the North Sea. In: *Characterization of Deep Marine Clastic Systems* (eds A.J. Hartley & D.J. Prosser), pp. 103–114. Geological Society of London Special Publication **94**.

Pemberton, S.G. & MacEachern, J.A. (1995) The sequence stratigraphic significance of trace fossils: examples from the Cretaceous foreland basin of Alberta, Canada. In: *Sequence Stratigraphy of Foreland Basin Deposits* (eds J.C. Van Wagoner & G.T. Betram), pp. 429–476. American Association of Petroleum Geologists Memoir **64**.

Pemberton, S.G., MacEachern, J.A. & Frey, R.W. (1992) Trace fossil facies models: environmental and allostratigraphic significance. In: *Facies Models: Response to Sea Level Change* (eds R.G. Walker & N.P. James), pp. 47–72. Geological Association of Canada, St Johns, Newfoundland.

Percival, C.J. (1986) Paleosols containing an albic horizon: examples from the Upper Carboniferous of northern England. In: *Paleosols: Their Recognition and Interpretation* (ed. V.P. Wright), pp. 87–111. Blackwell Scientific Publications, Oxford.

Perrier, R. & Quiblier, J. (1974) Thickness changes in sedimentary layers during compaction history; methods for quantitative evaluation. *American Association of Petroleum Geologists Bulletin* **58**, 507–520.

Pettijohn, F.J. (1975) *Sedimentary Rocks*, 3rd edn, 718 pp. Harper and Row, New York.

Pettijohn, F.J., Potter, P.E. & Siever, R. (1987) *Sand and Sandstone*, 553 pp. Springer-Verlag, New York.

Pickering, K.T., Hiscott, R.N. & Hein, F.J. (1989) *Deep Marine Environments; Clastic Sedimentation and Tectonics*, 416 pp. Unwin Hyman, London.

Pilkey, O.H. (1988) Basin plains: giant sedimentation events. In: *Sedimentologic Consequences of Convulsive Geologic Events* (ed. H.E. Clifton), pp. 93–99. Geological Society of America Special Paper **229**.

Pitman, W.C. (1978) Relationship between eustasy and stratigraphic sequences of passive margins. *Geological Society of America Bulletin* **89**, 1387–1403.

Plint, A.G. (1996) Marine and nonmarine systems tracts in fourth-order sequences in the Early–Middle Cenomanian, Dunvegan Alloformation, northeastern British Columbia, Canada. In: *High Resolution Sequence Stratigraphy:*

Innovations and Applications (eds J.A. Howell & J.F. Aitken), pp. 159–191. Geological Society of London Special Publication **104**.

Plint, A.G., Eyles, N., Eyles, C.H. & Walker, R.G. (1992) Control of sea level change. In: *Facies Models: Response to Sea Level Change* (eds R.G. Walker & N.P. James), pp. 15–26. Geological Association of Canada, St Johns, Newfoundland.

Porebski, S.J. & Gradzinski, R. (1990) Lava-fed Gilbert-type delta in the Plonez Cove Formation (Lower Oligocene), King George Island, West Antarctica. In: *Coarse-Grained Deltas* (eds A. Colella & D.B. Prior), pp. 335–351. International Association of Sedimentologists Special Publication **10**.

Posamentier, H.W., Jervey, M.T. & Vail, P.R. (1988) Eustatic controls on clastic deposition. I: conceptual framework. In: *Sea Level Changes: An Integrated Approach* (eds C.K. Wilgus, B.S. Hastings, C.G.St.C. Kendall, H.W. Posamentier, C.A. Ross & J.C. Van Wagoner), pp. 109–124. Society of Economic Paleontologists and Mineralogists Special Publication **42**.

Postma, G. (1984) Mass-flow conglomerates in a submarine canyon: Abrioja fan-delta, Pliocene, south-eastern Spain. In: *Sedimentology of Gravels and Conglomerates* (eds E.H. Koster & R.J. Steel), pp. 237–258. Canadian Society of Petroleum Geologists Memoir **10**.

Postma, G. (1990) Depositional architecture and facies of river and fan deltas: a synthesis. In: *Sedimentology of Gravels and Conglomerates* (eds E.H. Koster & R.J. Steel), pp. 13–27. Canadian Society of Petroleum Geologists Memoir **10**.

Potter, P.E. & Pettijohn, F.J. (1977) *Palaeocurrents and Basin Analysis*, 2nd edn, 425 pp. Springer-Verlag, Berlin.

Pratt, B.R., James, N.P. & Cowan, C.A. (1992) Peritidal carbonates. In: *Facies Models: Response to Sea Level Change* (eds R.G. Walker & N.P. James), pp. 303–322. Geological Association of Canada, St Johns, Newfoundland.

Press, F. & Siever, R. (1986) *Earth*, 2nd edn, 649 pp. W.H. Freeman, New York.

Puigdefabregas, C. & van Vleit, A. (1978) Meandering stream deposits from the Tertiary of the southern Pyrenees. In: *Fluvial Sedimentology* (ed. A.D. Miall), pp. 469–485. Canadian Society of Petroleum Geologists Memoir **5**.

Purser, B.H. (1973) *The Persian Gulf: Holocene Carbonate Sedimentation and Diagenesis in a Shallow Epicontinental Sea*, 417 pp. Springer-Verlag, Berlin.

Pye, K. (1987) *Aeolian Dust and Dust Deposits*, 334 pp. Academic Press, London.

Racey, A., Love, M.A., Bobolecki, R.M. & Walsh, J.N. (1995) The use of chemical analyses in the study of biostratigraphically barren sequences: an example from the Triassic of the central North Sea (UKCS). In: *Non-Biostratigraphical Methods of Dating and Correlation* (eds R.E. Dunay & E.A. Hailwood), pp. 69–105. Geological Society Special Publication **89**.

Rampino, M.R. & Caldeira, K.C. (1992) Major episodes of geological change: correlations, time structure and possible causes. *Earth and Planetary Science Letters* **114**, 101–111.

Read, J.F. (1982) Carbonate platforms of passive (extensional) continental margins: types, characteristics and evolution. *Tectonophysics* **81**, 195–212.

Read, J.F. (1985) Carbonate platform facies models. *American Association of Petroleum Geologists Bulletin* **69**, 1–21.

Reading, H.G. (1980) Characteristics and recognition of strike-slip systems. In: *Sedimentation in Oblique-Slip, Mobile Zones* (eds P.F. Ballance & H.G. Reading), pp. 7–26. International Association of Sedimentologists Special Publication **4**.

Reading, H.G. (1982) Sedimentary basins and global tectonics. *Proceedings of the Geological Association* **93**, 321–350.

Reading, H.G. (ed.) (1996) *Sedimentary Environments: Processes, Facies and Stratigraphy*, 688 pp. Blackwell Science, Oxford.

Reading, H.G. & Collinson, J.D. (1996) Clastic coasts. In: *Sedimentary Environments: Processes, Facies and Stratigraphy* (ed. H.G. Reading), pp. 154–231. Blackwell Science, Oxford.

Reading, H.G. & Levell, B.K. (1996) Controls on the sedimentary record. In: *Sedimentary Environments: Processes, Facies and Stratigraphy* (ed. H.G. Reading), pp. 5–36. Blackwell Science, Oxford.

Reid, I. & Frostick, L.E. (1985) Beach orientation, bar morphology and the concentration of metalliferous placer deposits: a case study, Lake Turkana, Kenya. *Journal of the Geological Society of London* **142**, 837–848.

Reineck, H.E. & Singh, I.B. (1973) Genesis of laminated sand and graded rhythmites in storm-sand layers of shelf mud. *Sedimentology* **18**, 123–128.

Reineck, H.E. & Singh, I.B. (1980) *Depositional Sedimentary Environments*, 2nd edn, 549 pp. Springer-Verlag, Berlin.

Reinson, G.E. (1992) Transgressive barrier island and estuarine systems. In: *Facies Models: Response to Sea Level Change* (eds R.G. Walker & N.P. James), pp. 179–194. Geological Association of Canada, St Johns, Newfoundland.

Ricci-Lucchi, F. (1986) The Oligocene to Recent foreland basins of the northern Apennines. In: *Foreland Basins* (eds P.A. Allen & P. Homewood), pp. 105–139. International Association of Sedimentologists Special Publication **8**.

Rider, M.H. (1986) *The Geological Interpretation of Well Logs*, 175 pp. Blackie, Glasgow.

Roy, P.S. (1994) Holocene estuary evolution—stratigraphic studies from south-eastern Australia. In: *Incised-Valley Systems: Origin and Sedimentary Sequences* (eds R.W. Dalrymple, R. Boyd & B.A. Zaitlin), pp. 241–263. SEPM (Society for Sedimentary Geology) Special Publication **61**.

Ruddiman, W.F. & Kutzbach, J.E. (1990) Late Cenozoic plateau uplift and climate change. *Transactions of the Royal Society of Edinburgh: Earth Sciences* **81**, 301–314.

Sarg, J.F. (1988) Carbonate sequence stratigraphy. In: *Sea*

Level Changes: An Integrated Approach (eds C.K. Wilgus, B.S. Hastings, C.G.St.C. Kendall, H.W. Posamentier, C.A. Ross & J.C. Van Wagoner), pp. 155–181. Society of Economic Paleontologists and Mineralogists Special Publication **42**.

Schlager, W. (1992) *Sedimentology and Sequence Stratigraphy of Reefs and Carbonate Platforms*, 71 pp. American Association of Petroleum Geologists Continuing Education Course Notes Series **34**.

Schminke, H.V., Fisher, R.V. & Waters, A.C. (1975) Antidune and chute-and-pool structures in the base surge deposits of the Laacher See area, Germany. *Sedimentology* **20**, 553–574.

Scholl, D.W., von Huene, R., Vallier, T.L. & Howell, D.G. (1980) Sedimentary masses and concepts about tectonic processes at under-thrust oceanic margins. *Geology* **8**, 564–568.

Scholle, P.A., Arthur, M.A. & Ekdale, A.A. (1983) Pelagic environments. In: *Carbonate Depositional Environments* (eds P.A. Scholle, D.G. Bebout & C.H. Moore), pp. 620–691. American Association of Petroleum Geologists Memoir **33**.

Schumm, S.A. (1968) Speculations concerning palaeohydraulic controls of terrestrial sedimentation. *Geological Society of America Bulletin* **79**, 1573–1588.

Schumm, S.A. (1981) Evolution and response of the fluvial system; sedimentologic implications. In: *Recent and Ancient Nonmarine Depositional Environments* (eds F.G. Ethridge & R.M. Flores), pp. 19–29. Society of Economic Paleontologists and Mineralogists Special Publication **31**.

Schutz, L. (1980) Long range transport of desert dust with special emphasis on the Sahara. *New York Academy of Science Annals* **338**, 515–532.

Schwartz, R.K. (1982) Bedforms and stratification characteristics of some modern small-scale washover sand bodies. *Sedimentology* **29**, 835–850.

Seilacher, A. (1953) Studien zur Palichnologie I. Über die Methoden den Palichnologie. *Neues Jahrbuch für Geologie und Palaeontologie Abhandlungen* **96**, 421–452.

Sellwood, B. & Jenkyns, H.C. (1975) Basins and swells and the evolution of an epeiric sea (Pleinsbachian–Bajocian of Great Britain). *Journal of the Geological Society of London* **131**, 373–388.

Sharp, R.P. (1963) Wind ripples. *Journal of Geology* **71**, 617–636.

Shaw, A.B. (1964) *Time in Stratigraphy*, 365 pp. McGraw-Hill, New York.

Sheridan, R.E. (1974) Atlantic continental margin of North America. In: *The Geology of Continental Margins* (eds C.A. Burk & C.L. Drake), pp. 391–407. Springer-Verlag, New York.

Shinn, E.A., Ginsburg, R.N. & Lloyd, R.M. (1965) Recent supratidal dolomite from Andros Island, Bahamas. In: *Dolomitization and Limestone Diagenesis, a Symposium* (eds L.C. Pray & R.D. Murray), pp. 112–123. Society of Economic Paleontologists and Mineralogists Special Publication **13**.

Simpson, S. (1975) Classification of trace fossils. In: *The Study of Trace Fossils* (ed. R.W. Frey), pp.39–45. Springer-Verlag, Berlin.

Skinner, B.J. & Porter, S.C. (1987) *Physical Geology*, 750 pp. Wiley, New York.

Sloss, L.L. (1963) Sequences in the cratonic interior of North America. *Geological Association of America Bulletin* **74**, 93–114.

Smith, D.G. (1983) Anastomosed fluvial deposits: examples from western Canada. In: *Modern and Ancient Fluvial Systems* (eds J.D. Collinson & J. Lewin), pp. 155–168. International Association of Sedimentologists Special Publication **6**.

Smith, D.G. & Smith, N.D. (1980) Sedimentation in anastomosing river systems: examples from alluvial valleys near Banff, Alberta. *Journal of Sedimentary Petrology* **50**, 157–164.

Smith, R.B. & Braile, L.W. (1994) The Yellowstone hotspot. *Journal of Volcanology and Geothermal Research* **61**, 121–187.

Sparks, R.S.J. & Walker, G.P.L. (1973) The ground surge deposit: a third type of pyroclastic rock. *Nature Physical Science* **241**, 62–64.

Spears, D.A. & Kanaris-Sotirious, R. (1979) A geochemical and mineralogical investigation of some British and other European tonsteins. *Sedimentology* **26**, 407–425.

Steckler, M.S., Watts, A.B. & Thorne, J.A. (1988) Subsidence and basin modelling at the US Atlantic passive margin. In: *The Geology of North America*. Vol. 1/2. *The Atlantic Continental Margin* (eds R.E. Sheridan & J.A. Grow), pp. 399–415. Geological Society of America, Boulder, CO.

Stow, D.A.V. (1979) Distinguishing between fine-grained turbidites and contourites on the Nova Scotia deep water margin. *Sedimentology* **26**, 371–387.

Stow, D.A.V. (1985) Deep-sea clastics: where are we and where are we going? In: *Sedimentology, Recent Developments and Applied Aspects* (eds P.J. Brenchley & B.P.J. Williams), pp. 67–94. Blackwell Scientific Publications, Oxford.

Stow, D.A.V. (1986) Deep clastic seas. In: *Sedimentary Environments and Facies* (ed. H.G. Reading), pp. 400–444. Blackwell Scientific Publications, Oxford.

Stow, D.A.V. (1994) Deep-sea processes of sediment transport and deposition. In: *Sediment Transport and Depositional Processes* (ed. K. Pye), pp. 257–291. Blackwell Scientific Publications, Oxford.

Stow, D.A.V. & Lovell, J.P.B. (1979) Contourites; their recognition in modern and ancient sediments. *Earth Science Reviews* **14**, 251–291.

Stow, D.A.V., Reading, H.G. & Collinson, J.D. (1996) Deep seas. In: *Sedimentary Environments: Processes, Facies and Stratigraphy* (ed. H.G. Reading), pp. 395–453. Blackwell Science, Oxford.

Sturm, M. & Matter, A. (1978) Turbidites and varves in Lake Brienz (Switzerland): deposition of clastic detritus by density currents. In: *Modern and Ancient Lake Sediments* (eds A. Matter & M.E. Tucker), pp. 147–168. International Association of Sedimentologists Special Publication **2**.

Sugden, D.D. & John, B.S. (1976) *Glaciers and Landscape*, 376 pp. Edward Arnold, London.

Suthren, R.J. (1985) Facies analysis of volcaniclastic sediments: a review. In: *Sedimentology, Recent Developments and Applied Aspects* (eds P.J. Brenchley & B.P.J. Williams), pp. 123–146. Blackwell Scientific Publications, Oxford.

Swan, A.R.H. & Sandilands, M. (1995) *Introduction to Geological Data Analysis*, 446 pp. Blackwell Science, Oxford.

Swift, D.J.P. (1968) Coastal erosion and transgressive stratigraphy. *Journal of Geology* **76**, 444–456.

Talbot, M.R. & Allen, P.A. (1996) Lakes. In: *Sedimentary Environments: Processes, Facies and Stratigraphy* (ed. H.G. Reading), pp. 83–124. Blackwell Science, Oxford.

Talling, P.J. & Burbank, D.W. (1993) Assessment of uncertainties in magnetostratigraphic dating of sedimentary strata. In: *Applications of Palaeomagnetism to Sedimentary Geology* (eds D.M. Aïssaoui, D.F. McNeill & N.F. Hurley), pp. 59–69. SEPM (Society for Sedimentary Geology) Special Publication **49**.

Tanner, L.H. & Hubert, J.F. (1991) Basalt breccias and conglomerates in the Lower Jurassic McCoy Brook Formation, Fundy Basin, Nova Scotia: differentiation between talus and debris-flow deposits. *Journal of Sedimentary Petrology* **61**, 15–27.

Tarbuck, E.J. & Lutgens, F.K. (1997) *Earth Science*, 8th edn, 638 pp. Prentice Hall, Hemel Hempstead.

Tarling, D.H. (1983) *Palaeomagnetism*, 186 pp. Chapman & Hall, London.

Taylor, C.M. (1990) Late Permian–Zechstein. In: *Introduction to the Petroleum Geology of the North Sea* (ed. K.W. Glennie), pp. 153–190. Blackwell Scientific Publications, Oxford.

Taylor, S.R. (1992) The origin of the Earth. In: *Understanding the Earth, A New Synthesis* (eds G.C. Brown, C.J. Hawkesworth & R.C.L. Wilson), pp. 25–43. Cambridge University Press, Cambridge.

Thomas, R.G., Smith, D.G., Wood, J.M., Visser, J., Calverley-Range, E.A. & Koster, E.H. (1987) Inclined heterolithic stratification — terminology, description, interpretation and significance. *Sedimentary Geology* **53**, 123–179.

Thornburg, T. & Kulm, L.D. (1981) Sedimentary basins of the Peru continental margin: structure, stratigraphy and Cenozoic tectonics from 6°S to 16°S latitude. *Geological Society of America Memoir* **154**, 393–422.

Till, R. (1974) *Statistical Methods for the Earth Scientist*, 154 pp. Macmillan, London.

Tipper, J.C. (1987) Estimating stratigraphic completeness. *Journal of Geology* **95**, 710–715.

Tissot, B.P. & Welte, D.H. (1984) *Petroleum Formation and Occurrence*, 699 pp. Springer-Verlag, Berlin.

Tucker, M.E. (1982) *Field Description of Sedimentary Rocks*,

128 pp. Open University Educational Enterprises, Milton Keynes.

Tucker, M.E. (ed.) (1988) *Techniques in Sedimentology*, 394 pp. Blackwell Scientific Publications, Oxford.

Tucker, M.E. (1991) *Sedimentary Petrology*, 2nd edn, 260 pp. Blackwell Scientific Publications, Oxford.

Tucker, M.E. (1992) Limestones through time. In: *Understanding the Earth, A New Synthesis* (eds G.C. Brown, C.J. Hawkesworth & R.C.L. Wilson), pp. 347–363. Cambridge University Press, Cambridge.

Tucker, M.E. (1996) *Sedimentary Rocks in the Field*, 2nd edn, 153 pp. Wiley, Chichester.

Tucker, M.E. & Wright, V.P. (1990) *Carbonate Sedimentology*, 482 pp. Blackwell Scientific Publications, Oxford.

Turner, P. (1980) *Continental Red Beds*, 562 pp. Elsevier, Amsterdam.

Udden, J.A. (1914) Mechanical composition of clastic sediments. *Geological Society of America Bulletin* **25**, 655–744.

Underwood, M.B. & Moore, G.F. (1995) Trenches and trench–slope basins. In: *Tectonics of Sedimentary Basins* (eds R.V. Ingersoll & C.J. Busby), pp. 179–220. Blackwell Science, Oxford.

Vail, P.R., Mitchum, R.M. & Thompson, S. (1977) Seismic stratigraphy and global changes of sea level. Part 4: global cycles of relative changes of sea level. In: *Seismic Stratigraphy, Applications to Hydrocarbon Exploration* (ed. C.E. Payton), 516 pp. American Association of Petroleum Geologists Memoir **26**.

Van Andel, T.H. (1994) *New Views on an Old Planet: A History of Global Change*, 439 pp. Cambridge University Press, Cambridge.

Van Wagoner, J.C. (1995) Overview of sequence stratigraphy of foreland basin deposits: terminology, summary of papers and glossary of sequence stratigraphy. In: *Sequence Stratigraphy of Foreland Basin Deposits* (eds J.C. Van Wagoner & G.T. Betram), pp. ix–xxi. American Association of Petroleum Geologists Memoir **64**.

Van Wagoner, J.C., Posamentier, H.W., Mitchum, R.M., Vail, P.R., Sard, J.F., Loutit, T.S. & Hardenbol, J. (1988) An overview of sequence stratigraphy and key definitions. In: *Sea Level Changes: An Integrated Approach* (eds C.K. Wilgus, B.S. Hastings, C.G.St.C. Kendall, H.W. Posamentier, C.A. Ross & J.C. Van Wagoner), pp. 39–45. Society of Economic Paleontologists and Mineralogists Special Publication **42**.

Van Wagoner, J.C., Mitchum, R.M., Campion, K.M. & Rahmanian, V.D. (1990) *Siliciclastic Sequence Stratigraphy in Well Logs, Cores and Outcrop: Concepts for High Resolution Correlation of Time and Facies*, 55 pp. Methods in Exploration Series **7**. American Association of Petroleum Geologists, Tulsa, OK.

Visser, M.J. (1980) Neap–spring cycles reflected in Holocene subtidal large-scale bedform deposits: a preliminary note. *Geology* **8**, 543–546.

Walker, G.P.L. (1973) Explosive volcanic eruptions — a new

classification scheme. *Geologische Rundschau* **62**, 431–446.

Walker, R.G. (1992a) Facies models. In: *Facies Models: Response to Sea Level Change* (eds R.G. Walker & N.P. James), pp. 1–14. Geological Association of Canada, St Johns, Newfoundland.

Walker, R.G. (1992b) Clastic sediments. In: *Understanding the Earth, A New Synthesis* (eds G.C. Brown, C.J. Hawkesworth & R.C.L. Wilson), pp. 327–346. Cambridge University Press, Cambridge.

Walker, R.G. & James, N.P. (eds) (1992) *Facies Models: Response to Sea Level Change*, 409 pp. Geological Association of Canada, St Johns, Newfoundland.

Walker, R.G. & Plint, A.G. (1992) Wave- and storm-dominated shallow marine systems. In: *Facies Models: Response to Sea Level Change* (eds R.G. Walker & N.P. James), pp. 219–238. Geological Association of Canada, St Johns, Newfoundland.

Walther, J. (1894) *Einleitung in die Geologisch Historische Wissenschaft*, Bd 3, *Lithogenesis der Gegenwart*, pp. 535–1055. Fischer-Verlag, Jena.

Ward, J.D. (1988) Aeolian, fluvial and pan (playa) facies of the Tertiary Tsondab Sandstone in the central Namib desert, Namibia. *Sedimentary Geology* **55**, 143–162.

Warren, J.K. & Kendall, C.G.St.C. (1985) Comparison of sequences formed in marine sabkha (subaerial) and salina (subaqueous) settings—modern and ancient. *American Association of Petroleum Geologists Bulletin* **69**, 1013–1023.

Warren, W.P. & Ashley, G.M. (1994) Origins of the ice-contact stratified ridges (eskers) of Ireland, *Journal of Sedimentary Research* **A64**, 433–449.

Wells, N.A. & Dorr, J.A., Jr (1987) A reconnaissance of sedimentation on the Kosi alluvial fan of India. In: *Recent Developments in Fluvial Sedimentology* (eds F.G. Ethridge, R.M. Flores & M.D. Harvey), pp. 51–61. Society of Economic Paleontologists and Mineralogists Special Publication **39**.

Wentworth, C.K. (1922) A scale of grade and class terms for clastic sediments. *Journal of Geology* **30**, 377–394.

Wescott, W.A. & Ethridge, F.G. (1990) Fan deltas—alluvial fans in coastal settings. In: *Alluvial Fans: A Field Approach* (eds A.H. Rachocki & M. Church), pp. 195–211. Wiley, Chichester.

Whittaker, A., Cope, J.C.W., Cowie, J.W., Gibbons, W., Hailwood, E.A., House, M.R., Jenkins, D.G., Rawson, P.F., Rushton, A.W.A., Smith, D.G., Thomas, A.T. & Wimbledon, W.A. (1991) A guide to stratigraphical procedure, *Journal of the Geological Society of London* **148**, 813–824.

Whittington, H.B. & Conway-Morris, S. (eds) (1985) Extraordinary fossil biotas: their ecological and evolutionary significance. *Philosophical Transactions of the Royal Society, London* **311B**, 1–192.

Wilson, I.G. (1972) Universal discontinuities in bedforms produced by the wind. *Journal of Sedimentary Petrology* **42**, 667–669.

Wilson, J.L. (1975) *Carbonate Facies in Geological History*, 471 pp. Springer-Verlag, Heidelberg.

Wilson, J.T. (1966) Did the Atlantic close and then re-open? *Nature* **211**, 676–681.

Woodburne, M.O. (1977) Definition and characterization in mammalian chronostratigraphy. *Journal of Paleontology* **51**, 220–234.

Worsley, T.W., Nance, D. & Moody, J.B. (1984) Global tectonics and eustasy for the past 2 billion years. *Marine Geology* **58**, 373–400.

Wright, V.P. (1986) Facies sequences on a carbonate ramp: the Carboniferous Limestone of South Wales. *Sedimentology* **33**, 221–241.

Wright, V.P. & Burchette, T.P. (1996) Shallow-water carbonate environments. In: *Sedimentary Environments: Processes, Facies and Stratigraphy* (ed. H.G. Reading), pp. 325–394. Blackwell Science, Oxford.

Yang, C.-S. & Nio, S.-D. (1985) The estimation of palaeohydrodynamic processes from subtidal deposits using time series analysis methods. *Sedimentology* **32**, 41–57.

Zak, I. & Freund, R. (1981) Asymmetry and basin migration in the Dead Sea rift. *Tectonophysics* **80**, 27–38.

Index